BODY, TEXT, AND SCIENCE

PHAENOMENOLOGICA
SERIES FOUNDED BY H.L. VAN BREDA AND PUBLISHED
UNDER THE AUSPICES OF THE HUSSERL-ARCHIVES

144

MARIANNE SAWICKI

BODY, TEXT, AND SCIENCE

THE LITERACY OF INVESTIGATIVE PRACTICES AND THE PHENOMENOLOGY OF EDITH STEIN

MARIANNE SAWICKI
Midway College,
Midway, Kentucky, U.S.A.

Body, Text, and Science

The Literacy of Investigative Practices and the Phenomenology of Edith Stein

KLUWER ACADEMIC PUBLISHERS
DORDRECHT / BOSTON / LONDON

A C.I.P. Catalogue record for this book is available from the Library of Congress

ISBN 0-7923-4759-5

Published by Kluwer Academic Publishers,
P.O. Box 17, 3300 AA Dordrecht, The Netherlands.

Sold and distributed in the U.S.A. and Canada
by Kluwer Academic Publishers,
101 Philip Drive, Norwell, MA 02061, U.S.A.

In all other countries, sold and distributed
by Kluwer Academic Publishers,
P.O. Box 322, 3300 AH Dordrecht, The Netherlands.

Printed on acid-free paper

Printed in the Netherlands.

For Helene

Table of Contents

Preface

What is "scientific" about the natural and human sciences? Precisely this: the legibility of our worlds and the distinctive reading strategies that they provoke. That account of the essence of science comes from Edith Stein, who as Husserl's assistant 1916-1918 labored in vain to bring his massive *Ideen* to publication, and then went on to propose her own solution to the problem of finding a unified foundation for the social and physical sciences.

Stein argued that human bodily life itself affords direct access to the interplay of natural causality, cultural motivation, and personal initiative in history and technology. She developed this line of approach to the sciences in her early scholarly publications, which too soon were overshadowed by her religious lectures and writings, and eventually were obscured by National Socialism's ideological attack on philosophies of empathy. Today, as her church prepares to declare Stein a saint, her secular philosophical achievements deserve another look.

I have chosen to place Stein's phenomenology of science into three contexts, each of which is appropriate in its own way. First, the historical context in which Stein wrote was formed by the learned conversations of turn-of-the-century German hermeneutics and of the Phenomenological Movement around the time of the First World War. Husserl was in mid-career and Heidegger was a tongue-tied young man. The first two chapters of this book reconstruct phenomenology as it was being done at that time, when the question of intersubjectivity had not yet received its familiar framing by the pronouns *we* and *thou*.

But the historical context is not the only possible place to begin studying Stein. Since her death at Auschwitz, Stein's writings have attracted controversial interpretations in various academic disciplines and popular media. There now exists a rapidly expanding literature addressing her life and thought. Stein's profound appeal to diverse constituencies--to working men and women as well as to academics of various stripes--is an intriguing part of the picture and must not be overlooked. I found that the various interpretations of Stein could not be dealt with in a perfunctory review of secondary literature, but required the in-depth consideration that I give them in chapter five. These interpretations form the second context for understanding Stein, and they offer readers an alternative avenue of approach to her thought.

The third context for grasping Stein's theory of interpersonal understanding is the present-day debate over constructionism and cultural relativism in the sciences. Feminist theorists and other critics, often drawing on some branch of materialist or psychoanalytic social theory, have argued that scientific data are socially determined to a very great extent. Stein concurs, in principle. Her works on empathy and on psychology establish that natural science is indeed a cultural achievement, for it rests on the ability to isolate *caused* data by recognizing and subtracting *motivated* data from raw data. This subtractive literacy is the most basic scientific competence, and it is fundamentally interpersonal. In the final chapter of this book, I suggest that the reality of data as the *illegible causal remainder* defeats the critiques of science recently offered by psychoanalytic and standpoint feminisms.

My presentation of Stein's own phenomenology of empathy comes in chapter three, which is addressed to Stein herself. I found that I had to speak my observations and criticisms directly to the person whose thought I was trying to follow. I beg the reader's indulgence for this unconventional second-person construction, and hope that its appropriateness will become apparent in light of the demands of Stein's own theory of empathy. In the fourth chapter I revert to the usual scholarly third-person style as I evaluate Stein's interpretive practices. Here readers will find an account of my most surprising discovery: that Stein composed significant portions of Husserl's manuscript for *Ideen II*.

This book began as a doctoral dissertation at the University of Kentucky under the direction of Professor Ronald Bruzina, for whose mentoring I am deeply grateful. During an all too brief visit to Germany in the rainy summer of 1993, I received guidance and encouragement from Professor Hugo Ott of the University of Freiburg and from Professor Elisabeth Ströker of the University of Cologne. I was cordially welcomed at Edith Stein's own religious community, Karmel Maria vom Frieden in Cologne, where Sister Maria Amata Neyer granted me access to the Edith Stein Archive and shared stories of the small details of Stein's life. In Tübingen, Sister Waltraud Herbstrith of the Edith-Stein-Karmel helped me to appreciate the theological dimensions of Stein's work. During my excursion to Louvain, Steven Spileers introduced me to the Husserl Archive and directed me toward the texts in which I found indications of the Husserl-Stein collaborations.

My first tentative formulations of this project benefited from conversations with the late Sister Mary Catharine Baseheart, Professor Emerita at Spalding University. Subsequently I received generous advice from Professors Alasdair MacIntyre and Ralph McInerny at the University of Notre Dame and Linda Lopez McAlister of the University of South Florida. I am indebted to Professors R. Philip Buckley of McGill University, Barry Smith of the State University of New York at Buffalo, and Karl Schuhmann of the University of Utrecht for their insightful comments on earlier drafts. Portions of this research were presented before the Society for Phenomenology and Existential Philosophy in October 1995 and the American Philosophical Association in December 1995. Earlier formulations of selected aspects of the work appear in my essays "Empathy Before and After Husserl," *Philosophy Today* 41/1: (Spring 1997) 123-7; and "Caste and Contact in the Galilee of Jesus," *Galilean Archaeology and the Historical Jesus*, edited by Richard A. Horsley and J. Andrew Overman (Philadelphia: Trinity Press International, 1997). My former teachers at Kentucky who will see their influence in these pages include Professors Monica Udvardy, Don Howard, and Ted Schatzki.

My philosophical studies at Kentucky, commencing as they did in the fifth decade of my life, are the fruit of my husband's enabling trust in me. The work is dedicated to an exemplary teacher of empathy, Helene Amanda Nelson Sawicki, who is my dear mother.

Chapter One
The Genesis of Phenomenology

Misunderstanding of Edith Stein begins, ironically, with "empathy." That word translates the technical term *Einfühlung* in the title of Stein's 1916 doctoral dissertation. Stein herself used the term with precision, for she appreciated the issues at stake in the turn-of-the-century academic debates out of which it came to her. For the contemporary reader, the term *Einfühlung* need not be a stumbling block. It can serve instead as a stepping stone back into the philosophical context in which Edith Stein began her academic career. The first two chapters of this work recover the state of the question of *Einfühlung* before the First World War when, as a student in her early twenties at Göttingen, Edith Stein confronted it.[1]

In the summer of 1913, after one semester of study with Edmund Husserl, Stein asked him to approve a specialized topic for her doctoral studies, a topic on which she would also write a preliminary essay to satisfy requirements for the *Staatsexamen pro facultate docendi*. Her choice of the topic of *Einfühlung* reflects her interest in psychology, the science that she had studied for four semesters at Breslau before coming to Husserl. But she recalls:

In seinem Kolleg über Natur und Geist hatte Husserl davon gesprochen, daß eine objektive Außenwelt nur intersubjektiv erfahren werden könne, d.h. durch eine Mehrheit erkennender Individuen, die in Wechselverständigung miteinander ständen. Demnach sei eine Erfahrung von anderen Individuen dafür vorausgestezt. Husserl nannte diese Erfahrung im Anschluß an die Arbeiten Theodor *Lipps* Einfühlung, aber er sprach nicht darüber aus, worin sie bestünde. Da war also eine Lücke, die es auszufüllen galt: ich wollte untersuchen, was Einfühlung sei. Das gefiel dem Meister nicht übel. (ESW 7: 238)	In his course on nature and [intellect], Husserl had said that an objective outer world could only be experienced intersubjectively, i.e., through a plurality of perceiving individuals who relate in a mutual exchange of information. Accordingly, an experience of other individuals is a prerequisite. To the experience, an application of the work of Theodore Lipps, Husserl gave the name *Einfühlung*. What it consists of, however, he nowhere detailed. Here was a lacuna to be filled; therefore, I wished to examine what empathy might be. The Master found this suggestion not bad at all. (CWES 1: 269)

[1]In order to recover the discussion of *Einfühlung* as Stein discovered it, one must examine the texts that she studied. Wherever possible, I cite original texts (although spelling and punctuation are adjusted to conform to the critical editions). Translations are provided in parallel columns. I cite the published English translation wherever there is a suitable one available, indicating any adjustments within brackets. Translations not followed by a reference (year and page) are my own. If the date when a manuscript was written is more significant than its publication date, then the composition date is given in brackets. References are listed by chapter at the end of this work.

Stein's divergence from Husserl's account of *Einfühlung* would hinge upon her creative misconstrual of what he meant by "an experience of other individuals." As we shall see, Husserl meant to specify that an objective world was a world available for other people to experience. For him, "experience of other individuals" refered to *their* experiencing of the world. Stein, however, was interested in *one's own* experiencing of other people as human beings; that is, how people are available to one another to be understood.

These two takes on "*von Anderen*" unfolded out of nineteenth-century advances in the natural sciences as well as in the social or cultural sciences, the so-called *Geisteswissenschaften*. Academic theorists in Germany thought big: their quest was for a unified account of these disciplines that would guarantee in one stroke the reality of the world and the reliability of knowledge. At that time the insulation of the humanities, including philosophy, from psychology and other sciences was not yet in place as we know it today. Husserl conceived phenomenology as philosophy of science, that is, the philosophy of the foundation of all the sciences.

Theodor Lipps's term *Einfühlung*, adapted by Husserl, was one of a cluster of terms coined for nineteenth-century explorations of the availability of the world to the mind and of the mind to other minds.[2] In the wake of Leibniz, the German hermeneutical tradition revered the monadic character of the mind. Leibniz had taught that each mental individual in some way contains or implies the knowledge of all other individuals comprising its universe. For such an individual or "monad," experience would consist not in absorbing information from others outside itself, but in unfolding within itself that of others which was already given in its own being as necessary. This notion of an in-built, pre-given (albeit rather minimal) mutual accessibility among individuals continued to spur the imaginations of German intellectuals, even after the Rationalism of Leibniz went into eclipse behind Romanticism in the humanities and Positivism in the sciences. Schleiermacher believed in the possibility of penetrating the mental processes of another and termed it *Divination*. Scheler wrote of *Mitfühlen*; Dilthey of *Nachfühlen, Hineinversetzen, Nacherleben*. As we shall see, their explorations of these experiences often explicitly associated such capacities with femininity. Stein herself was alert to the gendering of understanding, and preferred to discuss the human being as *die Person* rather than *der Mensch*.

Yet one must go beyond vocabulary in characterizing the legacy and the agenda of German hermeneutics as it stood at the turn of the nineteenth century--no simple task. On the one hand, there is a linear development from Schleiermacher to Dilthey to Husserl and Stein. On the other hand, more of the work of Schleiermacher is published today than was known to Dilthey, and more of Dilthey is published now than was accessible to Husserl and Stein in the 1910's, the period of their collaboration. Thus we face nice puzzles about the availability of earlier writers to later ones, before we can begin to investigate what the

---

[2]The term *Einfühlung* comes from Robert Vischer. See Mallgrave and Ikonomou (1994: 21).

respective writers had to say about the very issue of availability--interpersonal, intertextual, or otherwise. Recent commentaries base their interpretations on complete reviews of the Works of an author, but one must not project the possibility of such global familiarity backward in time. For example, we are justified in assuming only that Stein knew Dilthey's positions as stated in the published works that she cites, which are those of the 1870's, '80's and '90's. Beyond that, we can speak generally of issues and common agenda; but in many cases it will not be possible to establish direct dependence. About conversations between Dilthey and Husserl, little can be known.[3]

With those caveats in mind, this introductory chapter discusses how the term *Einfühlung* came to express a significant problem in phenomenology and the philosophy of science in the pre-war years. The survey will cover: (A) Schleiermacher's and Dilthey's hermeneutics; and (B) the Munich style of phenomenology as pursued by Lipps, Pfänder, Conrad-Martius, Scheler, and Reinach. Subsequent chapters will discuss Husserl's appropriation of that tradition and Edith Stein's own account of the mutual availability of individuals, which can be characterized as a *hermeneutic* or theory of understanding. My application of the term "hermeneutic" to Stein's theory of empathy follows the usage of Schleiermacher, for whom intimate personal conversation was the touchstone of communication and the model for theorization of the reading of texts. Later in the twentieth century, Hans-Georg Gadamer would characterize human existence itself as essentially hermeneutic (that is, interpretive) in its stance toward the natural world, the cultural world, other people, and texts. These inclusive uses of the term, which I favor, contrast with the narrower but more common designation of theories and the scholarly disciplines that they govern: *hermeneutics* positions persons (readers and writers) in relation to texts; *communications* positions persons in relation to persons or to electronic message systems (which are pseudo-personal); and *studies of intertextuality* position texts in relation to other texts. Common to those contrasting juxtapositions is the pretense that the act of understanding constitutes *an identical meaning* in multiple discrete locations--be they textual, mental, or electronic.

A. The Nineteenth-Century German Hermeneutical Tradition

1. Friedrich Schleiermacher

Friedrich D. E. Schleiermacher (1768-1834) was a theologian, preacher, and philologist as well as a professor at Berlin. He called for a general hermeneutics as the art of understanding, which would explicate the common principles underlying the specialized interpretive work of biblical exegesis, literary criticism, legal interpretation, and so forth. He made no practical distinction between understanding texts and understanding unwritten communications such as

[3]Husserl acknowledged the impact of these conversations in his correspondence. See, for example, Makkreel and Scanlon (1987: ix-x). Three letters from Husserl to Dilthey and one response, all dating from 1911, are published in Husserl's correspondence (1994, 6: 43-53).

those involving speech, gesture, or symptoms; indeed, he tended to assimilate the spoken word to the written. Schleiermacher regarded understanding as the reversal of composition, and held that the reader re-experiences the mental processes of the writer.[4] For him the possibility of access to the mind of another was given in texts.

Schleiermacher's work *Psychologie* was published posthumously in 1862, but more influential were his lectures on general hermeneutics. These are known to us, as they were to Dilthey, in an outline composed in 1819 but glossated thereafter by Schleiermacher and published in 1838 with lecture transcripts added by a student.[5] The paradox of the so-called hermeneutical circle underlies Schleiermacher's theorization of understanding.[6] What must be understood about understanding, he says, is that there is no logical place for it to begin. One understands the whole of a work only by considering its parts, yet the parts have their meaning only by virtue of their participation in the whole. This paradox did not daunt Schleiermacher or drive him to skepticism, for as an heir of Leibniz he took it for granted that access to other minds is indeed given a priori within the monadic mind itself. In fact, we *do* understand; therefore understanding must be possible. Philosophers need only show *how* it happens, *how* a priori understanding is rendered possible.

Working with these presuppositions, Schleiermacher pointed out that any given text is understood as a part of two entirely different kinds of whole. On the one hand, the text fits into a universe of language whose grammar, vocabulary, genres, and other formal features find instantiation in the text. On the other hand, the text fits into the individual psychic life of its author, as one expression of the author's aims. These two "wholes"--language, and individual psychic life--are to be understood through the same textual expressions, but in two different ways.

One of those ways, "grammatical interpretation," follows the relations between the text and language, using *comparative* method. Language if taken as a whole would be unknowable, for it is merely a sea of infinite uncertainty; but in text, certain of its possible constructions are actualized as they determine what is said. Grammatical interpretation shuttles between the whole (linguistic possibility) and the part (the actual text) and so finds meaning.

The other way, "psychological interpretation," follows the relations between the text and its author's ongoing psychic life, using what Schleiermacher

---

[4]For introductions to Schleiermacher's hermeneutics, see Palmer (1969: 84-97); Duke (1977); Niebuhr (1967); Mueller-Vollmer (1985: 72); Ormiston and Schrift (1990: 11-14); Gadamer (1960); and Redeker's (1966) introduction to Dilthey's *Schleiermachers System als Philosophie*.

[5]Schleiermacher's former student L. Lücke edited and published *Hermeneutik und Kritik* in 1838. In 1959 Heinz Kimmerle published Schleiermacher's earlier manuscripts on hermeneutics. Recent discussions of Schleiermacher have been able to take this evidence of his development into account, but one must not assume that these manuscripts were available to Dilthey.

[6]Schleiermacher found the hermeneutic circle described in a work published in 1808 by Friedrich Ast, according to Ormiston and Schrift (1990: 11-12).

terms *divinatory* method. The human being if taken as a whole would be unknowable, for personal perspectives are infinite in range and there is no end to the influences near and far that shape a human being. But the authoring of a text brings certain of the psyche's "motivating principles" into focus. Psychological interpretation then can shuttle between the whole (the author's psychic life) and the part (this text) and so can find another dimension of meaning in addition to the meaning recovered by grammatical interpretation.

The work itself, of course, is also a whole comprised of parts by virtue of some unifying principle. Schleiermacher writes:

Aber die Einheit des Werkes, das Thema wird hier angesehen als dass den Schreiber bewegende Princip, und die Grundzüge der Composition als seine in jener Bewegung sich offenbarende eigentümliche Natur. ([1838] 1974: 103)	The unity of the work, the theme, will be viewed here as the writer's motivating principle, and the foundation of the composition as his peculiar nature as it is manifested in each motif. ([1819] 1990: Part 2:1)

What one understands is the *Bewegung* of the work: literally its movement or motivation, for the parts of the work have meaning only in light of its overall movement. But this amounts to understanding something of the author that moved the author in the production of this work. To accomplish this understanding, Schleiermacher says, one must put oneself in the place of the writer, even transform oneself into the writer. Hence understanding is the re-enaction of creation, but with this difference: along the hermeneutic circle, understanding runs counterclockwise to the futureward expressive direction in which creation runs, inasmuch as understanding works back from text to author rather than forward from author to text.[7]

Schleiermacher terms this intuitive displacement of the author *Divination*, and he characterizes it as a feminine force.[8] However it is always employed in

[7]For a contrasting account of the hermeneutic circle and the interplay between the comparative and the divinatory moments see Ellison (1990: 68-89). Although unfortunately Ellison has been misled by a tendentious translation of Schleiermacher's *Hermeneutik*, she astutely describes the suppression of the divinatory moment by Gadamer, Ricoeur, and Habermas.

[8]The *divinatorische* is also *prophetische*. ([1862] 1967: 146) Kimmerle notes that "divinatorische steht als Korrektur über profetische" in the manuscript at one of the two occurrences. *Divination* balances and is balanced by *komparative* or historical method. They are feminine and masculine, respectively. Ibid. 153. Gadamer (1960: 166-167) takes a different meaning here: that comparison concerns what is common and familiar, while divination concerns what is alien and individual. But this is to obscure the point that the interpreter, too, is an individual and so is intimately acquainted with individuality. Moreover, while Gadamer himself overlooks the gendering, his reading would associate the masculine with the familiar and the feminine with the alien--contrary to Schleiermacher's apparent intention to coordinate the two in one act of understanding. Richardson (1991: 186) concludes that Schleiermacher advocated and practiced "psychological androgyny" owing to his collaboration with women friends during his early professional life. On gender in Schleiermacher see also Richardson (1992) and Briggs

tandem with the masculine force: comparative method, whose task it is to place the text in the context of the linguistic universe. Comparison yields knowledge of the genre or type to which the work belongs; yet without divination, comparison would continue endlessly and could never reach the unity of the type. For its part, divination is saved from getting lost in fantasy thanks to the control afforded by comparison. In other words, the recognition that the author is "of a type" (or comparable) with oneself is what constrains the interpreter to avoid attributing impossible motivations to the author and the work.

Schleiermacher's schematic remarks on hermeneutics are far from clear, much less self-evident. He asserted that the interpreter's task was to understand a discourse better than its creator had.[9] Dilthey took him at his word, and developed a rigorous "life philosophy" out of Schleiermacher's provocative but cryptic "art of understanding."

2. Wilhelm Dilthey

Wilhelm Dilthey (1833-1911) studied philosophy at Berlin two decades after Schleiermacher's death, and returned there to teach in 1882. As a young man Dilthey had written a prize-winning essay on Schleiermacher, and in 1870 he published his biography. The ambition of Dilthey's long career, only partly accomplished, was to establish a common epistemological foundation for all the human sciences, or *Geisteswissenschaften.* These would be sciences of *understanding*, not sciences of *explanation* like the natural sciences. Understanding is historical; that is, its object is something with a temporal structure, something that flows along in a futureward direction. Where Schleiermacher had seen that both language and the individual psyche were continually in flux yet unbounded by beginnings and endings, Dilthey saw the streaming of life. Where Schleiermacher had described the *Divination* through which the text allows the reader access to the writer's psychic motivation, Dilthey described *Nacherleben*: the living out in understanding of some experience already lived by another.

Dilthey brought the term *Erlebnis* into the German academic vocabulary.[10] An *Erlebnis* is a unit of living experience whose content includes its very occurrence. The *Geisteswissenschaften* study subject matters that have to do with the living-through of occurrences to which we can have a kind of internal access that is never brought into play in the world of the natural sciences. Physical events are *explained* in terms of the causes that produce them; they are not, strictly

(1992).

[9]For Schleiermacher, the task of understanding is infinite since each of the two wholes into which the text fits--language and the author's psychic life--is unbounded ([1838] 1967: 146). The implication is that an author cannot have known the significance that the text might take on in the future.

[10]See the discussions of *Erlebnis* in Gadamer (1960: 55-63); Müller-Vollmer (1963: 35-38); Plantinga (1980: 31-35 and 86-88); and Palmer (1969: 107-111). For overviews of Dilthey's thought, in addition to these see Betanzos (1988); Rickman (1967 and 1979); and Bulhof (1980).

speaking, *understood.* The "why's" of nature are differently available for knowledge than the "why's" of human action.

Like Schleiermacher, Dilthey looked to textual materials for the means of making an approach to understanding (although at first he did not recognize the significance of his doing so). In his *Poetics* of 1887, he proposed that European history had unfolded in three waves, each with its characteristic literary genre. First, feudalism and its continual warfare spawn epic poetry, in which the *plot* is the predominant feature. Dilthey regarded the literature of the Latin, Catholic nations as stalled in that stage. Second, strong monarchies open the way for developments in science, commerce, and industry, with playwrights like Shakespeare portraying *character as vividly as plot.* This Dilthey took to be a Protestant, Germanic accomplishment.[11] Finally, the French Revolution inaugurates a new age of rational industrial organization and historical consciousness, in which the literary portrayal of *heroic character* will be perfected.[12] Dilthey now looked for the advent of a new kind of dramatist, modeled after Goethe, who would "speak to us about our sufferings, our joys, and our struggles with life." This poetic genius is to do for the inner life what Shakespeare did for the outer world.[13]

Dilthey's psychological theory arises from his poetics. In bygone eras when plot had supplied the structure for the literary work, Dilthey argues, poetic theory had needed only to make comparisons between the infinite causal chains of actual events and the finite connections created within the work.[14] With the advance from plot to heroic characterization, however, the structuring of the work became more complex while its relation to life became more difficult to account for. The inner life of someone like Goethe was not a causal nexus of events. Rather, Dilthey believed, a poet was a genius who somehow comprised within himself, with great clarity and intensity, those identifying conditions of his age that everyone else was living through as well. The genius would reach beyond the mere plot-like sequence of live experiences--*Erlebnisse*--to bring out their deeper coherences, which might be emotional, ethical, and cultural rather than temporal. The poetic genius would personify and typify the motifs of cultural life, lifting them up for others to understand and thereby enabling others to understand themselves.

But how? Dilthey's poetics needed a psychology to account for *Motivation*, which is what gives coherence to mental life and to literary works (just as causality imparts coherence to physical events). There must be a descriptive psychology to address just this dimension of access to the other-than-sequential,

---

[11]Dilthey regarded William Shakespeare and Charles Dickens as Germanic authors.

[12]See Dilthey ([1887] 1958: 151, 166-73). In the tradition of Aristotle, Dilthey centered his poetic theory on drama, the fullest artistic representation of human life. Plays, epics, myths, modern novels, and even dreams are considered.

[13]See Dilthey ([1887] 1958: 173 and 264).

[14]See Dilthey ([1887] 1958: 149, 153-154).

other-than-causal *motivated* coherences of life as found, on the one hand, in the very living of the poetic genius and, on the other hand, in the structured imagery of his literary work.[15]

What sort of science would descriptive psychology be? The natural sciences are adapted to grasp the coherence of physical events, a coherence that one may know only *from the outside*, that is, without participating in it.[16] An entirely different kind of science was needed in order to grasp the coherence of streaming human life. This sort of coherence displays itself as purposive and as generated from within life itself. It is therefore knowable only immediately, internally, through participation. Dilthey held, in his psychological works of the 1890's, that introspection was essential to the methods of all the *Geisteswissenschaften*, although each would pursue the spiritual/cultural (*geistig*) coherences distinctive to its own particular subject matter. In other words, *geisteswissenschaftliche* understanding comes from *Nacherleben*: to live-along-after the lived experience of others, getting to know it from the inside out.

After 1900 Dilthey no longer insisted upon introspection and descriptive psychology as the foundation of the *Geisteswissenschaften*. He turned from psychology to hermeneutics, and brought the literary expression--the text itself--to the center of his analyses. He did not address the intersubjective constitution of the socio-historical world itself (as Husserl would do), confining himself instead to literary expressions occurring within that world.[17] The elder Dilthey seemed to mistake the canon for the world. Nevertheless Husserl and Stein were able to make use of the earlier, "psychological" Dilthey, whose conception of life as coherence itself afforded phenomenology some of its most basic insights.[18]

In addition, Stein appropriated the so-called methodological individualism

---

[15]Dilthey assumed that the poetic genius would be a male.

[16]To say that physical reality is known externally is to exclude the possibility of understanding physical processes through empathic insight, such as the biologist Barbara McClintock claims to have achieved with plants. Plantinga (1980: 158) wonders whether Dilthey would also exclude mathematical relationships from among the *Zusammenhänge* that are understood rather than explained.

[17]See Plantinga (1980: 116). For a periodization of Dilthey's long career, see pages 46 and 55; and see page 104 for the opinion that Dilthey remained focused on text and writing in his later, hermeneutical period. By comparison, where Dilthey turned to text, under the influence of Husserl's criticism of "psychologism" Edith Stein would turn from introspective analysis of *Erlebnisse* to analysis of *bodily* expression. Dilthey and Stein also made comparable moves from the recognition of shared human impulses to a theorization of community. For an account of how Dilthey makes this move, see Plantinga (1980: 90-92).

[18]Dilthey ([1894] 1957: 144) wrote that ". . . life exists everywhere only as *Zusammenhang*." See Plantinga (1980: 131). Plantinga says (157) that "Dilthey stopped short of an explicit identification of life and being," and he faults Dilthey for not having provided "an ontology which explains how it is possible that this entire world be familiar and not alien" (159). For comparisons of Husserl and Dilthey, see the essays in the volume edited by Makkreel and Scanlon (1987).

of Dilthey's *Ideen* (1894). Dilthey regarded the human individual as the entry point for historical investigation, and biography as the basic historical science. Moreover, he held that some individuals may be considered *types*: personalities especially representative of their age.[19] In the mid-1930's Stein would craft her autobiography to reveal herself as type of the German Jew and the academic woman--a defensive effort in the face of National Socialism. Stein's post-phenomenological work in educational theory also would echo Dilthey's doctrine of types and of sexual differentiation.[20]

While Husserl was pursuing the insights of Diltheyan psychology as he developed his phenomenology in Göttingen, another band of phenomenologists was at work in Munich, where Theodor Lipps had founded a psychological institute.

B. The Munich Phenomenologists

1. Theodor Lipps

The wide-ranging research of Theodor Lipps drew numerous able students to him in the last years of the nineteenth century and the first decade of the twentieth. What came to be known as the Munich Circle can be traced back to 1895, when some of Lipps's students began to meet weekly as the *Akademischer Verein für Psychologie*. This group read Husserl's *Logische Untersuchungen* soon after their publication in 1900 and 1901. Husserl himself visited Lipps in Munich in 1904, and the next year several of Lipps's students went to Göttingen to study with Husserl. They included Adolf Reinach, who became *Privatdozent* at Göttingen in 1909, and they formed the nucleus of the *Philosophische Gesellschaft*, the Göttingen Circle.[21] Husserl's *Jahrbuch für Philosophie und phänomenologische Forschung* was co-edited by Munich colleagues. Between Göttingen and Munich there was a two-way exchange of people and of ideas.

Theodor Lipps (1851-1914) published works on logic, aesthetics, and psychology that were widely read and went through several editions both during his lifetime and posthumously. Lipps's interest lay in the epistemological dimensions of those diverse disciplines; that is, in the differing ways in which knowledge arises. He also edited the translation of David Hume's *Treatise of Human Nature*, introducing the empiricist epistemological critique into the land of Leibniz.

Lipps wrote that there are three fields of knowledge: things, myself, and other human individuals. Knowledge of the first two comes by sense perception and by inner reflective perception, respectively; while other people are known by empathy (*Einfühlung*.)[22] However empathy resembles the inner reflective

---

[19]See Plantinga (1980: 111-112).

[20]See, for example, Dilthey ([1894] 1957: 235-236).

[21]See Spiegelberg (1982: 166-170) for an account of the two phenomenological circles.

[22]See Lipps (1903a: 193).

perception that lets me know myself; often the two are indistinguishable. *Einfühlung* is able to deliver knowledge of others as human individuals precisely because of this. In other words, my inner perception of my own life has that same distinctive quality of saturating inner coincidence (*Einfühlung*) that I sometimes experience when I am captivated by the distinctive individuality of another human being.[23] *Einfühlen*--while it is happening--allows no differentiation between myself and another individual. But as it concludes I can tell whether its content has belonged to my me or to another's me. This subtle play of overlay and emergence among individuals, lovingly described by Lipps, was his legacy to phenomenology.

In English, the word "empathy" is not a good equivalent for *Einfühlung*, but the alternatives are awkward. For Lipps the term meant something like "lived inner coincidence." *Fühlung* is a "feeling" in the sense of a touch or a quasi-tactile contact, but not necessarily in the sense of an emotional arousal. *Einfühlung* connotes inward awareness, awareness-within: literally, "in-feeling."[24] The "in" of *ein* can indicate either position inside, or motion toward, or both. Thus *Einfühlung* can occur "within" oneself or "into" another. It is the way in which one inhabits or coincides with one's own subjectivity--ablatively, so to speak--but it is also the way in which one overtakes, saturates, and receives someone or something else--accusatively, as it were. One cannot tell which kind of inward awareness has been going on until the experience is brought to termination, as we shall see.

Husserl's explorations of subjectivity owe a debt to Lipps's formulations. However, Husserl would transgress Lipps's rule that there could be no further reduction beyond the three sources of knowledge mentioned above, which for Lipps were ultimates.[25] Lipps regarded the outer world of things and the inner world of the psyche alike as objective realms, requiring empirical scientific investigation. He regarded logic as a psychological discipline, since it is "the theory of the forms and laws of thinking," and since thinking is a psychic happening and knowledge occurs only in the mind. However Lipps distinguished carefully between logic and psychology. The latter investigates how knowledge and error alike are to be made intelligible, but it neither accounts for their difference (as logic does) nor claims that knowledge and error are the same.[26] Logic informs us that what we have is knowledge when the chains of inference delivering it exhibit the proper pattern. Logic tells us that we have error when the pattern is improper.

But it is the patterns, inference chains, and thought-forms themselves that

---

[23]Lipps (1907: 674, 693) defines *Gefühle* as immediately experienced i-qualities or i-competencies, while "feeling" (*das Empfundene*) is no such thing.

[24]On the difficulty of translating *Einfühlung* and related terms, see Mallgrave and Ikonomou (1994: 22, 25). The choice of English "i" to translate German "Ich" is discussed below in note 32.

[25]See Lipps (1907b: 722).

[26]See Lipps (1893: 1-2). Lipps insists that nobody would wish to make psychology disappear into logic, for the two are distinct disciplines.

interest Lipps. They are what the mind understands. To think is to follow connections. The title of Lipps's psychology textbook is a pun: *Leitfaden* can mean a primer or guidebook; but the term literally means "guiding thread." The intelligible connections whose threads Lipps's logic pursues are of two kinds. *Causal* connections are recognized in the world of real things, while *motivated* connections are lived through (*erlebt*) in conscious life. Causality and motivation thus are equally empirical, and both have the character of something understood rather than something substantive.[27] Objects of consciousness are brought to understanding not so much individually as serially. Lipps writes:

»Zusammenhang zwischen Gegenständen« aber besagt jederzeit, daß ein Gegenstand eine auf einen anderen bezügliche Forderung stellt. So geschieht alles Motiviertsein am Leitfaden objektiver Zusammenhänge d.h. eben am Leitfaden der Forderungszusammenhänge. Motivation ist der subjektivierte Forderungszusammenhang oder ist das Erlebnis dieses Zusammenhanges, so wie es eben im individuallen Bewußtsein und unter den Bedingungen desselben stattfinden kann und tatsächlich stattfindet. (1903a: 30)	But "connection between objects" always means that one object places a claim concerning another one. Thus all being-motivated occurs along the guidelines of objective connections, i.e., precisely with respect to lead-threads of the claim-connections. Motivation is the subjectivated claim-connection, or is the live experience of that connection, in just the way that it can take place and actually does take place within the individual consciousness and under the stipulations of the latter.

When it understands, consciousness is following empirically along the serial connections among objects, whether they be the *causal* connections of its outer-world objects or the *motivated* connections of conscious life objectively considered. But there is another sort of psychic coherence that grounds such rational following. An *Icherlebnis* (or live experience of one's own subjective deployment) constantly accompanies the changing objects of consciousness, Lipps writes. I keep finding me engaged in whatever mental activity I happen to be pursuing. Whatever I do, I feel myself (*fühle mich ein*).[28]

[A brief historical digression may be in order here. These formulations of Lipps's restate and develop a line of thought that he had first presented in his psychological studies of the 1880's and his *Logik* of 1893; however they also respond to Husserl's *Logische Untersuchungen* of 1901.[29] A close comparison of the mutual engagement between Lipps and Husserl on issues of subjectivity would

---

[27]See Lipps (1903a: 34, 28). Lipps writes on page 28: "Kausal Beziehungen gibt es nur für den Verstand. Sie sind von ihm erschlossen, und sie gehören der von uns unabhängigen Außenwelt an." ("There are causal connections only for the understanding. They are deduced by it and they pertain to the outer world independent from us.")

[28]See Lipps (1903a: 27-28; 1906: 1-3). In English, we cannot say "I empathize myself," approximating something like the middle voice of classical Greek.

[29]See Stern (1898) for a detailed discussion of both Lipps's work on *Einfühlung* before the turn of the century and its historical antecedents.

lie beyond the scope of the present discussion, but its main points can be briefly mentioned here. Schuhmann (1973) and Marbach (1974) report that Husserl marked up his own copies of Lipps's books, and Marbach also comments on excerpts from Lipps's works that are found among Husserl's papers on intersubjectivity. Both Marbach and Mensch detail the development of Husserl's conception of subjectivity between 1901 and 1913. Mensch (1981: 6) points out that where the *Logische Untersuchungen* had declared the i or subject "to be an element in real being, in the *Ideen* it is taken out of this category." Marbach concludes that Husserl's theory of attention and his approach to the determination of the unity of consciousness both were worked out in close confrontation with parallel developments in the positions of Lipps. For example, Lipps's accounts of the streaming unity of consciousness and of the individuation of the i had preceded Husserl's formulations in *Ideen*. Husserl's notion of an "i pole" or egoic ray, and his noetic-noematic analysis of attention, both are indebted to Lipps's imaginative metaphors. Marbach writes:[30]

> Bildlich gesprochen führt Lipps auch aus, das Bewusstseinserlebnis sei eine „Linie mit zwei Endpunkten", dem Inhalt und dem Ich, das er besser als „Anfangspunkt" bezeichnen will. „Dieses Ich ist ein einziger Punkt. . . . Bewusstseinserlebnisse sind Linien, die von einem einzigen *Punkt*, dem *Ich*, ausgehen und am anderen Ende einen Inhalt tragen."

> To put it figuratively, Lipps also worked out that the live experience of consciousness would be a "line with two endpoints": the content and the i, which he preferred to designate as a "jumping-off point." "This i is a unique point. . . . Live experiences of being conscious are lines that go out from a unique point, the i, and lead toward a content at the other end."

Lipps--possibly responding to Husserl's criticisms in the Logische Untersuchungen --anticipated Husserl's later proposals by calling for a science of consciousness to investigate i-experiences (*Icherlebnisse*). As such, this science would stand in contrast to the empirical sciences that investigate experiences of objects, and it could therefore serve as a "first philosophy" for their foundation. In this connection Lipps distinguished between mediated experiences, which one has of objects, and unmediated experiences, which one has in regard to the i. Not individually conscious egoic experiences, but "the pure i" or consciousness as such, would be the concern of a radically new psychology intended as "*Wissenschaft von den Icherlebnissen*," according to Lipps.[31] Lipps's theory of empathy (*Einfühlungstheorie*) was to be be the key to such a science.]

But what is *Einfühlung*, and what is an *Ich*?[32] Lipps insists that the i is

[30]See Marbach (1974: 243), citing page 3 of the first (1903) edition of Lipps's *Leitfaden*; compare Husserl's 1913 *Ideen*, §§76-95, especially §92.

[31] See Marbach (1974: 226-227).

[32]In German the first person singular pronoun *ich* is not capitalized when appearing as the pronominal subject of a sentence. But phenomenologists sometimes turn the pronoun into a noun by capitalizing it: *das Ich, die Iche*. (These neuter forms serve as

activity, not substance. The i is conscious life prior to numerical differentiation into individuals pursuing individual conscious lives.[33] Psychology's proper subject matter is consciousness, but not individual consciousness and not the i's that are the substrates to which individual conscious experiences refer themselves. Such substrates are real i's, transcendent i's.[34]

A real i is a human individual; yet his or her individuality is known only reflectively, not immediately. *Prior to* reflective individuation, the i is immediately grasped as a dimension of every human activity. Or rather than *grasped*, it is *lived* as a midpoint between two contrasting dimensions of inner coincidence (*Einfühlung*). On one side, Lipps says, I feel me active, busy, and free in the willful realization of the comings and goings of the objects of consciousness for me. Those objects belong to me and depend on me. On the other side, I feel me bound, constrained, obliged, unfree, or passive in my presentation of objects of consciousness. Those objects are independent from me. On the one side is consciousness of subjectivity, on the other is consciousness of objectivity. However in this phase their hinge, the i, does not yet belong to anybody in particular.[35]

This early (1893) account of a "busy i" makes scant distinction between *living* the i and *feeling* the i--that is, between *Icherlebnis* and *Einfühlung*--for here the i is lived simply as hinge between creative and determined *Einfühlung*. (Comparable to this hinge-enactment are the live, immediately lived connectings between antecedents and consequents, or between independent and dependent variables--"causes" and "effects" so called. All are i-livings, *Icherlebnisse*.) Subsequently, in the second volume of his *Ästhetik*, Lipps would restrict the term *Einfühlung* to instances in which the activity of the i contributes something original: something optional and dependent upon the i such as a mood, a desire, a yearning, a choice. (This restricted sense accords more closely with the denotation of "empathy" in English.) The i is felt precisely as the source of such colorings. By contrast, in Lipps's later terminology the i is not said to be "felt" but rather is "lived" in instances of judging, meaning, and believing. These are instances of consciousness of truth and validity. The i is not flavoring them, for it is doing nothing original or distinctive. It is rather affirming a reality as independent of itself; indeed, the i experiences itself *as just this assenting understanding*. Lipps insists that truth is not an affair of feeling but of

---

both subjects and objects in sentences.) In English we do the opposite. We do not capitalize nouns as the Germans do, but we do capitalize the first-person nominative singular pronoun, I. Since the equivalent of *ich* is I, *das Ich* will be rendered here as "the i," *die Iche* as "the i's," and so forth. It is hoped that this inversion of the capitalization rules will serve to capture the distinctions that are important for the German text. For adjectival forms (*ichlich* and so forth), the conventional term "egoic" will be retained for want of an othographically viable alternative.

[33]See Lipps (1903a: 27-28, 33).

[34]See Lipps (1903a: 33; 1893: 5, 10).

[35]This is a paraphrase of #12 in Lipps (1893: 5).

understanding. Although in both cases--assenting *Icherlebnis* and constitutive *Einfühlung*--the i is busy, still these are two contrasting kinds of busy-ness.[36]

But Lipps himself never nailed this contrast down tight. The distinction between i-enactment (*Icherlebnis*) and empathy (*Einfühlung*) emerged in his work as part of the gradual elaboration of the concept of inward awareness itself, in which a certain equivocation was never resolved. The term plays along a spectrum of meanings. *In one direction*, empathy denotes an optional technique that can be taken up and practiced at will when an occasion arises, that is, whenever one confronts an aesthetic object and undertakes to understand it. Inward awareness can also occur as a spontaneous response to such an object; indeed, even nonaesthetic artifacts and things of nature can provoke this projective ensoulment (*Beseelung*). Natural sciences humanize (*vermenschlichen*) things by imputing connectedness to them, that is, by projecting causality into physical events. Thus science animates the outer world.[37] This *accusatively* projective and/or receptive awareness is properly termed "empathy."

In the other direction, inward awareness (*Einfühlung*) is the spontaneous, nonoptional, virtually continuous business of the i that grounds the reflective process that will yield individual i's. Only by reflection does one move *from* the i dissolved within the subject-object relation (i.e., dissolved within what will subsequently be recognized to have been such), *to* the individual self-aware i whose conscious object is his or her own individual and real i. Lipps says that the individual i is not immediately given but is rather thought-toward (*hinzugedacht*).[38] The individual i comes into consciousness *at and as* the termination of inner involvement by the nonindividuated i with the appearings of objects to consciousness. An individual substantive i emerges as *something*, a unity, when it is constituted as an object for consciousness, an object to which the series of its various live experiences (*Erlebnisse*) now is attributed. Self awareness concludes the inner contact (*Einfühlung*) in which the i was engaged with objects: i becomes

---

[36]See Lipps (1906: 2-3). Husserl will put this distinction to good use. The i busy with assent to truth or validity becomes for Husserl the "transcendental i." However Husserl's transcendental i is much busier and more creative than anything Lipps conceived. In the recognition of truth by that i, Husserl will find a conferral or *constitution* of sense for its objects; while in the recognition of validity by that i, Husserl will find a grounding origination of logical form. Husserl's transcendental i will have no passively receptive, unfree side, for it will be entirely self-determining in its rationality. Many of his contemporaries read Husserl's 1913 *Ideen* (along with the "Nachwort" to the second edition of the *Logische Untersuchungen*) as having parted ways with realism here, and perhaps also with reality.

[37]See Lipps (1903a: 27-28). Because projective *Einfühlung* tends to humanize the natural world, we are tempted to understand physical causality as if it were deliberate, motivated, goal-directed movement. Only when that temptation is successfully resisted do we achieve scientific understanding of causes. However the unidirectional connections that differentiate causes from effects lie within i as its own enactments (*Icherlebnisse*). Causal connections are live; they are not static items strewn about in the material world.

[38]See Lipps (1893: 10).

I, and I recall that i was so engaged. This *ablatively* deployed awareness, which the kindling of self-consciousness extinguishes, has no good name in English; so we'll call it ablative empathy.

Awareness of others emerges at/as the termination of ablative empathy whenever the i engaged in the concluded subjective activity turns out *not* to have been I. Lipps calls on the circus acrobat to illustrate his meaning. Watching the acrobat sail through the air, one is pulled into the experience of flying and forgets that one is still seated below. One goes along with the acrobat, sharing the kinesthetic consciousness of movement. Oddly, one does not see what the acrobat sees; rather, the optical impression of the performer "up there" overcomes the kinesthetic impression of resting "down here" and displaces it with the experience of sailing "up there." This goes on for as long as one passively rides with the acrobat. It terminates when one moves a hand or a foot and so is recalled to the awareness that one is numerically distinct from the performer after all. I realize that I have not been the one flying. Nevertheless *there was i* in the flying--for the flying was a subjective *Erlebnis*--and moreover *there is a real i* whose willing directed the flying and for whom the flying becomes something that "I did."[39]

For Lipps's aesthetics and perhaps also for his logic, the movement of the acrobat is the exemplar of all form. Form is what we can follow. We understand logical arguments, artworks, and our fellow human beings alike by moving along with them, quasi-kinesthetically. Thus Lipps's conceptualization of understanding is quite in keeping with Dilthey's insistence that the cultural world is structured through motivations that can be understood. But Lipps adds an account of how we are enabled to enter vicariously into experience. We do so *by instinct*, he says. Emotion-laden sounds excite similar emotions in us. Expressive gestures induce us to go along with the sentiments that are in them. Outwardly, imitation can be only imperfect because two bodies cannot materially coincide.[40] Inwardly, however, one can go along *completely* with the living (*Erlebnis*) of another. This happens involuntarily, thanks to two coordinated and inwardly felt instincts: an instinctual expression of inner conditions (*der Trieb der Äußerung*) and an instinctual internalization of expressed conditions (*der Nachahmungstrieb*).[41] These are presuppositions for certain live experiences that we recognize, at their conclusion, as having been shared with another individual. Lipps argues that such experiences cannot be accounted for adequately as having been brought on either by analogical inference or by habitual association.[42] Rather, an expression is a *symbol* for an i; in its expressions an i is *symbolized*.[43]

This instinctive imitation is an impulse or impetus (*Anstoß*) that drives me,

[39]See Lipps (1903b: 114-115). See also the discussion of instinctive imitation of yawns, Lipps (1907b: 716-17).

[40]See Lipps (1903b: 124).

[41]See Lipps (1907b: 713-721).

[42]See Lipps (1907b: 697-712).

[43]See Lipps (1907b: 721-2).

rather than an aspiration or endeavor (*Streben*).[44] Yet, it would seem, one can and does sometimes *deliberately initiate* empathy by approaching an artwork, by opening one's eyes for the optical impression that leads into kinesthetic oneness, or by tracking a syllogistic argument. This initiation amounts to an overlay "upon" or a penetration "into" the motivation "in which" subjectivity already has been operative.[45] Yet this initiation is merely a chronological beginning. The logically prior state, says Lipps, is the living (*Erlebnis*) of movement or motivation empathized (*eingefühlt*) by the i. Its conclusion yields objective knowledge of another human being.[46] That of the other that is known by me is twofold: the *motivation* inhering in the logical, bodily, and/or aesthetic moves that the other has made, as well as the *human being* of the other as an i that is not I. These are correlated, and both are known as real.

In short, *there are* other human beings. Knowledge of their existence is given to me simultaneously with knowledge of my own existence *as an individual*. Lipps says that the i that I originally know prior to knowing other i's is not "my" i nor is it even "an" i or "this" i. It cannot become "mine" unless and until there are others.[47] What I first experience is ablative i-at-work: busy i. My own individual substantive off-duty i is secondary; but *equally* secondary, so to speak, are the other i's. At the conclusion of any activity, reflection returns me to myself as the one who has been busy at that activity, deployed therein, feeling it from the inside out. With certain activities, this deployment has the character of a double tending: tending toward expression or tending toward imitation. Empathizing i *is* that tending. Subsequently the reflective act that concludes the ablative phase of such activities does so by catching subjectivity red-handed as i who *was* that

---

[44]See Lipps (1903b: 118). However as we shall see, Lipps's student Alexander Pfänder will use the term *Streben* more precisely to describe the involuntary--that is, *pre*voluntary--conditioning of a freely consented-to aspiration by a prior, spontaneously arising urge. Therefore when discussing Pfänder's work and Edith Stein's appropriation of it, we will have to translate *Streben* in that technical sense as "inclining" and *Strebung* as "inclination."

[45]A single content now is doubly *eingefühlt*: accusatively as well as ablatively, so to speak. Accusatively: empathy as feeling-into is the voluntary projection of oneself "into another's place" through the entry way offered in artwork, in logical argument, in expressive bodily gestures, or in ordinary conversation. But it presupposes the recognition that *there is* someone else. That prior recognition of the existence of others is given as the termination of ablative empathy. Lipps places the co-emergence of knowledge of oneself and knowledge of another individual at the conclusion of that lived inner coincidence. The primary feeling-within must be prior to any secondary projective feeling-into, logically as well as chronologically or biographically. The original emergence of other i's for me out of (the termination of) empathized experiences is the condition for any subsequent deliberate attempt to experience what someone else experiences, as well as for any spontaneous empathetic responses in which I feel pulled into the joy or pain of another, perhaps so deeply that I temporarily lose myself in it.

[46]See (Lipps 1903b: 125-6).

[47]See Lipps (1907b: 694; compare 1893: 34-37).

tending. Insofar as the i has been a tending *to imitation*, the i is my own: I.[48] But insofar as the i has been a tending *to expression*, that i belongs to someone else, and therefore I have knowledge of another i. That is, I have reflective, non-immediate knowledge that I have just concluded a live experience (*Erlebnis*) in which another i was given immediately. Thus Lipps holds that i as immediately experienced (*erlebt*) is neither mine nor thine, whereas the i as thoughtfully *known* must be either mine or thine or someone else's.[49]

Before examining two critical receptions of Lipps's doctrine of empathy, let's briefly consider his overall approach. Munich, as is often remarked, was a city of music and the arts, whereas Göttingen was imbued with mathematics and the sciences. Lipps took for granted that human being was the supreme value and the ultimate criterion against which other values were to be measured.[50] He wrote:

In Wahrheit ist der Mensch dem Menschen das Schönste, oder kann es sein, weil er eben Mensch ist. Der Mensch, so müssen wir sagen, ist nicht schön wegen seiner Formen, sondern die Formen sind schön, weil sie Formen des Menschen, und demnach für uns Träger menschlichen Lebens sind. (1903b: 105)	For the human being, truly the most beautiful thing there is or can be is the human being, precisely because he or she is human. Thus we have to say that the human being is not beautiful on account of his or her form; rather, the forms are beautiful because they are forms of the human being, and thus they are for us a carrier of human life.

The human was admired, but it was approached through nonhuman artifacts: sculpture, painting, orchestral music. Lipps grounded his doctrine of empathy in the bodily expressions of the emotion-laden voice and the nimble acrobat. Moreover, the instinctual genesis that he proposed for empathy was bodily based. Yet Lipps left the body behind in his applications of this theory in the fields of logic, epistemology, psychology, and aesthetics generally. He worked on texts, without taking it sufficiently into account that those textual representations were not human bodies themselves. Logical texts "move" only metaphorically. Pictures and statues arrest movement. An inference is not a move; formally considered, it has no before and after, and it changes nothing. There is a vast difference between

---

[48]See Lipps (1907b: 715-717). The distinction is easily ilustrated in the experience of losing oneself while reading. The reader follows the turns of an argument or of a dramatic plot vicariously, but may not recognize what had been happening until the reading concludes.

[49]In Lipps's terms, we do not ever get *knowledge* of deployed, busy, non-individuated i; we can only *experience* it, that is, have inner coincidence with the activity in which i abides deployed as executive subjectivity.

[50]The following quotation indicates the sense in which human being itself is the transcendent*al*, the necessary source from which all other meaning arises. The inclusive pronouns, "he and she," are justified in the translation because they reflect Lipps's own gender-inclusive intention. He deals at length with questions arising from the fact that the two basic human forms, the male and the female, cannot imitate one another's motions completely. See Lipps (1903b: 147-151).

the kind of *living* being from which Lipps derived his doctrine of empathy, and the *textual* being of the artifacts to which he and others sought to apply it.[51]

Yet Lipps's doctrine of empathy makes an important transposition of logical terms back into bodily ones. The contrast between necessity and contingency is reconfigured as the opposition between the instinctive and the voluntary (or, alternately, the immediate and the inferred). In the tradition of Leibniz, Lipps held that certain knowledge of whatever could be known of the universe would have to be internal to one's own "monad," since its necessity could be derived from one's own existence. All reality, including the reality of other human beings, was available (for knowing) internally in the monad. But how? With Lipps's transposition, it is *instinct* that initiates empathy, the inner coincidence that renders human individuals available to one another. The i that is prior to the termination of ablative empathy, and therefore prior to individuation, is a monad *of sorts*: a monad incarnate but skinless, one monad not yet faced off against "others." In principle, then, all individual human beings "start with" equal access to their common humanity, and through it to all other individuals who partake of that humanity. The human race is Lipps's monad.

Lipps died in 1914, but his doctrine of the priority of ablative empathy stood in the way of National Socialist ideology. His empathy theory is singled out by name in a 1943 philosophical dictionary as contradictory to Nazi racial doctrine.[52] Alfred Rosenberg, the ideologue of the Third Reich, raises two specific objections to Lipps's account of empathy. First, Lipps does not adequately distinguish between natural objects, on the one hand, and people and artworks, on the other. According to Lipps's account, one could undertake to project oneself into a mountain as well as into an artwork or another human being.[53] (This objection overlooks Lipps's insistence that pre-individuation experience is primary, so that the subsequent deliberate excursions of awareness toward an artwork or another person occur only on that basis.)

More seriously, Rosenberg objects that only the sharing of folk-racial characteristics can be the basis for a bridge between the known object and the knowing subject, and therefore for understanding among individuals. He asserts, quite correctly, that Lipps's doctrine of empathy undermines that key tenet of National Socialist ideology. According to Rosenberg, what is brought from inside an artist and expressed outwardly, especially in music, can be internalized by the "art-consumer" only because both producer and consumer already share a Nordic soul. That power, awakened by art, persists in the consumer even when the artwork no longer is present.[54]

---

[51]Lipps's theory could also be applied to realtime media of the later twentieth century: moving pictures, sound recording, interactive video.

[52]See Schingnitz and Schondorff (1943: 119-120).

[53]See Rosenberg (1930: 416). The argument is riddled with snide puns.

[54]See Rosenberg (1930: 417-8).

2. Alexander Pfänder

Rosenberg's perverse alternative to Lipps's doctrine of *Einfühlung* still lay far in the future during the productive years of the Munich Circle under the leadership of Lipps's most prominent disciple, Alexander Pfänder.[55] Pfänder (1870-1941) taught at the University of Munich from 1900 to 1935. Through his publications, his lectures, and his conversations with colleagues, Pfänder profoundly influenced the Phenomenological Movement. Indeed, he had titled his *Habilitationschrift* a "phenomenology" before Husserl adopted that term to characterize his own investigations in logic.[56] Pfänder later collaborated with Husserl on the editorial board of the *Jahrbuch für Philosophie und phänomenologische Forschung*, and published his *Logik* there in 1921. Ill health during retirement prevented Pfänder from consolidating a lifetime of lecture notes into a comprehensive and systematic write-up of his philosophy, as he had planned to do.[57] But the thought that was "in the air" in the 1910's, the thought that engaged and influenced both Husserl and Stein, was Pfänder's work on motivation. This component of Pfänder's philosophy was set out in his 1899 Habilitationschrift, *Phänomenologie des Wollens*, and especially in his 1911 essay "Motive und Motivation," a refinement of his argument written for the Lipps *Festschrift*.[58]

There are indications that Husserl had hoped to see Pfänder succeed him at Freiburg.[59] But during and after the war, as Husserl pursued his investigation of the transcendental constitution of the world for subjectivity, Pfänder sided instead with the realism of the Munich Circle and with the program that would come to be known as *Realontologie*. Pfänder's estrangement from Husserl's project was a consequence of Pfänder's critical advance beyond certain formulations of his own mentor, Theodor Lipps, concerning the structures of *Einfühlung*--formulations that would persist as tacit presuppositions within Husserl's phenomenology.

Munich realism is not a naively dogmatic claim that entities exist independently of human knowledge of them. It is rather a commitment meticulously to describe how entities are *meant as* real within conscious rational life. In this it resembles the better known phenomenology of Husserl, to which it is often contrasted. Lipps, as we have seen, described intersubjective

---

[55]Herbert Spiegelberg, Pfänder's interpreter to English readers, overstates the import of his teacher's advances when he insists that Pfänder was never the disciple of Lipps because he criticized Lipps's "psychologism." See Spiegelberg (1967: xvi).

[56]See Spiegelberg (1967: xvi). Husserl's copy of Pfänder's *Phänomenologie des Wollens* is extensively marked up. See Schuhmann (1973: 29-35) and Marbach (1974).

[57]Pfänder's *Nachlaß* is preserved at the Bavarian State Library, and one volume of his papers has been published posthumously. See the bibliography in Spiegelberg and Avé-Lallemant (1982: 359-370).

[58]As Melle (1988: xxx) points out, Husserl used Pfänder's essay as the basis for expanding his own 1914 lectures on fundamental questions in ethics and value theory. See Husserl's "phenomenology of will," Hua XXVIII: 102-125.

[59]This was Husserl's intention before Heidegger came on the scene, according to Spiegelberg (1967: xix).

understanding in terms of *Einfühlung*: the feeling of coincidence with another's lived coherences as they have been posturally, gesturally, verbally, textually, or artistically expressed. According to Lipps, such instinctive following is made possible by the logical unity of the executive i, prior to its individuation into "oneself" and "the others" when empathy ceases. The twin experiences of coming to oneself and recognizing the others, which mark the conclusion of ablative empathy and the transition from i, to i's and I, may also be characterized as a transition from *sameness* to *similarity*. The singular i becomes a plurality of i's; and *I* now am enabled to regard others as similar to myself in their capacity to live as i's that are "like me" but not numerically identical with me.[60]

On Lipps's view, then, it takes an act of projection in order to reverse that transition so as to recover the pre-individuated coincidence of subjectivity and thereby to attain empathic (*eingefühlt*) understanding of the living experiences of others. When I understand, I do so by feeling-into the other a similarity with myself. I recognize his or her expressions as similar to those that I myself have produced, and so I empathicly project into those expressions a meaning similar to the meaning that I myself have meant to express. This recognized similarity, Lipps would say, is the basis for a twofold *association* that yields the understood meaning: the other's expressions resemble my expressions, and therefore the other's feelings resemble the feelings that I expressed in those similar expressions. But all the while, similarity for Lipps is something projected from the knower.[61] Similarity belongs to the subjective side of the subject-object relation.[62]

In practice, Husserl tacitly accepts this Lippsian account of similarity while Pfänder rejects it. Husserl's phenomenology relies upon the predication of similarity at several important points, which can be briefly enumerated here. First for Husserl, the subjectivity of others is assumed to be similar to one's own, although numerically distinct. Second, in "ideation" one imaginatively reviews possible versions of an object that are similar to its current appearance in order to discern what is essential, that is, common to all possible instances of this kind of object. (See *Ideen* §3 and §4). Third, the *eidetic reduction* occurs as a reorientation of one's regard away from an individual material thing and toward the essence that the thing would share with others similar to itself. Fourth, objects transcendent to consciousness give themselves in a series of adumbrated appearings whose similarity and connection is recognized in perception. Fifth, noetic-noematic analysis remarks that the worldly object achieves selfsameness as a meant unity

---

[60]This enablement turns out to involve acts of association that will be called into question by Edith Stein, as we shall see.

[61]The significant role that the notion of similarity plays in the thought of Lipps and the Munich school is pointed out by Smid (1983). My remarks here are based on his investigations.

[62]Even though he theoretically rejects inference by analogy as a means of access to an alien i, characterizing it as "high sophism" (Hua XIII: 36-38), Husserl does not see the extent to which he begs the question of similarity in the diverse components of his own phenomenology mentioned below.

through an act of subjective meaning basing itself upon multiple similar appearances.[63] In all these instances, similarity is presupposed as something already constituted for subjectivity. Similarity is not simply perceived, but during perception is *projected* (or *hineingefühlt*, as Lipps might say). Or in Husserl's terms, similarity is constitutively determined by subjectivity rather than transcendent to it. This originary power of subjectivity is precisely what Husserl will pursue in the *phenomenological reduction*, in which the constitution of similarities is investigated.

Pfänder must refuse Husserl's phenomenological reduction, because unlike Husserl Pfänder rejects Lipps's axiom that like things are associated on the basis of an inwardly lived experience (*Erlebnis*) that originates in subjectivity. Pfänder instead regards similarity as an objective knowledge. The association of two like things happens in consciousness, says Pfänder, but it is determined by the conditions under which the objects appear. Smid (1983: 611) comments:

> . . . die Frage auftritt, was die Ähnlichkeit von Gegenständen ausmache. Ähnlichkeit ist in den Gegenständen fundiert und nicht mit dem rein subjektiven Bewußtsein der Ähnlichkeit zu verwechseln, was zum „Psychologismus" führen würde. Diese Argumentation richtet sich ebenso wie Pfänders Ablehnung eines Gefühls der Ähnlichkeit gegen Lipps.

> . . . the question arises, what constitutes the similarity of objects. Similarity is founded in the objects and not to be confused with the pure subjective consciousness of similarity, which would lead to "psychologism." This line of argument, just like Pfänder's refusal of a feeling of similarity, is leveled against Lipps.

For Pfänder, the likeness of two things is a result that registers in consciousness through a spontaneous act of association. But in order to find out how consciousness of similarity comes about, one can run that association backwards, as it were, in order to come upon the objective qualities that have conditioned it. These qualities originate elsewhere than in consciousness. Pfänder wrote already in 1899 that his phenomenology would be more retrospective than introspective.

> Diese sogen. „subjektive" Methode besteht nicht notwendig in einer direkten Beobachtung des unmittelbar Erlebten; denn eine solche ist meistens, wie gerade beim Wollen, unmöglich. Die „subjektive" Methode geht vielmehr notgedrungen meistens von der Festhaltung der unmittelbaren oder weiter zurüchliegenden Erinnerungsbilder aus. Also is diese „introspektive" oder Methode der „inneren Beobachtung", wie man sie irrtümlicherweise genannt hat, in Wahrheit keine „introspektive", sondern grösstenteils eine „retrospektive". (1899: 6)

> This so-called "subjective" method does not necessarily consist in a direct observation of what is immediately experienced; for this is mostly, and especially in the case of willing, impossible. Rather the "subjective" method starts by necessity from the fixation of the immediate or more remote images in memory. Hence, this "introspective" method or method of "inner observation," as it has been erroneously named, is actually not "introspective" but is mostly "retrospective." (1967: 8)

---

[63]Further examples could be given. Husserl will be discussed at greater length in the next chapter.

By contrast, with Husserl the backwards or regressive move, the phenomenological reduction, brings him to the transcendental i as the constituting source of the law-governed appearing of essences (the *eidē*). Similarity itself is not an *eidos*, but belongs rather among the regularities exhibited by the appearings of phenomena.[64]

Pfänder's principal contribution to the realist argument, however, is not his association-theory but his phenomenology of will and its guiding maxim: *That which is willed is always thereby meant as something realizable.* It is a phenomenological commonplace that consciousness is intentional; that is, consciousness always is "of" something, never indeterminate, even though the content of consciousness may or may not be something real. In this vein, Pfänder insists that willing, too, is always "of something," but of something *insofar as it can be brought into existence by one's own action.* To will, therefore, is to intend a future state of a real world.[65]

In Pfänder's account, to will (*Wollen*) is one kind of experience belonging to a larger class: experiences of inclining (*Streben*).[66] That class includes other varieties of being inclined: to wish, to fear, to love, to hate, to hope. All of these necessarily have a target (*Erstrebtes*), something meant as the target of the inclining. All register spontaneously in the i as unchosen pressures toward or away from that something. And all entail valuations; for example, somatic feelings of delight or distress attending the attraction or repulsion, or rational assignment of relative worth. However, willing is the only kind of inclining, in Pfänder's view, that engages the very core of the i. Willing is the subjective consent to go along with an impulse that has been felt as originating in the body of the i or perhaps altogether outside the i. Willing is to decide: yes, I choose to do something about the inclining--the desire or aversion--that currently registers in my consciousness. Thus willing essentially includes an act of meaning some state of affairs *as realizable.* The i can invest itself in an inclination already underway (and so transmute that *Strebung* into a *Wollen*) only when the corresponding targeted goal (*das Erstrebte*) *is meant as real.* This consenting act of the i thus transforms the

---

[64]*Eidē* themselves can be "similar" to one another only in being *eidē*--which is no similarity at all because *eidos* is not an *eidos.* In other words, there is no form "form" to which all forms belong by virtue of their similarity one to another.

[65]In his 1914 lectures on ethics, Husserl would discuss the reality intention as a formal-hypothetical imperative. See Hua XXVIII: 109.

[66]This is an odd and idiosyncratic use of the term. Pfänder selects the word *Streben* as the name for a general category comprising volitional and nonvolitional affective experiences. Herbert Spiegelberg, his translator, simply uses the English cognate: "striving." But "striving" cannot support the nuances in Pfänder's description of this phenomenon, which is an activity of registering attraction (or, revulsion) *prior to* the acts of decision and pursuit (or, flight). The lexical meanings "to tend" and "to gravitate" come closer to the special phenomenological sense of *streben.* I have chosen to render *Streben* as "inclining" and *Strebung* as "inclination," finding them better approximations of Pfänder's meaning. Edith Stein discusses *Streben* in the Pfänderian sense of "being inclined" throughout the portion of her *Habilitationsschrift* that deals with psychic causality.

intended goal into a *Motiv*.[67]

A motive therefore does not exert causal influence over a subject. Pfänder supports this assertion with two lines of argument. First, the subjective willing (*Wollen*) is not determined by an inclining (*Streben*); rather, it is determined by the i *on the basis of* an inclining. The inclining itself receives a determination from the i, precisely by being transformed into a willing. Second, a cause must precede its effect. But the motive arises subsequently to the willing and through the determination enacted in the latter. Thus the i determines itself as a willing on the basis of living inclinedly, and in the same stroke it makes a motive out of what had been a mere inclination. The transition from inclining to willing does something to the i; or rather, the i does something to itself by enacting the transition. The i actualizes itself in choice; it gives itself a determination. But this is not at all comparable to the transition that Lipps described as occurring at the conclusion of inward ablative empathy: the transition from i to i's (one of whom is I). One difference is that after individuation, I repeatedly reverse that transition so as to dissolve myself back into nonindividuated awareness. After decision, by contrast, I cannot un-determine myself again by dissolving my willing and reverting to merely being inclined. Unlike the conclusion of ablative empathy, the transition from inclining to willing is not reversible.[68] *Motivation* is not something that *happens to* a subject; motivation is rather the subject's initiative transforming the in-felt inclination and mobilizing itself for creative engagement with a world meant as real.

Pfänder framed these arguments to meet certain challenges that he had found in the work of Arthur Schopenhauer, whose philosophical agenda was inherited from Kant and Fichte--rather than from the hermeneutical tradition of Schleiermacher and Dilthey through which we have been unfolding the theme of *Motivation*. The Kantian way of deducing the reality of the world favored the route of action over that of perception. The universal experience of moral obligation, Kant reasoned, bespeaks the existence of a realm in which the required actions may be carried out. In short, what I am *obliged* to do must be *possible* for me to do, so there must be space and time enough for action. Fichte regarded the world both as the enabler of action and as the realm of hindrances to action. Human projects succeed through continual collisions with material worldly impediments to their success. These stumbling blocks, or *Anstoßen*, empower subjectivity as they press against it, but they do so from outside of subjectivity itself.

---

[67]This summarizes the argument of Pfänder (1899); see also Spiegelberg (1967). Paul Ricoeur, who drew upon Pfänder (1899 and 1911) in writing his own *Philosophie de la volonté* (1949), remarks that the correlation of inclining with its target (*Streben* with *Erstrebtes*) in Pfänder's work anticipates Husserl's correlation of noesis and noema. See Ricoeur 1975: 121; but see also Ricoeur (1952), which takes Husserl's work as its point of departure and neglects to mention Pfänder.

[68]This argument overlooks the possibility of repentance. While repentance cannot undo the factuality of past decisions, it alters the i's ongoing engagement with them. The repentant i purges itself of those decisions *now*.

Schopenhauer imported the pressures of reality into subjectivity but made them "blind"--that is, unmotivated, unreasonable, unknowable, goalless, and doomed to frustration. As blind striving, will (*Wille*) in its uncomplicated unity *was* reality for Schopenhauer, and as such it was the antithesis of reason. By contrast, motivation lay on the side of reason, as one of the four varieties of reasonableness governing the reality-deficient realm of presentations (*Vorstellungen*); while will was the ungoverned, undetermined reality, objectified in one special presentation, the human body.[69] In other words, the body was where reality and its antithesis--that is, *will* and *reason*--conducted their hopeless struggle.

Schopenhauer's arbitrary alienation of the will from reason is called into question by Pfänder, and the alleged "blindness" of the will is placed under suspension. According to Pfänder:

> Mag nun das menschliche Wollen die Grundfunktion des menschichen psychischen Lebens sein oder nicht; mögen alle psychischen Vorgänge nichts weiter sein als Äusserungsweisen des Willens, mag also alles psychisches Geschehen im Grunde ein Wollen sein oder nicht; jedenfalls müssen alle derartigen allgemeinen Behauptungen zunächst suspendiert werden. Denn eine Untersuchung des menschlichen Wollens muss immer von dem sprachgebräuchlichen Sinne des Wortes „Wollen" ausgehen. Und dieser schliesst die Annahme in sich, dass nicht jedes beliebige psychische Geschehen ein Wollen ist, dass vielmehr nur bestimmt geartete psychische Thatbestände den Namen eines „Wollen" verdienen. (1899: 4)

> Whether or not human willing is the basic function of human psychic life, whether or not all psychic processes are nothing but modes of expressing the will, whether or not all psychic events are fundamentally cases of willing--all such general assertions must at first be suspended. For an investigation of human willing must always start from the ordinary usage of the verbal noun "willing." And this usage includes the assumption not that every conceivable psychic event is a willing, but rather that only specifically qualified psychic facts deserve the name of "willing." (1967: 6-7)

Pfänder then shows that willing is but one kind of inclining: specifically, an inclining transformed into a rational motive through the designation of its object (*Erstrebte*) as realizable and to-be-realized. So like Schopenhauer's will, Pfänder's willing is an index of reality. But Pfänder's conception departs from Schopenhauer's in two significant ways. First, not every "blind" impulse becomes a willing, but only those impulses that can "see their way" to a realizable goal. Second, besides being realizable the goal must in fact be embraced. The subject-sided self-investment of i in *Wollen* imparts rational "vision" to it. Thus human freedom for Pfänder is rational and goal-directed, not blindly driven as for Schopenhauer.

---

[69]Schopenhauer had in mind the male body in a permanent state of frustrated sexual arousal. In relation to the male-gendered will, reason was feminine, which for Schopenhauer meant controlling and limiting the freedom of the male will. This metaphor is apparent throughout his major work, *Welt als Wille und Vorstellung*.

In his 1911 essay, Pfänder elaborates this conception by distinguishing between "blind" and rational moments and between causal and grounding factors within the phenomenon of motivation. He positions the i as a center, surrounded by a subjectively indwelt human body (*Ich-Leib*) and receptive to inclinations arising either within itself (*zentral*) or from outside (*exzentrisch*). From this core a "centrifugally" directed act of consciousness first establishes an object, its target. This conception may be schematized as in Figure 1.1

FIGURE 1.1: PFÄNDER'S MODEL OF INCLINATION

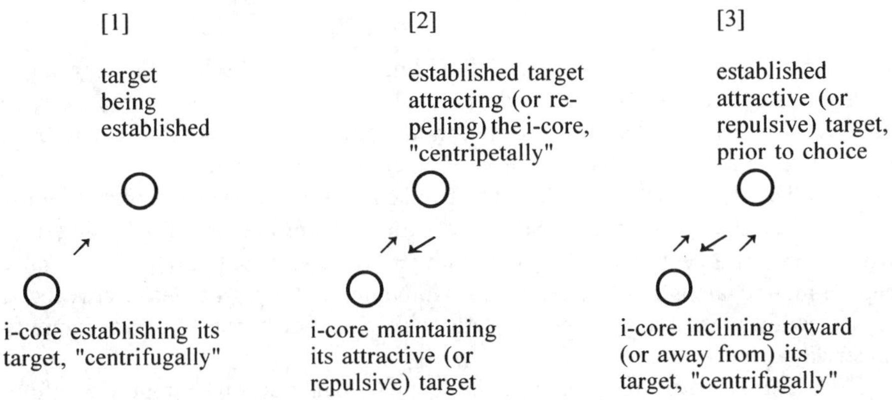

Second, its object seems to affect the i centripetally. Third, another centrifugal tendency arises, an inclination for which the object has become the target (*das Erstrebten*).

So far, the inclination is "blind" concerning the reality of the object.[70] One may say that the i is affected by its target, in the sense that the subject-object relation in which the i participates takes on the character an inclination *of* i. The i in-feels that inclining within itself; it registers there. However, the inclining in no way causes or compels the i, nor may it be said to "motivate" the i. The condition of being inclined is merely the warrant for a possible willing. Yet the i remains quite capable of a willing that runs contrary to its inclination, and even of willing in the absence of any relevant inclination at all. Pfänder writes that

. . . *Willensbegründung* ist keine *Verursachung des Wollens*, und Motive sind keine phänomenalen Ursachen des

. . . the grounding of the will is no causation of willing, and motives are no phenomenal causes of willing. It is a

[70]This account is given in Pfänder (1911: 139-140; 1967: 17-18). The inclination may be any sort of self-involving orientation toward the object, whether attraction and desire, as diagramed here, or repulsion, fear, hatred, and so forth. Husserl heavily annotated this passage in his copy of Pfänder's essay, according to Schuhmann (1973: 96).

Wollens. Es widerspricht daher den Tatsachen, wenn Schopenhauer behauptet, die Motivation sei die Kausalität von innen gesehen. („Über die vierfache Wurzel des Satzes vom . . . zureichenden Grunde." Kap. VII §43.) Denn was bei der Bestimmung des Wollens durch Motive „von innen gesehen" wird, ist völlig verschieden von einer Verursachung. . . . Das Wollen ist seinem Wesen nach phänomenal immer frei, d.h. nicht durch etwas vom Ich-Zentrum Verschiedenes verursacht. (1911: 157-8)

contradiction of the facts, therefore, when Schopenhauer affirms that motivation is causality seen from within. For what is seen "from within" in the case of the determination of the will by motives is completely different from causation. . . . According to its essential nature, willing is always free phenomenologically; it is not caused by something different from the ego-center. (1967: 34)

While motives do not "cause" acts of willing, they do disclose their grounds. Thus, motives are what one can understand about human actions. Actions can be interpreted precisely because they are "motivated," which is to say that what is intelligible about an action is that it is the realization of a willing.

The term *Motivation*, then, has two senses for Pfänder. Motivation in the *future-directed* or realizing sense denotes the transformation of inclining into willing, and of a centripetally attractive target (*das Erstrebte*) into a centrifugally chosen motive, along the way to action. Motivation in the *retrospective* or realized sense denotes the comprehensible rational goal-directedness of an activity underway or now completed.[71]

Pfänder thus has argued in favor of human freedom within a carefully restricted scope. The i is free to decline to will any of the inclinings that it experiences. Nevertheless it is impossible for the i to will something without at the same time meaning it to be something *real*. Willing must will a realization; and therefore it must intend a real world.

This does not insure the i against making a mistake or even deceiving itself.[72] In any particular instance of willing, the i may very well be wrong about the reality-status of the goal it has in view. But taken as a whole, a life that includes acts of willing is a life during which the real existence of the world cannot rationally and coherently be denied. Or, as Hedwig Conrad-Martius will argue, reality is not a phenomenon alongside other phenomena, constituted by a transcendental i and subject to "reduction."[73] Reality itself is necessarily intended,

---

[71]The metaphor of motion and moving will be adapted by Heidegger to characterize the non-necessity of human existence. For him, human lives are not rationally motivated but merely thrown and sailing along arbitrary trajectories toward worldly death.

[72]Pfänder (1911: 16; 1967: 37) briefly mentions involuntary repression, pretending, and concealing from oneself the motivations for one's actions. Yet even in such cases, the motivations would be there and could, by some judicial or therapeutic interpretive procedure perhaps, be brought out into the open for understanding.

[73]This point is made by Kuhn (1975: 3-5). Hedwig Conrad-Martius first formulated this criticism of Husserl's transcendental program in 1916, before Husserl had modified and nuanced his earlier call for a bracketing of being. Heidegger's plan of attack in his

even within the will to suspend belief in the reality of this or that particular item.

Although Pfänder did not subscribe fully to a *Realontologie* like that of his student Conrad-Martius, he demanded scrupulous observance of the *epochē* or suspension of judgment about the reality-status of beings prior to phenomenological examination of their appearances.[74] Writing in the *Deutsche Literaturzeitung* in 1929, Pfänder endorsed the charge that Husserl's transcendental idealism depended upon the decidedly unphenomenological prejudgment that there could be no being independent of consciousness. That thesis was argued by a former student of Husserl's whose book Pfänder favorably reviewed. In effect, Pfänder signed on to an anathema excommunicating Husserl from the phenomenological movement.[75] His remarks summarized the case against Husserl's transcendental phenomenology as follows:

Die Phänomenologie, als Wissenschaft vom reinen Bewußtsein, verbietet sich ihrem Wesen nach und ausdrücklich jedes Urteil über alles Bewußtseinstranszen-dente. Der H.sche Idealismus vollziehe aber ein solches Urteil, indem er behaupte, die bewußtseinstranszendente	Phenomenology, as science of pure consciousness, forbids itself, explicitly and according to its essence, every judgment about whatever may be transcendent to consciousness. Yet Husserlian idealism executes just such a judgment, in that it declares that the

1928 *Sein und Zeit* is comparable. Against Husserl's program of postponing any decision to believe in the real existence of something until after its essence has been recovered, Heidegger insists that the essence of being human consists precisely in its distinctive manner of existing--and so would become obscured at the outset by the Husserlian technique of turning off the predication of its existence.

[74]See Spiegelberg (1973: 11). Spiegelberg holds that Pfänder followed Husserl's first or eidetic reduction, which came to be called the *epochē*, but that Pfänder did not accept the transcendental or phenomenological reduction, which Husserl formally presented in his *Ideen* of 1913. According to Geiger (1933: 14-15), Husserl had indicated as early as the *Logische Untersuchungen* that he was more interested in analyzing "the mode of being-given of the given" than in analyzing the givens themselves, which occupied Pfänder.

[75]Commenting on a work by Theodor Celms, Pfänder (1929) guardedly concluded: "Wenn man die Resultate betrachtet, die der Verf. in überzeugender Weise begründet, so wird man verstehen, daß er trotz seines ausdrücklichen hohen Respektes vor den Leistungen, die sein Lehrer in der Phänomenologie erreicht hat, sich zu dem ziemlich schweren Vorwurf schmerzlich genötigt sieht, H. habe sich bei der Grundlegung seines Idealismus mit erstaunlicher Leichtigheit einfach der Tradition überlassen. Das Buch ist viel zu ernst und zu gediegen, als daß man es einfach mit der Erklärung abtun dürfte, sein Darlegungen beruhten auf Mißverständnissen und auf Unkenntnis der weiteren Entwicklung der Phänomenologie." ("If one examines the results, which the author establshes in a persuasive manner, then one will understand that he sees himself painfully obliged to a fairly weighty reproach: Husserl has simply forsaken the tradition with astonishing ease by laying the foundation of his idealism. The book is much too serious and too massive for one to be permitted to lay it aside with the explanation that its expositions rest on misunderstandings and on ignorance of the wider development of phenomenology.") These words must have been disappointing to Husserl, who had retired the year before. Spiegelberg (1973: 11 n15) reports that Husserl had responded favorably to Celms's study when it was published in 1928.

physische Welt habe kein Sein an sich, sondern nur ein Sein für ein Bewußtsein, sie sei darüber hinaus ein Nichts. Diese Idealismus folge nicht notwendig und könne nicht folgen aus der Phänomenologie, sondern er verlasse ihre notwendige Grundlage. (1929: 2049)	physical world transcendent to consciousness has no being in itself, but only a being for a consciousness; beyond that, it is a Nothing. This idealism does not necessarily follow and cannot follow from phenomenology; on the contrary, it forsakes its necessary fundamentals.

The Munich phenomenologists insisted that prejudice against reality ruins phenomenological rigor as surely as does prejudice in its favor.

Husserl's 1913 introduction of the transcendental reduction as a methodological requirement for phenomenology was the first and most important of four events that eventually splintered and redirected the Phenomenological Movement. Two more followed quickly: a World War in which many young colleagues suffered and died; and Husserl's departure from the congenial circle in Göttingen to start anew at Freiburg in 1916. The fourth but decisive event was Heidegger.[76]

3. Hedwig Conrad-Martius

This splintering was first described by Hedwig Martius. Hedwig Conrad-Martius (1888-1966) was Husserl's student in Göttingen and under his direction wrote an award-winning essay on the epistemological grounding of positivism.[77] However she could not take the doctorate with him because Göttingen required doctorands to demonstrate competence in the Greek classics; girls did not receive instruction in that literature in their segregated secondary schools. But Munich had no such requirement, so Conrad-Martius went there to Pfänder, who accepted the prize-winning essay for the doctorate. Subsequently she tended an apple orchard with her husband and extended the hospitality of their farm home to many German intellectuals, including Edith Stein, whose baptismal sponsor she became in 1922. Conrad-Martius contributed to Husserl's *Jahrbuch* in the 1920's and published several volumes on philosophy of science and nature. She taught at the University of Munich from 1949 until just before her death, and remained a guiding influence over the later Munich Phenomenological Movement. Her papers are preserved at the Bavarian State Library.

As early as 1916, Conrad-Martius contrasted her own plans for a "real-ontological" phenomenology with two other programmatic strands: Husserl's strict method, which she characterized as epistemologically oriented, and the more lax general method, according to which the suspension of judgment about the reality

[76]The "event of Heidegger" would include his inversion of Husserlian transcendental subjectivity in the 1928 publication of *Sein und Zeit*, his appointment to the chair of the retiring Husserl also in 1928, his cooperation with Nazi influences in German intellectual life in the early and mid 1930's, and the development of French and German existentialism, which his work inspired.

[77]See Avé-Lallemant (1982) for a survey of Conrad-Martius' life and work. Further biographical details come from Hart (1972) and from Edith Stein's autobiography (Stein 1986).

of the world was optional. Conrad-Martius would *require* the execution of the general thesis that the world is real--the very thesis that, in her view, Husserl criticized as a naive unfounded belief. Real-ontology would be just as rigorous in requiring this reality-thesis as Husserl was in insisting upon its suspension.[78] In Conrad-Martius' view, the two "rigorous" branches of phenomenology--Husserl's, and her own!--outlasted the lax general variety and continued as live philosophical options alongside Heidegger's newer existentialist version until the early 1930's.

"Munich phenomenology" thus came to denote not only the tradition of Lipps and Pfänder, but also the inspiration of Conrad-Martius. She insisted that reality not be "reduced" and that the *epochē* not be applied wholesale to every noematic sense. Her argument is recapitulated by a former student:

Die Husserlsche Urteilsenthaltung (Epoché) beziehe sich nicht auf die zu irgendeinem noematischen Sinn gehörige Seinsvermeintheit, sondern nur darauf, ob solcher wie auch immer gearteter Seinsvermeintheit ein wirkliches Sein, eine wirkliche Wirklichkeit entspreche. Auf der einem Seite stehe hier „Wirklichkeit“ als noematische Bestand; auf der anderen stehe die bewußtseins-transzendente Wirklichkeit des noematischen Wirklichkeitsmomentes: eben die wirkliche Wirklichkeit. Diese könne niemals zum noematisch-phänomenalen Gesamtbestand der Welt gehören, weil es sich in ihr um das faktische „Auf-sich-selber-Stehen“ der Welt und aller ihre Bestände handele. In diesem Moment liege die wirkliche Wirklichkeit der Welt, ob es sie faktisch geben möge oder nicht. Bei Husserl komme das nirgends klar zum Ausdruck. (Avé-Lallemant 1975a: 33).

The Husserlian abstention from judgment does not cover the meant-to-be-ness belonging to just any old noematic sense at all. It covers only this: whether such meant-to-be-ness, no matter how conditioned, matches up with a real being, an actual actuality. Here on the one side stands "reality" as noematic preserve; on the other side stands the consciousness-transcending *reality of* the noematic import of reality: precisely the real reality. *That* reality could never belong to the noematic-phenomenal common fund of the world, because what is going on in it [i.e. in reality's noematic import] is the factual "standing on its own two feet" of the world and all its preserves. In this momentousness lies the real reality of the world, whether the world may be there factually or not. With Husserl this never comes clearly to expression.

This formulation enforces a certain subtle distinction between two senses of the term "reality" itself. On one hand, I am permitted to suspend judgment about whether there are any X's, so that I might examine whether and how the noematic sense of X includes specification that X's belong to the real world--the world in which I am carrying out the investigation just as anyone at all may do so. (X's might, for example, be meant instead as essences or as fictions or as hypothetical situations--none of which are meant as having real existence. In such cases there

[78]See Avé-Lallemant (1975: 34), who cites the 1916 manuscript. Although there were more publishing phenomenologists of the Munich-realist persuasion than Husserlian transcendental phenomenologists in those days, the views of the latter have attained canonical status while the former are footnoted as those who simply misunderstood Husserl. Edith Stein, as we shall see, sided with real-ontology but she escaped the silencing of the footnotes by way of the silencing of the cloister.

would be no reality meant within the noema of X.) On the other hand, I am *not* permitted to suspend judgment about whether there is any reality to the world at all. Conrad-Martius thought that because Husserl overlooked this distinction, he was willing to do away with the world hypothetically in pursuit of methodological rigor.

Where Husserl was too stingy with reality, she thought, Heidegger was too generous and would profligately multiply worlds without end. The number of real worlds that there are is the question that most clearly distinguishes the three divergent branches of phenomenological philosophy as identified by Conrad-Martius: the transcendental, the existential, and the real-ontological.[79] These represent contrasting positions on the reality-status of entities. The unity that constitutes a world consists in the coherence of its elements, however that coherence might be construed: logically, spatially, historically, epistemologically, ethnically, and so on. A world is *meant as real* when it is meant as having coherence of some kind; that is, when its parts are understood to connect to one another and eventually to some element known immediately to be real--oneself, for example. (Motivation and causation, in the senses discussed above, are varieties of coherence for real worlds.) There is *more than one* real world if connections cannot be completed among all acknowledged realities. There is *less than one* real world if no element can be established through immediate experience to be real. From the standpoint of real-ontology, neither Husserl nor Heidegger came up with a satisfactory value for the number of real worlds. Husserl establishes only that the number is either less than or equal to one, while Heidegger puts the number of realities at one or greater. In other words, Husserl's phenomenology cannot rule out the possibility that there *might* be no reality at all; Heidegger's phenomenology allows that there *might* be many.[80] Conrad-Martius' real-ontological phenomenology is meant to demonstrate that the number of worlds is one. There is exactly one reality.

4. Max Scheler

Among the Munich phenomenologists, two more names must be mentioned in order to round out this account of Edith Stein's philosophical milieu: Max Scheler and Adolf Reinach. Both had ties to Göttingen as well as Munich, both were committed realists, both were personal acquaintances of Edith Stein, and like her, both men were apostate Jews. Beyond that, their personalities and professional activities stood sharply in contrast.

Max Scheler (1878-1928) was trained in moral theory at Jena, where he

---

[79]These three are juxtaposed by Avé-Lallemant (1975: 21).

[80]In *Ideen I*, §48, Husserl argues that there can be no more than one real world; that is, that all that is real is continuous. In §49 he argues that the possibility that there is no real world cannot be excluded. Heidegger in *Sein und Zeit* construes the world as constructed through human projects as their current context as well as their future target. The multiplicity of human cares, projects, and life trajectories indicates the plurality of worlds. There is no way to bring congruence to them, that is, to insure the unity of reality.

took his doctorate in 1897 and began teaching in 1899. However scandals attending his three marriages, his wartime hyperpatriotism, and his other personal betrayals ruined his academic career at Jena and later at Munich. Subsequently he earned his living by free-lance lecturing and writing. Scheler's brilliant insights made him welcome among the members of the Munich phenomenological circle. He followed the tide of Müncheners emigrating to Göttingen, but stayed only for brief periods since he was unable to establish himself there.[81] Stein recalls that the Göttingen Philosophical Society selected Scheler's *Ethik* for study in the summer semester of 1913 and sponsored a lecture series by the author during the fall.[82]

Scheler admired Husserl and freely acknowledged his intellectual debt to him; nevertheless, Scheler's criticisms of Husserl's transcendental turn were quite caustic. Scheler also is indebted to Pfänder, his Munich colleague, although that debt remains unacknowledged in the texts to be discussed here.[83] In particular, Scheler appropriated three elements of Pfänder's phenomenology of motivation: the notion of the *correlation* between inclinations and their targets (*Streben* and *Erstrebtes*), the notion that to will always is to will *a realization*, and the notion that inclinings and willings entail *directional involvement of the i.* Equipped with this realist egology, Scheler gradually built from his earlier work in moral theory to a critique of Kantian formal ethics; on that basis he expanded his work into a systematic metaphysics of the human person.[84] Yet his philosophy was, and is, both thrilling and frustrating to read. On the one hand, there are passages of sublime insight and devilishly clever critique. On the other hand, writerly sins abound. Formal contradictions are not hard to find in Scheler's arguments. Technical terms such as "motivation" and "causality" are used imprecisely; "the psychic" is variously synonymous with "the spiritual," "the sentient," or "the mental." Positions of some authors, such as Freud, are fairly presented before being demolished, while those of others, such as Lipps and Husserl, appear more

---

[81]Biographical information and surveys of Scheler's work may be found in Shuster (1942), Bershady (1992), Lachterman (1973), Avé-Lallemant (1975b), and Spiegelberg (1982a: 268-305).

[82]See Stein (1986: 258). This choice by the students may have nettled Husserl, inasmuch as the first part of Scheler's *Ethik* had been bound together with the first book of Husserl's *Ideen* in the first volume of the *Jahrbuch für Philosophie und phänomenologische Forschung* in 1913. Husserl was discussing the *Ideen* in his own seminar, according to Stein (1986: 250).

[83]Since Pfänder's works are not cited by Scheler, it may be the case that Pfänder's vocabulary and approach had simply become the unattributed common currency of discussion within the Munich phenomenological circle. One of the delicious ironies of Scheler's literary achievement is his insistence that we often are deceived in considering the thoughts of another to be our own; see below.

[84]Much of Scheler's mature work lies beyond the scope of the present investigation. We must leave out of connsideration Scheler's contribution to the philosophy of religion, for which he is perhaps best known. Nor can we explore here the possible reciprocal influence of Scheler's writings upon Pfänder's mature work on the human soul.

or less in caricature.[85] Philosophical readers of Scheler today must tolerate these faults if we are to recover something of the excitement and inspiration that his work evoked among his contemporaries.

Moreover, once again some detective work is required in order to follow the unfolding of Scheler's influence during the first two decades of the twentieth century. Commentators typically interpret his work as a whole. For example, Barber (1993) finds in Scheler's phenomenology a coordinated critique of three institutional structural features of Western culture--namely, science, capitalism, and social formations--insofar as these features "reverberate" within three areas of philosophy--namely, epistemology, value theory, and theory of intersubjectivity. Yet those components had not yet fallen tidily into place at the time when our interest focuses. What is more, Scheler published his work piecemeal and in progressively revised editions. The texts that came into the hands of his first readers were substantially different from today's critical editions, upon which contemporary Scheler scholarship is based.[86]

At the time of Scheler's greatest influence in Göttingen, just before the First World War, he was working out a philosophical psychology designed with a twofold purpose in mind: constructively, to serve as the foundation of the cultural sciences, or *Geisteswissenschaften*,[87] and destructively, to overcome both the illusion of a Husserlian transcendental phenomenology and the folly of psychological theories and practices that were reductively modeled after those of the natural sciences. Scheler's new "*verstehende Psychologie*," or psychology of understanding, would be a phenomenology of sympathy (*Sympathie* or *Mitgefühl*) as an intersubjective function irreducible to anything more basic.

One branch of Scheler's investigations led to a refutation of the Kantian position that value has only a purely formal, ideal being. Against the idealist claim that "the only good thing is a good will," Scheler argued that the will is never empty. Willing isn't willing unless it wills *something* and also wills that something *as realizable*. Thus goodness lies not in will but rather in values, whose real existence is given immediately in volitional experience and can be exhibited phenomenologically. This case for a non-formal (*materiale*) ethics of value was introduced by Scheler in the first issue of Husserl's *Jahrbuch* in 1913 as *Der Formalismus in der Ethik und die materiale Wertethik*, part one.[88] Thus it came

---

[85]See, for example, how Husserl's *Ideen* with its "principle of principles" becomes an *Idol* in Scheler 1915, "Die Idole der Selbsterkenntnis." Edith Stein recalled Scheler as being insolent in person and half-baked in his published works. See Stein (1986: 259-260).

[86]Fortunately, the earlier editions can sometimes still be found, or can be reconstructed with the help of the careful notations available in the critical editions. See, for example, Frings (1973).

[87]This was, of course, Dilthey's old ambition. While Dilthey's work is not explicitly taken into account by Scheler, the influence is unmistakable.

[88]The second part of the work appeared in the next issue of the *Jahrbuch* in 1916. Scheler, Pfänder, and Reinach were co-editors of the *Jahrbuch* along with Husserl and

to the attention of the Göttingen Philosophical Society, and was discussed by that group in the summer semester of 1913--Edith Stein's first term in Göttingen.

Another branch of Scheler's investigations produced a study of the communication of thoughts, sentiments, and passions among human beings. Scheler rejected Lipps's theory of empathy; that is, he rejected the *version of* that theory that was conversationally current in the Munich circle, for there is no evidence that Scheler took the trouble to read and understand what Lipps had written. In regard to communication, the issue that Scheler seized upon arises from his recognition of the polarity between subjective contents and the subjective acts with which they are grasped; for example, between knowledge and knowing, between willed purpose and act of willing, or between attractive target and inclination. Scheler asks, in effect: What is it that is shared when human beings communicate? Content is shared, indeed; but in communication do we also achieve an inner coincidence with the very activity in which someone else has been intending that content? Do i's coincide in their activity (noetically) as well as in their targets (noematically)? In pursuit of this issue of interpersonal access, Scheler developed both a detailed account of conscious life and a thoroughgoing critique of the sciences of consciousness of his day. The former, his phenomenology of intersubjectivity, appeared in another 1913 publication, *Zur Phänomenologie und Theorie der Sympathiegefühle und von Liebe und Hass.*[89] The latter, his critique of contemporary psychological and phenomenological sciences of human being, appeared first in a 1912 journal article and subsequently in expanded form as "Die Idole der Selbsterkenntnis" in a 1915 collection.

These three works of Scheler--his *Ethik*, his so-called *Sympathiebuch*, and the *Idolenlehre*--all were carefully studied by Edith Stein as she prepared her doctoral thesis on *Einfühlung*, which depends substantially upon them. It will be instructive, then, to consider synoptically some of the positions and arguments that they present.

"Nothing," Scheler insists, "has harmed psychology as much as the thesis that it must be pursued by way of analogy with natural science" (1915: 89; 1973: 54; cf. Husserl 1911). Mechanical causality governs the interactions among elements of the physical universe, says Scheler, and causal sequences are the space-time connections that the natural sciences follow. But an entirely different sort of coherence obtains in human affairs, a coherence that is to be *understood* rather than *explained.* While physical events can figure into human affairs, both socially and inside the individual human organism itself, such interactions involve "causality" of a decidedly different sort. The i is activity, not an actor. The i can

---

Moritz Geiger.

[89]This work, revised and expanded, was brought out again in 1923 under the title *Wesen und Formen der Sympathie*. The 1913 version is focused along ethical lines, while the 1923 version enlarges Scheler's program into a comprehensive account of metaphysics and epistemology. The expansion appears to have come in response to Edith Stein's distinction between *Einfühlung* and *Einsfühlung*, or feeling-into and identification. This influence will be discussed in a later chapter.

accomplish effects at the various levels of its engagement with the world because at its core, *the i is effectivity itself.* Its engagement with things is not "caused" in a mechanical way. Rather, this engagement

. . . *folgt den in bezug auf alle psychische Kausalität »freien« Akten der Person* und dem Maße und der Art ihrer »Selbststellung«. Psychische Kausalität ist also in letzter Linie immer Ichkausalität, d.h. erlebte Wirksamkeit des einheitlichen Ich. Sie ist als solche wesenhaft individuelle Kausalität, d.h. ein solche, in der keine »gleichen Ursachen« und »Wirkungen« wiederkehren, also jede Ichänderung abhängt von dem Ganzen der Erlebnisreihe des Ich bis zu dieser Änderung. Diese *reine psychischen Kausalität* ist es, die wir auch Motivationskausalität nennen können und die nach allen Richtungen zu erforschen Aufgabe der *verstehende Psychologie* ist -- der Grundlage der Geisteswissenschaften. Sie »erklärt« nicht, sondern »versteht« alle Einzelvorgänge individueller oder typischer psychicher Einheiten auf Grund von deren individuellem oder typischem Gehalt; sie sieht also *nicht* ab von der »Individualität« oder den »Typen« der jeweiligen Ichtotalität, sondern hält gerade *diese fest* und macht sie zu dem besonderen Gegengstand, aus dem heraus sie »versteht«. (1913b: 438)	[... *conforms to the person's acts, which are "free" as far as psychic causality is concerned,* and to the extent and kind of her self-positioning.] In the final analysis, therefore, psychic causality is always [i]-causality, i.e., the experienced efficacy of a uniform [i]. As such it is by essence *individual* causality, i.e., one in which "the same causes and effects" never recur. Hence, every change in the [i] is dependent upon all experiences of the [i] up to this change. We can call this *purely psychic causality* the *causality of motivation.* It forms the foundation of the humanities. [The *psychology of understanding*] must investigate this area from all angles. This psychology does not "explain" but . . . "understand[s]" all singular processes of individual [or] typical psychic units [on grounds] of their individual or typical contents. It does *not* set aside the "individuality" or the "types" of [i]-totalities; rather it *sticks to them,* making them special objects from which it derives an "understanding." (1973b: 422)

Scientific explanation, which posits "like causes for like effects," is impossible with regard to human actions, Scheler is saying, because every act is determined by a unique series of past experiences and future expectations that is necessarily *un*like any other series. On this basis Scheler rules out "*erklärende Psychologie*"; that is, any psychological practice claiming competence to diagnose how human behavior is conditioned by external factors, and then to prescribe how disordered behavior can be modified by therapeutic manipulation of such factors.[90] He pokes fun at psychotherapy in general and Freudian psychoanalysis in particular. Unfortunately, says Scheler, the psychoanalysts have mixed up the two possible ways of intervening in someone's on-going psychic experiences: the "Socratic way" and the "surgical way." Like Socrates, they seek to lead patients to new insight into themselves. But like surgeons, they rely upon a mechanistic associationism to produce desired results. Thus even when the therapeutic goal is to heal

[90]Factors external to the i itself would include the constraints of the physical environment and physical interactions with other people, as well as physiological processes and neurological events occurring within the human body.

consciousness, that goal is pursued inappropriately by mechanistic interventions.[91]

Yet Scheler also finds corroboration in Freud's work for his own conception of psychic (*seelisch*) causality. For Freud, every experience is in some way stored up within psychic life and helps to shape every subsequent experience. Moreover, similar experiences "cause" *different* "effects" in consciousness, depending upon when in one's lifetime they happen to occur. Mental health and vitality depend upon having had experiences unfold in a certain sequence so as to permit the gradual maturing and strengthening of the i.[92] Yet for Scheler, because this i is not a substantial thing, it is not separated in time from what we commonly call its "past"; nor do so-called past experiences need to be recorded by some physical neurological mechanism in order to endure effectively into the present. The i itself is their record, their interwoven unity. The efficacy of past experiences in the present is owing to the essential character of the i itself.

[D]ie vergangenen Erlebnisse haben durchaus nicht zu sein und zu wirken aufgehört: Sie *existieren* im Ich und *»im« Ich*. . . . [Es gibt] eine psychische Mannigfaltigkeit -- sowie Verbindungsarten in dieser --, die ein echtes *»Ineinander im Ich«* darstellt; ein echtes Sein des Erlebnisses »im« Ich und nicht außerhalb seiner. Fragt man also: Inwiefern »sind« die vergangenen Erlebnisse, so sage ich: Sie *»sind« in meinem Ich*, das in jedem seiner Erlebnisse anders »wird« -- ohne sich doch dabei wie ein Ding zu »verändern«. Sie sind also nicht in einem mystischen Raume der Vergangenheit -- gleichsam blutlose Schatten, die zuweilen an meine Gegenwart pochen, um hier vom Blute meines Lebens zu trinken. (1913b: 434, 435)

"[P]ast" experiences do not have to cease to exist or cease to have effects. They *exist* in the [i] and *"in"* the [i].... [T]here is a psychic manifold, as well as types of combinations, which represents a genuine *"interwovenness in the [i]"* -- a genuine being of experiences "in" the [i] and not outside it. If one were to ask how past experiences "are," I would answer that they "are" in my [i], which *"becomes" different* in all experiences, though without "changing" as a thing would change. Therefore they are not in a mystic area of the past, like bloodless shadows that sometimes knock at the door of my present to taste the blood of my life. (1973b: 418, 419)

Past experiences, woven into the i, need be neither consciously remembered nor even accessible to memory at all. Their being is their efficacy, for they factor into and shape the current living experience of the i. Here, too, Scheler finds corroboration from Freud, who held that one's early childhood experiences, although forgotten and unrecoverable, continue to determine the i. Scheler adds that a disposition such as love or hate also will make its effects felt even while it is not being consciously entertained. There is a difference, he points out, between a love that has gone out of existence, and one that is merely not acknowledged or brought into awareness at the moment. Therefore *sub*conscious realities form part of the i itself. Scheler insists on the distinction, however, between the

[91]See Scheler (1915: 16-18; 1973a: 10-11).

[92]See Scheler (1913a: 100-104; 1954: 196-200). This is termed Freud's *Ontogenie* or ontogenetic theory.

subconscious "part" of the i, and the so-called *un*conscious which, according to Freud, supposedly exerts just the sort of mechanistic causality upon the i that Scheler means to reject.[93]

One essential difference between mechanistic physical causality and i-causality is that the latter does not involve a temporal sequence. Thus Scheler faults psychology for having imported into the i the structures of sequential time that govern occurrences within the physical world. Colors and melodies, he says, are perceived instantly as wholes and not by gradually assembling their components. Recognition comes all at once, simultaneously with the giving of sense impressions, and not as the last item in a series of neurological registrations. This view does away with the bogus problem of constructing some kind of temporal bridge from causal stimulations to an effected conscious perception that would lie subsequent to them in time; for the i itself effects that span by *being* that span.[94] As with psychic transactions, so with those of the soul. There are no atoms of thought corresponding to the bits of language in a sentence; rather, meaning occurs all at once.

At this point it is somewhat confusing to find Scheler suddenly contrasting "the psychic" with "the soul," because elsewhere the terms are synonymous. Nevertheless they are conceived now as two distinct yet correlated levels of egoic function. At the *psychic* level, physiological events register in consciousness. At the level of *soul*, events of meaning register. "Below" the psychic level (also called the vital level) is the *physical* level. "Above" the level of soul is the *geistig* or *spiritual-cultural* level, where Scheler places the individual human person who loves and acts freely.

FIGURE 1.2: SCHELER'S LEVELS OF PERSONHOOD

	levels of i-function	**levels of i-hood**	
understand:	free acts, love, by *person*	the spiritual/cultural	(*geistig*)
understand:	meaning, thought, logic	the soul	(*seelisch*)
explain causally:	sense perceptions	the psychic/vital	(*psychisch*)
explain causally:	sensations	the sensible/physical	(*sinnlich*)

What goes on at the lower levels can be scientifically "explained," while activities at the upper levels can only be "understood." (One would expect, then, that Scheler would speak of "causality" only where there is a possibility of cause-and-effect explanation; but in fact he speaks--perhaps metaphorically--of soul and spirit causality as well.) Unfortunately Scheler is inconsistent in his application of

[93]See Scheler (1915: 73, 77; 1973a: 43, 45).

[94]See Scheler (1915: 139-142; 1973a: 80-82). As we have seen, the insistence that the whole is understood before the parts goes back at least to Schleiermacher. Scheler invokes it both in his argument that the human body is perceived before its parts and gestures, and in his claim that traditions and "what everybody knows" are prior to the emergence of individual knowledge. For the latter, see Scheler (1913a: 142; 1954: 258).

this very suggestive schema of the levels of i-function; nowhere does he offer a systematic account of it.[95]

This brings us back to the question of human communication. As Scheler wrestles with it, the question is simply *how the i shares what it does*--for i-efficacy at all functional levels is peculiarly replicant. In other words, among the unities that the i achieves--and that any theory of "psychic causality" must account for--is the unity of i's themselves while they are sharing in some informational content, emotional state, or decisional target. At such times, i's *seem to* coincide both noematically and noetically. Therefore any theory designed to guide the investigation of how such coincidence is achieved, Scheler insists, must not be modeled arbitrarily after the natural sciences; it must not reduce all efficacy to physical cause-and-effect efficacy, but should instead be tailored to fit the psychic and spiritual realities that are to come under investigation.

In this light, Scheler critically evaluates two theories that have been advanced to account for how and what i's can and do know of one another.[96] According to the first theory, *analogical inference*, I am constrained to deduce the existence of another human being from the sense impressions I receive of bodily parts, gestures, and expressions that resemble my own. I spontaneously associate such features with the presence of a human being. Scheler refutes this theory on grounds that parts are not given piecemeal in perception, and moreover that the chain of production invoked here is an invalid importation of physical causality into the realms of the psychic and the spiritual.

According to the second theory, projective empathy, I send out my own awareness to inhabit another's body and I imaginatively pretend to know, want, and feel whatever and however the other knows, wants, and feels.[97] Scheler objects that this theory presupposes the existence of the other human being, which is what it was supposed to account for. Moreover, this theory provides no way to check on whether the body into which I project my feelings is in fact a human body, and whether the feelings that I imagine to be his or hers really are so. Scheler's own description of sympathy or *Mitgefühl*--"fellow feeling"--is meant to overcome such objections by attending very carefully to the ways in which the various functions of the i, on the various levels of the i, *appear*.

---

[95]Scheler asserts that an i does not belong to the essence of mind (*Geist*), although mind is essentially personal. In this way he seeks to detour around the problem of connecting an i with an outer world. Personal mind already is with the world. Thus Scheler rejects the exclusivity of "the old Cartesian alternative" that assigns all realities to either the psychic or the physical. See Scheler (1913b: 404; 1973b: 389). See also Barber (1993: 123).

[96]This critique, "Über den Grund zur Annahme der Existenz des fremden Ich" ("On the Ground for the Supposition of the Existence of the Alien I"), was offered in an appendix to the 1913 edition of the *Sympathiebuch*. See Scheler (1913a: 118-148; 1954: 238-264). For the 1923 edition the title was changed to "Die Fremdwahrnehmung" ("Alien-Perception").

[97]Although Scheler attributes this account to Theodor Lipps, he has overlooked some important components of Lipps's *Einfühlungslehre*.

Appearances are (noematic) content for the i, and they are to be distinguished from the (noetic) functions or activities of the i. Both are multi-layered. As Scheler remarks, the Starnberg Sea as he remembers it still has a physical appearance (even though he now is registering that appearance in a psychic act).[98] By comparison, psychic and spiritual appearances occur as well; for example, the appearance of my own anger to myself, or the appearance of another person's anger to me. I know these "from the inside," that is, by partaking of that which *lives them through*: the i, which is intrinsically "*Ineinandersein*," a being-interwoven.

Was immer im Außereinander von Raum und Zeit erscheint, ist eine physische Erscheinung; was immer erscheint in einer unmittelbaren Ichbeziehung überhaupt, in einem »Zusammen«, dem dieses raumlich-zeitliche Außereinander fremd ist und das eine auf jene niemals reduzible Mannigfaltigkeit erfüllt, ist eine psychische Erscheinung. Beide Erscheinungsarten sind *gleich* unmittelbar gegeben, und im Aufbau der »Gegenstände«, die in beiden Bereichen des Erkennes das unmittelbar und mittelbar identifizierende Erkennen und Denken erfaßt, gibt es *dieselben Stufen* der Mittelbarkeit. Es ist daher ein grundlegender Irrtum, das Gebeit des Phänomenalen, überhaupt der unmittelbar und anschaulich gegebenen Erscheinung, mit dem Gebeit der »Bewußtseinerscheinung« oder der »psychischen« Erscheinung gleichzusetzen und das Physische erst als auf Denkakten (sei es bloß ein identifizierender Akt oder gar ein Schluß aus seelischen Wahrnehmungen oder aus der Empfindung auf eine »Außenwelt«) beruhend anzusehen. (1915: 56-57)

Whatever appears in the mutual externality of space and time is a physical appearance; whatever appears in an immediate relation to the [i] in general, in a "togetherness" which is foreign to this spatiotemporal externality and fills a manifold irreducible to the latter, is a [psychic] appearance. Both kinds of appearance are given with *equal* immediacy. Furthermore, the very same degrees of mediacy are present in the composition of the "objects" which cognition and thought identify, directly or indirectly, in both these domains of knowledge. Therefore, it is a fundamental mistake to equate the domain of the phenomenal, or, in general, the immediately and intuitively given appearance, with the domain of "appearance in consciousness" or ["psychic"] appearance. It is equally a mistake to regard the physical as resting on acts of thought (whether it is simply an identifying act or an inference from mental perceptions or from sensations to an "external world"). (1973a: 33)

What is more, neither physical nor mental appearances are self-validating. Scheler complains that in "today's fashionable philosophy," people assume that inner perception of something is equivalent to its reality. A lengthy criticism of Husserl follows in a footnote.[99] In fact, Scheler has framed the entire essay in which his

[98]See Scheler (1915: 50; 1973a: 29). A physical appearance is not necessarily a causal appearance.

[99]See Scheler (1915: 71-73; 1973a: 41-43). This note appears already in the 1912 version of the essay (see pages 109-111). It juxtaposes citations from the 1900-1 *Logische Untersuchungen* and the 1911 *Logos* article, and asks: "Has Husserl changed his opinion?" However the following indictment of *Ideen* §24 appears in the new introductory paragraph written for the 1915 version of the essay. Cf. §54 of *Ideen*.

Idolenlehre is presented as a challenge to Husserl's "principle of principles" (§24 of *Ideen*).[100] Scheler writes:

Es gibt vielleicht nichts, was für alle Art von Erkenntnis der seelische Welt ein so prinzipielles Hindernis darstellt, als die von vielen Forschern und Philosophen der Gegenwart und jüngsten Vergangenheit angenommene Meinung, daß innere Wahrnehmung im Gegensatze zur äußeren Wahrnehmung der Natur nicht täuschen könne, ja daß hier die Erlebnisse selbst mit evidentem und adäquatem Wissen von den Erlebnissen zusammenfielen. (1915: 5)	There is perhaps no more fundamental obstacle to any kind of knowledge of the psychic world than the position of many scientists and philosophers, both now and in the most recent past, that inner perception, as opposed to the external perception of nature, can never deceive, that here the lived experiences themselves coincide with self-evident and adequate knowledge of lived experiences. (1973a: 3)

In fact, Scheler observes, knowledge of one's own feelings comes about only through long and strenuous effort. To know oneself is at least as difficult as to know another. Or rather, the truly challenging task is to sort out individual feelings, thoughts, and desires from the flood of general opinion that washes over all of us in society.[101]

This indicates the direction of the "learning curve," as it were, that Scheler finds within all live experiences (*Erlebnisse*). These are experiences whose content is easily available to me; for example, emotions and prejudices. I first grab them by their content-pole, so to speak. They come to me as i-drenched feelings or knowledges or desires; but they do not disclose to me right away *whose* i it is that is operative in them. At first I get only a kind of general i-hood deployed diffusely in these experiential contents. I have to work along "against their grain" in order to grasp their egoic function-pole and discover whether they are originally *my own* feelings or someone else's.[102] Instances of emotional contagion illustrate

[100]Husserl wrote in §24 of *Ideen* that the "principle of all principles" for phenomenology was this: "daß jede originär gebende Anschauung eine Rechtsquelle der Erkenntnis sei, daß alles, was sich uns in der „Intuition“ originär, (sozusagen in seiner leibhaften Wirklichkeit) darbietet, einfach hinzunehmen sei, als was er sich gibt, aber auch nur in den Schranken, in denen es sich gibt, kann uns keine erdenkliche Theorie irre machen" (emphasis deleted) [. . . that any originarily oncoming disclosure would be a source of accuracy for cognition; that anything that tenders itself to us in originary "intuition" (in its incarnate reality, so to speak) would simply be acquired as what it comes on as. No conceivable theory can make us go wrong--but only within the channel of that oncoming.]

[101]See Scheler (1915: 112; 1973a: 65). Scheler holds that the capacity for feeling with someone else--*Sympathie* or *Mitgefühl*--is innate. It is not implanted from without, and it is not caused or engendered by anything extrinsic to itself. However, it does grow and become more discerning through practice. See Scheler (1913a: 31-37; 1954: 130-134).

[102]Thus, individual i's are disentangled by reflecting upon the "whose?" question implicit in any live experience. Scheler's account of this disentanglement resembles an account published many years earlier by Theodor Lipps in a passage discussed above but

this well.[103] But it occurs just as readily in regard to intellectual content. Scheler lays out the possibilities:

Nichts ist dann gewisser als dies, daß wir sowohl unsere Gedanken als die Gedanken *Anderer* denken, unsere Gefühle wie die *Anderer* fühlen können. Reden wir denn nicht Tag für Tag davon? Unterscheiden wir nicht fortwährend z. B. „unsere" Gedanken und diejenigen, die wir gelesen haben oder die man uns mitteilte? „Unsere" Gefühle von solchen, die wir nur „nachfühlten" oder von denen wir (unbewußt) angesteckt waren? . . . . Es kann sein, daß unser Gedanke uns auch „als" unser Gedanke gegeben ist; der Gedanke eines Anderen „als" der Gedanke eines Anderen z.B. beim bloßen Verstehen einer Mitteilung. Das ist der normal Fall. Es kann aber auch sein, daß der Gedanke eines Anderen nicht „als" solcher, sondern „als" *unser* Gedanke gegeben ist. . . . Es kann auch sein, daß ein Gedanke oder ein Gefühl, das unser ist, uns „als" Gedanke oder Gefühl eines „Anderen" gegeben ist. (1913a: 125-126)

For nothing is more certain than that we can think the thoughts of others as well as our own, and can feel their feelings . . . as we do our own. Are we not for ever distinguishing 'our own' thoughts from those we have read or which have been told to us? 'Our own' feelings from those we merely reproduce, or by which we have been infected (unconsciously)? . . . It may well be that our thoughts are presented 'as' our own, and those of others as theirs, e.g. in merely understanding a piece of information. That is the normal case. But it may also happen that the thought of another is not presented as such, but as a thought of ours. . . . It may also happen that one of our own thoughts or feelings is presented as belonging to someone else. (1954: 245)

In Scheler's view, the fact of communication can be taken for granted, since it is so evidently occurring. The reality of i-hood, too, is given with evidence *within* the givenness of that ubiquitous communication. Scheler insists that this datum is i-hood in general, but is not any sort of "transcendental i."[104] Anything like a Husserlian transcendental ego is ruled out. The possibility of i-hood is founded in the reality of the inimitable particularity of the actors who initiate the i-functions. Even the general i-hood that drenches all feelings and thoughts that infect us by social contagion is *given as* the "whose?" question; that is, as an invitation reflectively to discover whether a particular experience has been one that originates with myself or with another--and if the latter, then with whom. In fact, Scheler holds that experiences are not even fully known until one knows *of precisely which* individual i they are the experiences.[105] The quest for individuality, then, seems

apparently overlooked by Scheler: the emergence of multiple i's--exactly one of whom is I--from nondifferentiated i-hood at the conclusion of an acrobatic performance.

[103]See Scheler (1915: 31, 85-86; 1973a: 18, 50-51).

[104]See Scheler (1913b: 392; 1973b: 378). Scheler also allows for a "social i" almost as an afterthought to his *Idolenlehre*, but this term seems to mean simply the mask one wears in society; see Scheler (1915: 167; 1973a: 96). In his later work, "we" (*wir*) appears as a possible subject of live experiences.

[105]See Scheler (1915: 167; 1973a: 96).

to lead from the level of psychic function (where the operative i-hood may still be a generalized or anonymous one) to the level of *spiritual/cultural (geistig)* function, which is the realm of person. If i has been the weaver of unities, then "person" for Scheler seems to be *the unity of* those unities; that is, not their sum but their source. That elusive term receives the following "essential definition":

> *Person ist die konkrete, selbst wesenhafte Seinseinheit von Akten verschiedenartigen Wesens,* die an sich (nicht also πρὸς ἡμᾶς) allen wesenhaften Aktdifferenzen (insbesondere auch der Differenz äußerer und innerer Wahrnehmung, äußerem und innerem Wollen, äußerem und innerem Fühlen und Lieben, Hassen usw.) vorhergeht. *Das Sein der Person »fundiert« alle wesenhaft verschiedenen Akte.* (1913b: 397-398)

> *[P]erson is the concrete and essential unity of being of acts of different essences* which in itself (and therefore not πρὸς ἡμᾶς [*pros hēmas*, for us]) precedes all essential act-differences (especially the difference between inner and outer perception, inner and outer willing, inner and outer feeling, loving and hating, etc.). *The being of the person is therefore the "foundation" of all essentially different acts.* (1973b: 383)

The significance of "person" within Scheler's anti-transcendental program lies in the fact that persons are not interchangeable. The act of one person would not be essentially "the same" act if someone else tried to do it; there would be merely mimicry or imitation. In contrast, a logical operation *is* the same no matter who executes it. Assent to the correctness of a formal logical structure is the same, no matter who renders it.[106] Logical validity can be certified by any i equally well. Kant's formalization of ethics was an attempt to eradicate personal particularities from the conception of value. Scheler objects that life is such that it is impossible for someone to stand in another's shoes. Human beings cannot *personally* coincide, not even if they think the same *thoughts*, not even if they *think* the same. Noematic contents can coincide; noetic activity can clone itself; but only Cézanne can paint Cézanne's apples. A book has many readers but one author only.

Scheler's position would be contested by some literary theorists today; but in its time it offered a coherent account of exactly how far human communication can reach. For Scheler, persons could not "feel into" other persons as such; they could only "feel with." Personhood could not be shared but only companioned.[107]

In this connection, brief mention must now be made of Scheler's contribution to the philosophy of the human body. As we have seen, Pfänder had described a rudimentary directionality within experiences of inclining (*Streben*):

---

[106]Scheler (1913b: 397; 1973b: 382) conjectures that beings whose egoic functions were only those of knowing or willing would be logical subjects, but not persons.

[107]Spiegelberg (1982: 288) remarks that the doctrine of the person as an ontic unity of acts is the very center of Scheler's philosophy. Thus it is with regret that Spiegelberg complains that he "cannot discover sufficient phenomenological foundation for it, especially since Scheler does little if anything to substantiate his sweeping and often astonishing pronouncements in this area. Here Scheler's eagerness to reach metaphysical conclusions and to derive practical applications from them seems to have gotten the better of his phenomenological caution."

"centripetal" attraction seeming to move from the target toward the core of the i within the i-body, "centrifugal" desire running in the opposite direction. Scheler now elaborates Pfänder's spherical, gravitational model of the i. Things are not attractive or repulsive on their own, or in direct relation to the i, he says; rather they are experienced as having such effects only through the lived body. The living body (*Leib*, in distinction from *Körper*) gives location both to sensations and to what Scheler terms *Lebensgefühle*, vital feelings. Both physical sensations and vital feelings are lived-through as *extended* bodily. But vital feelings are "everywhere" throughout the body, while sensations are more focused, that is, here or there within the body. Thus I may experience simultaneously both the sensation of pain in my left wrist, and a general feeling of good health and alertness. Both levels of feeling are located in the body. But unlike sensations and vital feelings, soul feelings are not localized: these soul feelings (*seelischen Gefühle*) belong more properly to the i even though they may register in the body. (One says, "*My hand* hurts," but, "*I* feel sorrow.") Soul feelings are general non-localized states of the i; they are not extended by virtue of having some particular location in the body, nor are they primarily bodily at all like the vital feelings.[108]

Scheler finds yet another level of feelings beyond the i altogether, and he terms these *geistigen Gefühle*, spiritual feelings. They are never states. In Scheler's estimation there seem to be only two alternative spiritual feelings: bliss and despair. These are by no means responses elicited by something else; they simply flow forth from the person down across all levels of egoic experience, either bathing the i in sunlight or entombing it in a snowdrift of despondency. They can be neither explained scientifically, nor understood, nor shared in any way. The self-value of the person herself founds bliss or despair.[109]

In the original 1913 edition of his *Sympathiebuch*, then, Scheler admitted three varieties of human communication of feelings: (1) *Mitfühlen*, in which the same feeling is felt by several people; (2) *Mitgefühl* or *Sympathie*, in which one's feeling arises as a response to a similar feeling in another; and (3) *Gefühlsansteckung*, emotional contagion.[110] In the second edition of 1923, however, he added a fourth variety and made it foundational for the other three:

[108]One remarks that Scheler has not completely succeeded in differentiating these four levels of feeling. For example, a general state of shock can accompany a broken leg, and this general physical state can mask whatever joy or sorrow one might otherwise be experiencing in one's soul. It is perhaps more useful to think of Scheler's steps as a continuum connecting an i with its body.

[109]See Scheler (1913b: 253 and 350-357; 1973b: 246-247 and 338-344). See also the discussion of the body within Scheler's treatment of the person, (1913b: 413-431; 1973b: 398-415); and in the Idolenlehre, (1915: 64-69; 1973a: 37-40), where the the *Körperleib* and the *Leibseele* are said to be the same reality appearing on different levels of i-function. Scheler also mentions a *Leibich* or i-body as something given whenever an individual i is given. This need not be a living human body, apparently, but may be a cultural artifact that still expresses something of the experiences of that i. See (1913b: 392; 1973b: 378).

[110]See Scheler (1913a: 9; 1954: 12).

Einsfühlung, emotional identification. This new concept served as the principle for the expansion of the book from its original ethical focus into questions of epistemology and the metaphysics of the person.[111]

5. Adolf Reinach

The last of the Munich phenomenologists to be considered here is Adolf Reinach (1883-1917), although he may also rightly be called the first of the Göttingen circle. Reinach received his doctorate in Munich under Theodore Lipps late in 1904, but he went on to study with Husserl and habilitated under him at Göttingen in 1909. As *Privatdozent*, Reinach initiated students into phenomenology. Edith Stein took his course on "Introduction to Philosophy" and his seminar on motion in 1913. It was Reinach who trained many of those whom history remembers as Husserl's students: Stein, Roman Ingarden, Alexander Koyré, Hans Lipps, Jean Hering, Fritz Kaufmann. Reinach enlisted in the army in 1914, was baptized during a furlough in 1916, and was killed in action the next year.[112]

Reinach published several essays during his lifetime and co-edited Husserl's *Jahrbuch.* After his death, Edith Stein and other students collated his papers and prepared an edition of his *Gesammelte Werke*, which appeared in 1921. These convey an impression of a precise, careful thinker. Reinach "was the very opposite of Scheler, who published his ideas before he had fully digested them and thought them through."[113] His writings often are cited as an exemplar of perfected phenomenological analysis. Students in Reinach's seminar recall his manner as collaborative, directive, and encouraging. Stein remarked that it was even possible for students to convince Reinach to change his mind.[114] Reinach envisioned philosophy as a team effort by generations of relay workers. In a lecture at Marburg intended to introduce phenomenology to neo-Kantians, he said:

Spätere Geschlechter werden es nicht verstehen, daß ein einzelner Philosophien entwerfen konnte, so wenig, wie ein	To future generations it will be just as unintelligible that an individual should project a philosophy as today it is that an

[111]See below for the suggestion that Scheler reached this insight through reading Edith Stein's dissertation, where the term *Einsfühlung* is coined to denote what *Einfühlung* cannot be.)

[112]Reinach's patriotism, enlistment, and conversion from Judaism to Christianity all seem to be manifestations of his enthusiasm for Germanness. See Schuhmann and Smith (1987: 25). Smith 1982 provides an overview of Reinach's intellectual contributions. For further biographical details, see Husserl (1919), Brettler (1973), Spiegelberg (1982a), Crosby and Seifert (1981), Crosby (1983), and Schuhmann (1987).

[113]Crosby and Seifert (1981: 12). Reinach left instructions that his lecture notes and unfinished manuscripts be burnt in the event of his death. His widow, however, allowed Edith Stein to include some of the lectures in the *Gesammelte Werke*. Anna Reinach destroyed many of her husband's papers shortly after his death, and the rest of them when she fled Nazi Germany in 1942.

[114]See Stein (1986: 194). Stein elsewhere remarks that one could never convince Husserl to change his mind about anything.

einzelner heute die Naturwissenschaft entwirft. ([1914] 1921: 405)	individual should project natural science. (1969: 221)

Although Reinach revered the Greek classics and had loved Plato since boyhood, he espoused a materialism (that is, an anti-formalism) akin to that of Scheler and the other Munichers. *Essences* for Reinach were not to be conceived exclusively as empty forms or eternal ideals, in the manner of numbers. Rather, he said, there are sets of *material* essences systematically clustered within domains corresponding to the various sorts of objects of consciousness. It is these essences which can be brought to disclosure through phenomenological investigation.[115] For example, human emotions constitute a domain of objects whose essences can be exhibited to inner intuition. In this connection, like Scheler Reinach severly criticized the empirical psychology of his day and called for a new kind, an "a priori psychology."[116]

Reinach himself worked on the domain of judgments, particularly those (like promising and pardoning) that he termed "social acts." Trained in legal theory and thoroughly familiar with the German criminal code, Reinach had chosen the concept of causality in the law as the topic of his dissertation under Lipps. He argued that in the penal code, causal relations are not states of affairs (*Sachverhalte*) but rather are *relations of* states of affairs.[117] The term *Sachverhalt*, which became common in phenomenology, apparently was taken over from the legal jargon of the Imperial German Code of Civil Procedure.[118]

Reinach's 1909 *Habilitationsschrift* examined the essence of judgment. While that text no longer survives, portions of its argument reappeared in his 1911 essay for the Lipps *Festschrift*, "Zur Theorie des negativen Urteils" ("On the Theory of the Negative Judgment"). Reinach there discusses at length what is meant by a "state of affairs," in connection with a key question that would have important implications for Husserlian noetic-noematic analysis. Where, Reinach asks, does the negativity lie when one makes a negative judgment: in the state of affairs to which the judgment refers, or in one's activity of asserting it? (For

[115]Smith (1982: 297) writes: "The fundamental principle of Reinach's philosophy may be expressed as follows: for every domain of objects, whether psychological or material, mathematical or grammatical, a determinately structured family of essences can be discovered, standing in *a priori* relations to each other, as a reflection of which corresponding *a priori* laws hold of the objects in question."

[116]See Reinach ([1914] 1969: 217). This new psychology would resemble history in its focus upon rational motivations. Reinach's proposal compares with Scheler's call for a "*verstehende Psychologie*," and both are indebted to Dilthey's efforts to establish a unified basis for the *Geisteswissenschaften*. Thus, Joseph J. Kockelmans (1967) was mistaken in dating the origin of phenomenological psychology to Husserl's lectures in the mid 1920's.

[117]See Brettler (1973: 46).

[118]See Schuhmann and Smith (1987: 10). But Smith (1982: 311) quotes a 1913 work by Otto Selz to the effect that it "was Stumpf who introduced the expression 'Sachverhalt' as a technical term." Smith also remarks on the parallel uses of the term by Wittgenstein.

example, if I say that there are no cookies in the cookie jar, am I affirming a circumstance of cookie-jar-emptiness, or am I disaffirming a circumstance of cookie-jar-fullness?) While there may be no *practical* difference, the *ontological* stakes are enormous. To sort them out, Reinach distinguishes two acts that both are taken for judgments: belief, and assertion. While belief has also its negative form--disbelief--there is no corresponding negative form of assertion, he argues. Assertion is always a positive act. Therefore the negativity of negative judgment must lie with the state of affairs.[119] Or in Husserlian terms: denials are noetically positive and noematically negative--although Husserl himself fails to find this solution.[120] (With the example of the cookie jar, then, cookielessness inheres in the state of the jar and not in the act of speaking about it.)

Reinach thus accords to states of affairs an a priori status with respect to conscious acts. The reality of circumstances is independent of the activities that grasp and describe them. Reinach says that

Sachverhalte aber bestehen, gleichgültig, welches Bewußtsein sie erfaßt und ob überhaupt ein Bewußtsein sie erfaßt. ([1914] 1921: 397)	"states of affairs" obtain indifferently of what consciousness apprehends them, and of whether they are apprehended by any consciousness at all. (1969: 213)

Granted that fact, he says, one may then proceed to inquire how the states of affairs come to givenness. In the course of such an inquiry, certain essential structures will disclose themselves. For example, one sees that only states of affairs are numerically determinable, that is, have countable elements. Such elements are perceived as unities persisting over time and appearing under different aspects; yet each presents itself both as a "selfsame" (*selbig*) unity and therefore as one instance that can be counted into a group along with other ones that resemble it in some respect.

Selfsameness is not identity; and therein lies the distinction that Reinach sees between *states of affairs*, on the one hand, and *ideal objects*, on the other. An ideal object (for example, the number five) is identical with itself, but need not and does not achieve selfsameness through many perspectival appearings as do the

---

[119]Here Reinach disagrees with the positions of Wittgenstein and Pfänder, who held that all states of affairs are positive. Ingarden would criticize the ontological implications of Reinach's argument. See Smith (1982: 295-6). Reinach did not overlook the capacities of speech to bring states of affairs into being. His philosophy of "social acts" has been compared with the philosophy of "speech acts" of J.L. Austin, which it anticipated by some fifty years. See Smith (1982: 297-303); Mulligan (1987); Schuhmann and Smith (1987: 19).

[120]See Crosby and Seifert (1981: 11-12), who remark: "Reinach's position looks toward . . . the position which Pfänder subsequently elaborated in his *Logik* (1919) according to which the negativity of the negative judgment does not lie in the asserting function of the copula but rather in the referring function." See also the detailed discussion by Smith (1982: 294-297). For Husserl's view, see particularly §105, §106 and §108 of *Ideen*.

elements of states of affairs.[121] We recognize the latter *as real* in recognizing their possibility of moving, changing, and sitting alongside others of their kind. In the perception that *this particular* thing is moving or changing, its selfsameness already is given. This fact--that selfsameness of a *one* is given along with the perception of its movement--comprises an essential law governing the realm of states of affairs.[122]

Multiplicity, then, is the hallmark of the real world. Reinach says:

> Es ist eine allgemeine logische Ansicht, daß Selbigkeit und Identität wohl zu scheiden sind. Sicher ist, daß bei realen Dingen gewisse Schichten von Andersheiten sind, in denen sich Selbigkeit konstituiert, die bei anderen Gebilden, z.B. Zahlen, nicht möglich sind. Es fragt sich aber, ob das sich konstituierende selbst etwas ist, das sich wesenhaft von Identität unterscheidet. *Mann könnte sagen*: Jegliches Etwas ist als dieses erfaßbar, ist erfaßbar in seinem Eigensein. Ist es aber erfaßbar, so ist es beliebig oft erfaßbar und damit als Selbiges konstituierbar. ([1913-14] 1921: 432)

> It is a common logical view that selfsameness and identity are to be separated. It is clear that with real things there are certain layers of variance in which selfsameness constitutes itself--layers that are not possible with other formations such as numbers. However the question arises whether the self-constituting itself is something that differs essentially from identity. One could say: Every something is comprehensible as *this one*, is comprehensible in its proper being. But if it is comprehensible, then it is comprehensible as often as you please and thereby as a constitutable selfsame.

To this objection, Reinach answers that we recognize particular things through their peculiarities, not through what "all somethings" must have simply because they are "somethings." However *to be peculiar*--that is, to be recognizable in a variety of presentations--*is not itself a peculiarity* of "this something" while not of "that something." "All somethings" are peculiar in some way or other; and with this formulation we capture another essential law of the appearing of real things to consciousness. Moreover, Reinach writes, peculiarity is not a product of our having recognized it. The possibility of knowledge is rooted in the independent and a priori distinctiveness of real things.[123]

Perhaps the most fruitful and the most subtle distinction that Reinach makes is that between knowing and saying. Mere knowing changes nothing. But social acts of saying can indeed bring into existence such things as rights, obligations, and knowledges for other people as well as for oneself. The existence of legal claims and obligations depends upon the acts of promising, contracting, pardoning, and so forth through which they arise and are extinguished. When one undertakes phenomenological investigation of such creative speech acts, one

[121]See Reinach (1969: 207). The ontological implications are drawn out in Reinach's seminar on the essence of motion (Reinach [1913-14] 1921: 427 and 432).

[122]Reinach does not clarify whether the copulative verb, to be, asserts identity or selfsameness. Conversely, it is unclear whether identity or selfsameness is being denied when one asserts that identity "is not" selfsameness.

[123]See Reinach ([1913-14] 1921: 432-3).

discovers their essential temporal and social structures. These essential structures are "the a priori foundations of the civil law."[124] Furthermore, an *obligation* requires the existence of someone to bear the obligation, and it also requires a future time in which its content lies. In the same vein, a social act of *imparting information* requires (according to its essence) that there be two or more subjects "reticulated together in a specific way."[125] Acts of communication, then, offer occasion for the direct intuition of the existence of other i's.

Reinach discussed the problem of other i's in his course "Introduction to Philosophy" in 1913, which Edith Stein attended. In that course, Reinach presented Lipps's doctrine of empathy and reviewed the standard criticisms of it.[126] He suggested that this doctrine was designed to answer three questions: (1) How we catch on to (*erfassen*) alien experiences; that is, analysis of alien perception; (2) How we have developed the ability to catch on in this way; and (3) Whether our perceptions are valid and justified. The second and third are easily dealt with, says Reinach. The weightiest question is the first, the analysis of alien perception; and Lipps's theory does not have the resources adequately to deal with it.[127] Reinach's class notes read as follows (with bracketed material in the German supplied by the editor):

[Die] Frage [ist] dann, was faktisch vorgeht, wenn wir fremde Erlebnisse wahrnehmen. 1. Zuerst konstatieren wir: [Das] fremde Erlebnis hat hier [seine] Selbsterscheinung. *Vergegenwärtigung der Trauer kann z.B. übergehen in Selbsterscheinung; aber beides [ist] verschieden.* . . . 2. Es ist nicht nötig, [daß] die Sympathiegefühle meinerseits [vorhanden wären]. [Ein] Trauererlebnis auf meiner Seite braucht nicht vorzuliegen. . . . Fremde Trauer [ist] schon erfaßt, wenn ich dadurch angesteckt werde. 3. *Fremde Erlebnisse [sind] nicht so direkt zu erfassen wie eigene.* [Der] Gegensatz zwischen Fremdheit und Eigenheit von Erlebnissen [wird hier] verwechselt mit [dem] Gegensatz zwischen Mittelbarkeit und Unmittelbarkeit ihrer Erfassung. Ich erfasse	The question is, what factually goes on whenever we perceive alien live experiences? 1. First we establish: that an alien live experience is having its self-manifestation here. Representations of grief can fade into self-manifestations, for example; but the two are different. . . . 2. It is not necessary that sympathetic feelings be present on my side. A live experience of grief need not be put forward on my side. . . . Alien grief is already comprehended if I am infected with it. 3. Foreign live experiences are not so readily comprehended as one's own. The contrast between alienness and ownness in experiences is here [not to be?] mistaken for the contrast between mediateness and immediateness of their comprehension. I comprehend alien grief through something (demeanor or the like)

[124]This was the title of an essay by Reinach in the first volume of Husserl's *Jahrbuch* in 1913. See Smith (1982: 298-9).

[125]See Smith (1982: 302). In Reinach's phenomenology, then, the key terms were "doing and time"--*Tun und Zeit*, rather than *Sein und Zeit*, which later would occupy Heidegger.

[126]See Reinach ([1913] 1989: 389-393).

[127]Namely, Lipps's hypothesis of an imitative instinct (*Nachahmungstendenz*) does not do the job, in Reinach's estimation. See Reinach ([1913] 1989: 390-391).

fremde Trauer durch etwas (Gebärde o. dgl.) hindurch, aber [als] ein Selbstgegebenes. *[Das fremde] Erlebnis erscheint in etwas anderem, durch etwas anderes hindurch: indirekte Selbsterscheinung, durch Physisches vermittelt. Aber [ein] Schluß liegt nicht vor.* ([1913] 1989: 391-2)	--but I comprehend it *as something self-stated.* The alien live experience makes its appearance in something other, through something elsewhere: indirect self-statement, transmitted through something physical. But there isn't any logical inference here.

What, then, do we share when we share someone else's feelings? Reinach answers by distinguishing between the *content* of a feeling and the *activity* of feeling it. There may be as many different acts of feeling as there are people, he says, while the content of all those acts could be the same. Feelings are analogous to judgments, he says. Many people will make many executions of a syllogistic inference; but the judgment that they reach is identical. "A = B" represents an ideal object, not a state of affiars; so in the case of coincident judgments, there is identity (not selfsameness). By the choice of this analogy, Reinach seems to be saying that feeling-contents, too, are ideal objects. Thus there can be identity of grief, even while several people are sharing the feeling of grief in their discrete and separate actualizations of it.[128] An equivalent statement in Husserlian terminology would be this: when feelings are shared there is noematic identity but noetic multiplicity.

Reinach's line of reasoning thus establishes the essential possibility of alien subjects *in the realm of ideal objects*, even though it neglects to investigate the "material essence" of other people and their manner of self-presentation *in the realm of states of affairs*. Nevertheless Reinach stands within the realist tradition of Munich phenomenology. In this chapter, we have seen that the hermeneutical issues framed first by Schleiermacher and Dilthey were taken up by Theodor Lipps and brought into focus by means of three key concepts: motivation, the mutual availability of human beings through empathy, and the i. Pfänder, Scheler, Conrad-Martius, and Reinach advanced the agenda of phenomenology by developing those key concepts, particularly in connection with questions of how states of affairs are *realized* and how individuals *communicate* with one another. Their discussions enunciated the problems to which the best-known solutions today are those of Husserl. In fact, it has taken no small effort in this chapter to hold Husserl's interpretations at bay long enough to permit a reconstruction of the conversations to which he sought to contribute. Husserl's engagement with the questions of Munich phenomenology will be explored in the next chapter. This chapter concludes with a reminder that phenomenology as Edith Stein knew it was still very much a program undergoing construction and contestation.

[128]See Reinach ([1913] 1989: 393). As she took notes during Reinach's lecture, Stein could not have guessed that she would be editing those words just a few years later as she grieved for her teacher.

Chapter Two

Husserl's Early Treatments of Intersubjectivity

The agenda of the Phenomenological Movement, as we saw in the last chapter, came together in Munich around the investigation of three nested phenomena: motivation, empathy, and the i. Research inspired by Theodor Lipps gave Husserl the questions and the terminology upon which he would build his early investigations of intersubjectivity. Although he did not reside in Munich, Husserl was in close contact with colleagues there before the First World War. Reinach came from Munich to work as *Privatdozent* at Göttingen under the sponsorship of Husserl, whom the younger phenomenologists called *der Meister*.[1] Husserl was productive in his publications, prolific in his private *Forschungs-mansckripte*, and prodigally generous with his junior colleagues. First at Göttingen and later at Freiburg, Husserl freely shared the fruits of his research with his advanced students and he delighted in their original implementations of the phenomenological methods that he had pioneered.

Edmund Husserl (1859-1938) came from a Jewish family in Moravia but he received a German education. He studied mathematics at Berlin and Vienna and earned his doctorate in 1882 with a dissertation on the theory of the calculus of variations. Afterwards Husserl attended Brentano's lectures on philosophy, went to Halle to study psychology, and habilitated there with a thesis on the concept of number. At Halle he was baptized and he married a woman from his hometown community who also renounced her Judaism. Husserl's *Habilitationsschrift* was reworked into the first part of his *Philosophie der Arithmetik*, published in 1891. He made his reputation, however, with the two volumes of the *Logische Untersuchungen*, which came out in 1900 and 1901. In 1901 Husserl joined the faculty at Göttingen, where he would teach for 16 years. There he wrote his 1911 manifesto on "Philosophy as Rigorous Science" for the first issue of *Logos*, and worked out the comprehensive formulations of his phenomenology that are presented in *Ideen zu einer reinen Phänomenologie und phänomenologischen Philosophie*.

The first book of the *Ideen* appeared in the premier volume of Husserl's *Jahrbuch für Philosophie und Phänomenologische Forschung* in 1913. Shortly afterwards the World War disrupted Husserl's academic routine. His students dispersed. His older son, Gerhart, was wounded in February 1915 and sent to a sanitarium; and his younger son, Wolfgang, died in battle. Husserl himself was hospitalized for "indigestion and nicotine poisoning" in the fall of 1915.[2]

---

[1]Elements of respect, affection, and sarcasm mingle in this term, which often appears in the correspondence of Edith Stein. It is unlikely that anyone called Husserl *Meister* to his face.

[2]Schuhmann's *Husserl-Chronik* (1977: 196) places this illness between September 25

Materials for a second volume of *Ideen* had been prepared in 1912, immediately after the first book. Husserl used and augmented those materials in preparing his lectures and seminars on "Natur und Geist," offered at various times from 1913 through 1916. His assistant Edith Stein then arranged them for publication, but Husserl apparently was not satisfied with the result. Only after his death would the second and third books of the *Ideen* appear.[3] In 1916 Husserl accepted appointment to a professorship at Freiburg im Breisgau, the position from which he would retire in 1928. Landmark publications of his later years include *Formale und transzendentale Logik* in the *Jahrbuch* of 1929, and the *Méditations cartésiennes* of 1931; but those lie beyond the concerns of the present study. Husserl's last years were saddened by the escalation of National Socialism's racist policies against Jews. He died of pleurisy on Good Friday of 1938.

Husserl left behind some 45,000 pages of stenographic notes: lectures, essays, correspondence, transcriptions, and private reflections. Since he wrote episodically and in shorthand, Husserl's work needed to be transcribed, set in order, and edited before he could publish it. Manuscript preparation was called *Ausarbeitung*: elaboration or filling in the details.[4] During his lifetime Husserl entrusted this work to a series of assistants: Edith Stein, Ludwig Landgrebe, Martin Heidegger, Eugen Fink. To a wider circle of colleagues and students he offered access to research papers that might have bearing on topics of their interest. Husserl generously shared his "intellectual property."[5] Those papers remain a rich legacy. They are archived at the Catholic University of Louvain, and are still being published.[6] These materials make it possible to identify stages in the development of Husserl's thought, and therefore to reconstruct the contours of his phenomenology at the time when Edith Stein worked with him.

A reconstruction of Husserl's philosophy in the era of the First World War cannot rely uncritically upon what commentators have written about his work taken as a whole. Obviously, his work was not yet "a whole" when it attracted Stein to him first as a student and then as a junior collaborator. There are five considerations that require particular caution. (1) The version of the *Logische Untersuchungen* familiar to Stein was the original edition of 1900-01, not the

---

and October 20 of 1915, on the basis of correspondence with Frau Malvine Husserl and others. However these letters are not included in the index of the ten-volume *Briefwechsel* published in 1994. Husserl's mother died in July of 1917, while Edith Stein was editing his papers on temporality. Further biographical details come from Kockelmans (1967: 17-21), Herbstrith ([1985] 1992: 139), Edith Stein (1986), and Bell (1990: 3-4).

[3]In 1952, edited by Marly Biemel, as the fourth and fifth volumes of Husserliana, the collected works of Edmund Husserl.

[4]Ingarden (1962) attests to this practice.

[5]As Gibson (1971: 68) reports. Stein uses the same remarkable term, *geistiges Eigentum*, in the Foreword to her dissertation.

[6]For a personal memoir of the foundation and early years of the Husserl Archive in Louvain, see Biemel (1989).

revised edition brought out in 1913 and 1921 to adjust the earlier work to the new transcendental orientation of the *Ideen*. While the second edition was known to Stein, she agreed with those who criticized Husserl's transcendental turn as an abandonment of his original direction.[7] (2) Stein tells us that she did not see the relevant materials in *Ideen II* until after she had completed her own dissertation;[8] even so, she would have heard them discussed in Husserl's classes and seminars. (3) The overall arrangement and many of the section headings for *Ideen II* are known to be Stein's work.[9] She also composed introductory and transitional paragraphs, fulfilling her responsibilities to produce an "elaboration" of Husserl's raw manuscripts. (4) Husserl's mature work on intersubjectivity appears in his Fifth Cartesian Meditation, published in 1931, and commentators typically take this to be his definitive statement.[10] Earlier manuscripts, however, display a rather different approach to questions of interpersonal communication; and in the 1920's Husserl seems to have drawn closer to formulations attempted by Stein in her 1916 dissertation.[11] (5) Husserl's Lectures on Phenomenological Psychology from the summer of 1925 should not be read in isolation, but as the fruit of conversations whose milestones are Husserl's 1917 manuscripts on psychology and epistemology (to which Stein contributed) and Stein's 1919 *Habilitation* essay on "Philosophical Grounding of Psychology," (which Husserl published in the 1922 *Jahrbuch*).[12]

With these caveats in mind, we must prefer primary texts over commentaries. Husserl's early work on intersubjectivity should be read in the context of the hermeneutical phenomenological tradition. From that tradition he adopted the key terms *Motivation* and *Einfühlung*, whose career we have been tracing. The following texts offer relevant passages:

---

[7]See Société Thomiste (1933: 43-45).

[8]See E. Stein ([1917] 1980: vi; 1989: 1-2). The translator's introduction to E. Stein's dissertation, stressing its originality, overlooks the fact that materials included in *Ideen II* had been discussed by Husserl in classes and seminars that E. Stein attended. See W. Stein ([1962] 1989).

[9]See Rojcewicz and Schuwer's introduction to their translation of *Ideen II*, pages xii-xiii. Stein's contributions to the second book of the *Ideen* will be discussed in the last section of this chapter and in chapters four and six, below.

[10]See Carr (1974: 82-109) and Mensch (1988).

[11]This is the view of Ingarden (1967: 49-50). Kern (1973) shows that intersubjectivity was a sustained interest of Husserl's throughout his career.

[12]Stein's "Beiträge zur philosophischen Begründung der Psychologie und der Geisteswissenschaften" will be discussed below in chapter six. Stein edited the manuscripts that would become Books 2 and 3 of Husserl's *Ideas*, as well as the 1917 manuscripts on psychology and epistemology that would appear in Husserliana XXV in 1987. Husserl's 1925 lectures were published in Hua IX in 1962. That Husserl should accept and employ aspects of Stein's philosophy seems natural enough. The scandal is not that he adopted her ideas, but that having done so he denied that women could have academic careers in philosophy. See chapter five, below, for Husserl's lukewarm letter of recommendation for Stein.

- The First Logical Investigation (1901)
- Correspondence with Theodor Lipps (1904, *Briefwechsel* 2)
- Research papers on intersubjectivity (1905 and 1909-10, Hua XIII)
- The *Logos* article (1911)
- Correspondence with Wilhelm Dilthey (1911)
- *Ideen* (1913)
- Research papers on intersubjectivity (1911-1914, Husserliana XIII)
- Lectures and seminars on "Nature and Culture" (1913-1916, in *Ideen II*)

A few general remarks will serve to orient the discussion. We have seen that Dilthey, Scheler, and Reinach were calling for a new kind of psychology that would serve as the foundation for the human sciences, setting them apart from the natural sciences. In their view, the teleological *motivations* that impart intelligible coherences to occurrences in the cultural world (and to that world itself as a *world*) are quite different from the genetic *causes* that impart intelligible coherences to occurrences in the natural world (and to that world itself). This new "psychology of understanding" would grasp the distinctive coherences structuring the world of *Geist*, that is, culture--just as empirical sciences grasped the coherences of the natural world. This new kind of psychology would stand in contrast to the kind of empirical psychology being pursued in the laboratories of the day. That empirical psychology, insofar as it sought physical causes for mental events, really belonged among the natural sciences.

Husserl's quest was different. He wanted to investigate how it happens that one understands coherences and has worlds *at all*. Husserl noticed that one's *understanding of* causality and motivation is neither caused nor motivated. Furthermore, something underlies both the human sciences and the natural sciences that itself is neither kind of science. Husserl would struggle to formulate his investigations of this "pure logical" source in spatial terms ("realm," "field," "foundation") and also in terms of cognitive activity ("i," "ego pole," "constitution"). At the outset, he needed to insist that logic was not a product of thought, nor was it the tried-and-true, habitual patterning built up by correct thinking.[13] Thus he agreed with the Munich phenomenologists and Dilthey that one must consign a certain brand of psychological research, with its causal claims,

[13]It should be clear to the reader that the present account is a *history*, that is, an attempt at cultural science, an attempt to follow the motivational coherences exhibited by events comprising the early Phenomenological Movement. I am not attempting a phenomenology of those events. In this context, my claims of necessity are meant to imply conditional necessity in light of goals and outcomes, not absolute logical necessities. Thus, for example, Husserl can insist that the phenomenological reduction is "unmotivated" (Hua XIII: 156) yet it remains possible for the historian of philosophy to lay out the motivated coherences comprising the intelligibility of the occurrence of that claim in its particular context. This historical task differs also from the task of understanding why individual human beings might wish to study phenomenology, or might wish to apply it in other sciences. Husserl allows for such personal decisions to be "motivated," that is, for phenomenologists to have cognitive careers lived out apart from the fresh start of the phenomenological reduction.

to the realm of the *natural* sciences. Husserl alone, however, would come to regard the *cultural* sciences as equally problematic.

The new science that Husserl proposed--phenomenology--would investigate how other sciences, both *Geisteswissenschaften* and natural sciences, could exhibit their respective varieties of coherence: motivation on the one hand, and causality on the other. This new science would be a *seeing of essences*, whereas the other sciences were *followings of coherences*. Essences are unwavering, whole, simple, and given completely to the understanding (although painstaking work of some duration may be required to prepare for this event).[14] In contrast, as we shall see, both motivation and causality involve chains, segments, sequence, articulation. Thus the all-or-nothing *seeing* of essences contrasts with the quasi-kinesthetic step-by-step *following* of motivation and causation. In writings from the early 1900's which we are about to examine, Husserl uses the (metaphorical) term "insight" to characterize the way in which one grasps logical law, and he uses the term "felt" (*empfunden*) to characterize the way in which one follows an articulated chain, whether that chain be a process of physical reactions, a sequence of emotional responses, a national history, a biography, or a logical proof. In Husserl's formulations before the *Ideen*, these seem to be parallel but contrasting kinds of cognitive access: in-sight (*Einsicht*), yielding knowledge of logical laws, and in-feeling (*Einfühlung*), yielding empirical knowledge. The problem then becomes: How may the relationship between the two be understood?

In-sight and in-feeling are not simply opposed, set side-by-side as alternatives; rather they are in some way integral to one another. Insight seeks the logical "why's" that connect the steps through which the understander flows--by progressively in-feeling them--as she picks out causally related physical events or relives cultural events, constituting their coherence. These logical "why's" register as *recognitions affirming the rational coherence of* the directional flow of felt understandings of the empirical (physical or psychological) why's. Insight ratifies what has already been experienced in ablative and accusative empathy. In Husserl's work before the Armistice, two distinct attempts were made to establish the correlation between phenomenology and the other sciences--that is, between the intellectual seeing of essences and the quasi-kinaesthetic following of coherences. The first is the 1911 *Logos* article, "Philosophy as Rigorous Science." Its tactic is of a piece with the Prolegomena to the *Logische Untersuchungen*: the relation must be one of logical foundation. The second attempt is the arrangement of the second book of the *Ideen*, produced by Stein but never approved for publication by Husserl. There, the relation would be owing to the human person's direct access to other persons; all necessity is founded in that.

These introductory comments have previewed the gist of Husserl's project. In actuality, his use of terminology was not so tidy, and he tried out numerous

---

[14]In the Prolegomena to the *Logische Untersuchungen*, Husserl distinguishes between the intuitive grasp of a logical law, which one either has or does not have, and the multiple psychological steps through which one may have to pass in order to clinch that grasp; see §24.

approaches that did not bear fruit. With this in mind, we can now turn to a more detailed consideration of the texts.

A. From the *Logische Untersuchungen* to Seefeld

In December of 1903, Theodor Lipps sent Husserl a copy of his newly published *Leitfaden*, in hopes that Husserl would comment upon it.[15] In the cover letter, Lipps tells Husserl that their positions are in substantial agreement although he finds Husserl's terminology too psychologistic. This gentle teasing apparently did not amuse Husserl, who earnestly desired to distance himself from Lipps's entire psychological project. Husserl's response to Lipps is lost; however there is an extract of that letter made by Lipps's student Johannes Daubert when he visited Husserl on January 18 of 1904.[16] It bears the title "My Position on Psychologism --by Husserl," and it characterizes Husserl's project as follows:

Was heißt: Einsicht „erfaßt“ ein Sein, das ihr gegeben ist, oder Gesetze, die über jedes einzelne Sein hinausgehen, und Gesetze, die zu jedem Sein überhaupt gehören und ihm als a priori vorangehen? (1994, 2: 123)	What does this mean: Insight "comprehends" a being that is given to it, or [comprehends] laws that transcend each particular instance of being, and laws that appertain generally to every instance of being and precede them all as a priori?

This "in-sight" or discernment that is a grasping of a priori laws needs to be scientifically investigated. Yet psychology is not the science that can investigate it; at least not psychology carried out in the usual natural-scientific way. Up until now, Daubert copies, either epistemology has hinted at some mystical distinction between apriori norms of thought and natural laws, or it has collapsed the two. But if the laws of logic be merely "tendencies," analogous to gravitation and electrical forces, then there remains a question that cannot be answered:

Warum soll das Denken beim Verfolg logischer „Tendenzen“ als „Norm des Richtigen“ gelten? (1994, 2: 126)	*Why should* thinking along with a sequence of logical "tendencies" count as a "standard of what is accurate"?

If it indeed happens that certain thoughts usually occur together or follow one another, these associations and processes may very well be described by *natural laws* that will reliably predict the occurrence of one on the basis of another. But *logical laws* are not predictors of that sort. Nor are they the "tendencies" themselves. As Daubert writes, logic is not a force that works in combination with other psychic forces. It is not a movement, nor does it move anything else.

---

[15]Husserl reported on new publications in logic for the journal *Archiv für systematische Philosophie*. See Schuhmann and Schuhmann (1994, 2: 121).

[16]The letter roughly summarizes the case against "psychologism" in the Prolegomena. See Schuhmann and Schuhmann (1994, 2: 35, 122-127). This visit, and the divergence between Lipps's and Husserl's positions that apparently was discussed during it, marked the beginning of the migration of Munich phenomenologists to Göttingen.

Rather, there are essential laws intrinsic to acts of thought. These laws of pure logic have absolute validity.

Die *rein logischen* Gesetze drücken *Wesens*gesetze der „eigentlichen" Denkakte, d.i. zum spezifischen Wesen dieser Akte gehörige, also von ihm unberaubbare Möglichkeiten bzw. Unmoglichkeiten, sie in gewissen Konstellationen zu vollziehen, aus. (1994, 2: 126)	The *pure logical* laws express *essence*-laws of "intrinsic" thought-acts, that is, laws belonging to the specific essence of these acts, hence the respective possibilities and impossibilities, inseparable from them, of being consummated in certain constellations.

Essences are "seen" in their entirety, not followed sequentially.[17] The forms of valid inferences have no parts--even though when inference actually occurs, it must occur as a step-wise movement along a prescribed path. The validity of the path is not owing to "correct" connections between steps, much less to how well worn the path happens to be, but rather to its essential lawful form.

These notes of Daubert, drawn from the 1904 letter to Lipps, reprise the long and careful arguments set forth in the Prolegomena to the *Logische Untersuchungen*. Based upon lectures delivered in 1896 in Halle, the Prolegomena try to establish that "pure logic" is a field of essences distinct from but foundational for the various enterprises that ordinarily go by the name of logic. Pure logic is not a technique employed in the sciences, a duty operative in the intellectual disciplines, or a record of typical psychological occurrences.[18]

Man muß durchaus unterscheiden: Gesetze, welche *zur Normierung* der Erkenntnistätigkeiten *dienen*, und Regeln, welche *den Gedanken dieser Normierung selbst enthalten* und sie als allgemein verpflichtend *aussagen*. . . . Auch hier wandelt sich allererst durch die Einführung des normativen Gedankens das Gesetz in die Regel, die seine selbstverständliche apodiktische Folge, jedoch nach dem Gedankengehalt von ihm verschieden ist. (Hua XVIII:159, 161)	Throughout one must distinguish: laws, which happen to be useful for regulating the activities of knowledge, and rules, which include the very notion of this regularization itself and assert it as generally binding. . . . (It is) through the introduction of the normative notion (that) the law first of all converts itself into the rule that is its self-evident apodictic consequence yet is different from it as9 regards notional content.

Husserl is interested in the lawfulness of rules. He wants to know how correctness pertains to what scientists recognize as correct procedures and inferences. Therefore Lipps's attempt to place logic within psychology is among several

[17]The transition *from* feeling the motivated flow of sequential appearings *to* grasping the unity of an object is what Husserl would later describe as object constitution. The flow itself he would explicate as internal time consciousness. Husserl's early manuscripts on internal time were edited by Edith Stein with a view toward supporting his later, transcendental phenomenology.

[18]The 1970 Findlay translation of the *Logische Untersuchungen* is not cited here because it is not precise enough to support close analysis of Husserl's argument.

approaches that come in for criticism in the Prolegomena (although in the preface to the 1913 edition, Husserl acknowledges that Lipps has altered his position).[19]

At the beginning of the First Logical Investigation, Husserl elaborates his distinction between insight and the feeling of being pulled along by motivation. These are contrasting ways of perceiving the connections among elements of thought, and they pertain to two contrasting varieties of a relation whose common name is "indication" (*Anzeige*). In one sort of indication, Husserl says, the perception of a sign or a symptom calls into the mind the thought of something else. This connection is a "motivating" (*Motivierung*). The pull from one to the other is felt (*empfunden*) -- but felt nonperspicaciously, without insight (*nicht-einsichtig*). The unity between sign and signified is lived as the unity between two judgments: the judgment that the sign obtains, and the judgment that the signified must therefore obtain as well. Husserl says that this felt, motivated, two-step connection is expressed in the word "since" (*weil*).[20]

On the other hand, there is a different sort of unity between premises and the conclusions that are deduced from them by the laws of logical inference. In a proof, we say "that the relation of consequence could be inspected or seen-into" (*daß die Konsequenz eingesehen werden könne*).[21] There is an objective relation between ground and result, which Husserl calls an ideal unity. Such unities are not judgment-experiences (*Urtheilserlebnisse*); they are *the ideal contents of* judgment experiences. This is so, regardless of the fact that the actual working out of proofs usually occurs in a "motivated" way, that is, by following procedures that one has been authoritatively taught. Husserl calls this mere technique, while insisting that insight into the logical necessity inhering in the form of the proof is something else altogether.

Husserl's thinking on unity, wholes, and parts derives from the "mereological" theory of Franz Brentano, and it is reminiscent of Schleiermacher's "divination" as well.[22] Brentano held that whatever is present to consciousness is in some way complex, yet because of its *presence* this complexity cannot be sequential or syntactic. Thus the essential structure of the *whole* inference is not present *in any of the parts* while the proof-pursuing mind is working its way

---

[19]Husserl quotes Lipps to the effect that the laws of logic are natural laws and are directed toward the accomplishment of goals. He fleshes out this "psychologistic" contention by suggesting that it means to use the term "law" to denote what governs the real causal connections among mental events occurring in succession. See Hua XVIII: 65-67; Husserl (1970: 93-94).

[20]See Hua XIX/1: 31-32. The nonperspicacity or eidetic blindness of motivation as one inwardly coincides with the flow of events, asserted by Husserl, should be compared to the blindness of the will asserted by Schopenhauer and also to the thrownness of human existence described by Heidegger. Against these kindred notions stands Pfänder's countervailing claim, discussed above: that willing is not blind but rather sees the possibility of a future reality.

[21]See Hua XIX/1: 33.

[22]On Brentano, see Bell (1991: 17-28). On Schleiermacher, see above, chapter one.

through the articulated steps of inference one by one. The premises do not causally produce the conclusion, nor does the conclusion motivationally arrange the premises. In other words, when parts are *intrinsically* related to the whole, their relation is neither motivated nor caused, but essential. By contrast, motivation and causation are two ways of understanding the integrity of wholes whose parts are *not* intrinsic to them: motivated series and causal series, respectively. One "sees" essential unities but "feels" motivations and causations.

Although both are feelable (in his early terminology), Husserl asserts that these two kinds of series are distinct. Motivation is not causation.[23] It is easy to overlook this remark, and Bell apparently has done so, for he asserts that Husserl has written the *Logische Untersuchungen* from a naturalistic standpoint. Bell writes:

> A philosophical theory is naturalistic to the extent that it is committed to the view that the universe contains nothing but natural phenomena--a natural phenomenon being any object, event, property, fact, or the like, whose explanation can in principle be couched exclusively and without remainder in terms acceptable within the natural sciences.[24]

In other words, for naturalism the coherence of the universe is accounted for in terms of connections that are causal in character. But if this were Husserl's position, he would be contradicting himself when he asserted that the essential laws of logic are not self-enforcing, (that is, they do not "cause" us to make correct inferences). In the naturalist's universe, there would be no principle other than causality to account for why scientists do indeed follow the rules of logic. Yet it is fair to say that Husserl's *Logische Untersuchungen*, even if not absolutely naturalistic, are overwhelmingly concerned with the foundational principles inherent in the practice of the natural sciences. Phenomenology is meant to establish the possibility of precisely those practices; not until the 1911 *Logos* article will it claim to secure the foundation for *both* the causal chains explored in the natural sciences *and* the motivated coherences explored in the historical-cultural sciences.

Throughout these arguments, Husserl has not yet addressed the status of other people.[25] Bell has some warrant for describing Husserl's methodological

---

[23]Or more exactly, to perceive something as motivated is to perceive it as not caused (Hua XIX/1: 35). However Husserl will later expand the denotation of the term "motivation" to cover both causal and volitional sequences, as mentioned above.

[24]See Bell (1991: 154). Bell's argument will be that Husserl turned from naturalism to transcendental idealism after writing the *Logische Untersuchungen*. This is a helpful general periodization of Husserl's career. However we must be more exact in detecting hints that already in 1901 Husserl admitted non-causal sequential coherences as well as causal ones. Non-causal motivations are what impart structure to the cultural world.

[25]In the *Logische Untersuchungen* Husserl sometimes presupposes the rule that thinkers are plural: not as an unacknowledged first premise, but certainly as an unreflected determinant of method (if one may apply to intersubjectivity a distinction that Husserl makes in his critique of psychologism in §19 of the *Prolegomena*; compare his admission, in note 2 of §6 of the *Logische Untersuchungen*, that his phenomenology follows a "zig

stance in the *Logische Untersuchungen* as solipsistic "in a weak sense": while it stops short of denying the possibility of knowing other minds, it confidently assumes that the conscious experience of an individual can be treated adequately "in total isolation not only from facts about the inanimate environment of the individual, but also from facts about other psychological subjects."[26] Husserl's brief comments on interpersonal communication in the First Logical Investigation bear this out.[27] A speaker, he says, is able to impart the sense of her psychic acts to a hearer if and only if the hearer understands *that* she is trying to communicate--and that *that* requires that the hearer take the speaker to be a person (*er den Sprechenden als eine Person auffaßt*) who is not just making noise but rather is *speaking to him.*[28] However, to *understand* information is not at all to have precise conceptual knowledge of it. Understanding consists merely in the hearer's intuitive comprehension of the speaker as a person who is expressing this and that; he *perceives* her as such. Nevertheless, Husserl says,

> Die Kundgabe nimmt der Hörende in demselben Sinne wahr, in dem er die kundgebende Person selbst wahrnimmt--obschon doch die psychischen Phänomene, die sie zur Person machen als das, was sie sind, in eines [A]nderen Anschauung nicht fallen können. (Hua XIX/1: 40 [restoring "Anderen" as in Husserl 1901: 34])

> The hearer perceives the information in the same sense in which he perceives the speaker herself who is doing the informing--although after all the psychic phenomena that make her into a person cannot, as that which they are, drop into another person's intuition.

Thus the individual is insulated from sharing intimately the live experiences of another person, thanks to a fine distinction that Husserl draws between full-out adequate intuition, and a relatively intuitive presentation. Since Stein will challenge this quibble, it is worth citing it in full.

> Der Hörende nimmt wahr, daß der Redende gewisse psychische Erlebnisse äußert, und insofern nimmt er auch diese Erlebnisse wahr; aber er selbst erlebt sie nicht, er hat von ihnen keine „innere“, sondern [nur] eine „äußere" Wahrneh-

> The hearer perceives *that* the talker is expressing certain psychic experiences, and to that extent he also perceives these experiences. But he himself does not experience them live. He has of them only an "outer" perception, not an "inner"

---

zag" course to found the presuppositions by which it operates).

[26]See Bell (1991: 156). For a contrasting view and the argument that Husserl ultimately overcame his "transcendental solipsism," see Ströker (1993: 117-145, especially page 139). Bernet *et al.* (1993: 206-211) distinguish two incompatible and unreconciled approaches to i-hood in Husserl's work: through intersubjectivity, and through eidetic analysis. On the former account, the pure i can be differentiated into "I" and "you." On the latter account, the pure i is the universal form of consciousness and is logically prior to any such personal differentiation. See p. 206.

[27]Kern discusses this passage as background to Husserl's treatment of intersubjectivity. See Kern (1973: xxiii-xxiv).

[28]See Hua XIX/1: 39.

mung. Es ist der große Unterschied zwischen dem wirklichen Erfassen eines Seins in adäquater Anschauung und dem vermeintlichen Erfassen eines solchen auf Grund einer anschaulichen, aber inadäquaten Vorstellung. Im ersten Falle erlebtes, in letzteren Falle *supponiertes* Sein, dem Wahrheit überhaupt nicht entspricht. Das wechselseitige Verständnis erfordert eben eine gewisse Korrelation der beiderseitigen in Kundgabe und Kundnahme sich entfaltenden psychischen Acte, aber keineswegs ihre volle Gleichheit. (Hua XIX/1: 41 [restoring "nur" as in Husserl 1901: 34-35])

one. There's a big difference between the actual seizing of a being in adequate intuition, and the virtual seizing of such a being on the basis of an intuitive but inadequate presentation. In the former case [we have] a being that is something experienced live; in the latter case an *alleged* being, which ultimately falls short of displaying its reality. Reciprocal understanding calls for precisely a steady correlation of psychic acts deploying themselves in information-giving and information-taking on both sides. But in no way does it call for their full equality.

If this be solipsism, it is a fragile, subtle, yet stubbornly persistent variety. It is not overcome in the *Ideen*, where Husserl tries to dissolve the exclusivity of phenomenology and open up its practice to anyone at all. Husserl declines to show that *there is* anyone at all besides himself. As we shall see, Husserl will deduce the *possibility* of other human individuals, but he will deduce it *from* the data of individual experience. The *actuality* of other human beings does not enter into Husserl's phenomenology until the 1930's. Nevertheless the question of other people begins to be treated in Husserl's research manuscripts during the first decade of the twentieth century--if only from a standpoint of "weak solipsism."[29] To these we now turn.

In 1905 Husserl vacationed in Seefeld with his new friends Daubert and Pfänder, and they talked about human relationships.[30] Husserl's tentative jottings during those pleasant days contain the seeds of his critical alternative to Lipps's account of how one understands some of the items in the surrounding world to be individuals like oneself. How, Husserl asks, do I recognize my own identity as continuing throughout the flux of my perceptions? I am the selfsame in the continuous comprehending of them: appearances, presentations, judgments; in doubt, feeling, and will. If the principle of my sameness were "the form of the connection of all these moments and [if it] achieved the connection *empirical unity* through 'association and habit'," then it too would be an item in the flow. In that case it would be something creat*ed* rather than creat*ive*; and then its own creation

---

[29]Interestingly, Husserl's concern with the problem of other people began in earnest just after Lipps's students began to flock to Göttingen. As Kern remarks, Lipps's theory of empathy is the horizon against which Husserl works on the problem of other people's experience. See Kern (1973: xxv). Kern also notes (p. xxvi) the possibility that Husserl may have taken the term *Einfühlung* from Meinong. Nevertheless Husserl returns often to reconsideration of Lipps's theory, with which he disagrees. Kern's reading of Husserl's manuscripts on intersubjectivity has guided my interpretation of them here.

[30]The fruits of their discussions also are reflected in the so-called Seefelder manuscripts, Hua X: 237-268. Although this early phenomenology of interpersonal communication was superceded in Husserl's mature work, it was still an important component of Husserl's thought as Edith Stein encountered it in 1913.

and ensoulment (*beseelen*) would still remain to be accounted for. But personality is not something findable phenomenologically as a discrete datum, Husserl continues; rather, it is the unity acknowledged (*Geltungseinheit*) amid the plurality perceived--like the substance of a thing.[31]

In the case of other people, Husserl muses, I arrive at their individuality through analogy, just as extrapolation by analogy gives me the constitutive qualities of things that I do not and cannot possibly perceive, such as the interior of the earth, the sun, or the stars. That is, I see something other than myself and I assume it's another person--without seeing a person as such, inasmuch as I do not perceive those intimate intrapersonal phenomena that are and can be given only to her. Instead, I *think my way* to the affirmation of an i-coherence (*Ichzusammenhang*) to whom such phenomena would belong. The alien i is a conclusion, not an intuition for me. All that is "evident" to me is that a phenomenon can belong only to one i. That "evidence" perhaps is owing to the anchoring of different i's (*der verschiedenen Ich*) to different physical bodies (*Körpern*) that belong to a space and a regularity of spatial appearing. The appearance available to me at my present location is not available to me at any other location; however, at another time I could have access to a different appearance if I were to occupy a different location. Likewise an "other" can have access to that appearance right now if he is in that spot. The possibility of changing places is required for attaining consensus about empirical observations.[32]

Thus by 1905 the qualifying notions of "here or there" and "now or later" modify the "pure logic" and "immediate insight" that Husserl's earlier phenomenology had pursued. These notions surface in the early research manuscripts as the themes of *memory* and of *body*--body considered first as *Körper* permitting the spatial arrangement of appearances, and then as *Leib,* sentient in its own right. (The plurality of bodies eventually will prompt Husserl to dislocate i-hood from any one of them and make it "transcendental" as that guarantor of rational certitude and constitutor of essences that is not the property of any particular embodied i.)

The sentience of bodies, plural, leads into the problem of experiences that are had by other people but somehow understood also by oneself. Husserl later will apply the Lippsian aesthetic theory of empathy to this problem. But his primary use of the term *Einfühlung* before 1910 is logical, not aesthetic.[33]

---

[31]See Hua XIII: 1-2.

[32]See Hua XIII: 2-3.

[33]Thus I take exception to Kern (1973: xxviii) when he writes: "Das Charakteristische dieser frühen Beschäftigung mit der Problematik der Fremderfahrung besteht wohl darin, dass es in ihr *primär* um die ‚Einfühlung' von *Empfindungen und Empfindungsfeldern* in den fremden Leib geht, um die Einfühlung dessen, was Husserl die ‚ästhesiologische Schicht' nennt." ("What is characteristic of this early engagement with the problematic of other people's experiences consists entirely in this, that it *primarily* concerns the 'empathy' of *feelings and sensory fields* in the alien live body, that is, the empathy of what Husserl called the 'aesthesiological level'.") On the contrary, before 1910 this sentience interests Husserl chiefly insofar as it bears upon questions of logic, science, and shared understanding.

Excerpts from an unpublished 1905 lecture[34] indicate that Husserl first took up the term *Einfühlung* as a tool to reinforce the distinction between the seeing of essences and the feeling of motivations (a distinction already introduced in the First Logical Investigation, as we have seen). The point at issue is: to what degree is sharing possible when two people communicate? Husserl holds that what is communicated registers in the receiving consciousness only as a simulation, not as the live experience that the communicator has had.[35]

[Ergibt sich die Spaltung] zwischen aktueller Frage und Einfühlung in die Frage, der aktuellen Freude und der Einfühlung in die Freude. . . . Statt ‚Einfühlung' können wir auch sagen ‚sich hineindenken', ‚sich hineinphantasieren'. (Hua XIII: xxvi)	[There results the separation] between an actual query and empathizing into the query, between the actual gladness and empathizing into the gladness. . . . Instead of "empathizing" we could also say "thinking yourself into," "imagining yourself into."

Oddly, Husserl's way of modeling the event of communication invests the initiative on the side of the one receiving the information: the receiver goes out and gets the information for himself by intruding imaginatively into the subjectivity of the other. Husserl accomplishes this modeling by suppressing one side of the double sense of *ein-* established by Lipps. According to Lipps's usage, I feel subjective involvement in every activity while I am doing it; only later do I distinguish whether I have been inhabiting my own self (ablatively, monadically), or visiting the subjectivity of someone else (having entered it accusatively to retrieve something for export and import elsewhere). In this passage, Husserl is retaining only the accusative dimension of *ein-*, construing it as *hinein*. He drops the ablative side of empathy, in which several people might share one experience--*hier*.

The suppressed *Hier* of *Einfühlung* would correspond to those immediate and adequate intuitions of essences that I have inwardly and that according to Husserl I can get from no one else. But for describing my access to those intuitions Husserl has chosen another term: insight (*Einsicht*). He contends that the *content* of judgments about essences can be shared, but *insight into* the judging acts can never be shared.

Wir können uns auch in das Urteil *hineinfühlen*, ohne selbst zu urteilen . . . . Die blosse Vorstellung ist also die Einfühlungsmodifikation des Urteils. (Hua XIII: xxvi, emphasis added)	We can also *feel ourselves into* the judgment without executing judgment ourselves . . . . The bare presentation thus is the empathized version of the judgment.

Empathy in logic, then, is for Husserl a nonperspicacious "feeling one's way" through the steps of inference. Empathy is how one blindly pursues logical

---

[34]Cited by Kern (1973: xxvi).

[35]One notes here again the use of the term *Einfühlung* to name a cognitive process (questioning) as well as an emotion (gladness).

techniques of the sort that Husserl criticized in the Prolegomena.[36]

B. The 1910 Lectures on Basic Problems of Phenomenology

Things change, however, when Husserl sets his sights upon a grander role for phenomenology than that staked out in the *Logische Untersuchungen*. By 1910, phenomenology becomes the science grounding both the natural sciences and the human sciences. Phenomenology now aspires to account for the coherences of motivated historical events as well as the coherences of caused physical reactions. Before, the task was to resist assimilation into the human sciences while vying for primacy over the natural sciences. Now, the task will become one of coordinating the processive projects of both the human and the natural sciences by establishing a single non-processive foundation for both.

This grand undertaking is announced in the *Logos* article of 1911, but it is foreshadowed in the intersubjectivity manuscripts dating from 1909-1910. They develop the two new themes already intimated at Seefeld: the role of the human *body*, and the parallel between *memory* (of one's own experiences) and empathy (of someone else's).[37] Husserl puts his finger on the weakest point of Lipps's theory, a point which as we have seen would later become the target of ridicule in Nazi ideology. If understanding originates in my projection of my own sensibilities into other human bodies and their artifacts, what prevents my projecting (inappropriately) into trees and rocks as well? Lipps for his part had accorded priority to the original unity of experiencing and understanding within the flow of nondifferentiated, ecstatic, engaged subjectivity; this was the assumed basis for the possibility of subsequent "returns" to communion in discrete instances of engagement with persons and artworks. Lipps also had attributed understanding to the coordination of two instincts: a drive to express, and a drive to submerge oneself in expressions and follow along with them. Presumably, these drives operate as natural physical causes.

But Husserl points out what is left implicit in this account of Lipps's: before commencing to understand an expression, I must somehow first take it to be the configuration of a living body. I do not go from "expression" to "soul"; I must first recognize soul before I can recognize expression. The body (*Leib*) must already be given as such, and constituted differently from other items in the

---

[36]"Logical empathy" sounds absurd. This is owing to the incongruity of the sort of mechanical rote performance of logical techniques that Husserl criticized. Such followings do not deserve the name of logic, in his estimation. Accordingly, the designation of *Einfühlung* as a cognitive process seems discordant with Husserl's later and better known phenomenology. Nevertheless the early texts clearly show that Husserl was invoking the contrast between in-sight and in-feeling that I have summarized here as I reconstruct the growing edges of phenomenology as Edith Stein encountered it in Göttingen in 1913.

[37]In the Fifth Cartesian Meditation Husserl would further elaborate the comparison between memory of one's own experience and empathy of someone else's.

physical world.[38] Givenness and constitution are to be investigated phenomenologically, that is, on the basis of what one intuits with full adequacy, in evidence. Evidence cannot be imported into subjectivity. One imports, through simulation, only the sequential flow of a series of events. Husserl notices that one needs a body in order to do this, for sequences are followed quasi-kinesthetically and only a body can go from one location to another. But the *logic of* the flow is not imported; it is seen intellectually with insight. Furthermore, the logic is seen to have been either that of causes or that of motivations. Causal coherence is grasped as the kind of movement that one could live through insofar as one is a physical body (*Körper*), and motivated coherence is grasped as the kind of movement that one could live through insofar as one is a sentient body (*Leib*) as well. The human body is the principle of access to the two kinds of coherences followed in the natural and cultural sciences, respectively.

While never formulated so neatly by Husserl, this will be the gist of his framing of the manifesto "Philosophy as Rigorous Science," the *Logos* article of 1911. Spatialization and temporality become increasingly important considerations in Husserl's research. The earlier notion that I *think my way* to the affirmation of a conscious personal coherence in another human being despite my lacking access to the phenomena of her consciousness is revised in a 1909 discussion of human hands.[39] My own hand is known to me directly as a living thing because, in addition to its spatial extension and movability, Husserl says, it localizes a sensory field for me and it goes where I make it go. A human hand that is not one of mine withholds from me the primary evidence of its life; nevertheless I "apprehend" (*auffassen*) it as living--that is, as sentient and self-moving--on the basis of its similarity to my own hands. That is, I *remember* my hands when I perceive the alien hand. By 1909 Husserl no longer deems such apprehension to be an articulated inference.

Eine Auffassung ist kein Schluss, sonst wäre schliesslich jede Assoziation ein Schluss. Es ist eben eine Auffassung, und zwar eine apprehensive; das Gesehene, das hier nicht bloss Empfundene, sondern als Körper Wahrgenommene ist Träger einer „beseelenden *Deutung*", einer beseelenden Auffassung, und diese bringt apprehensive Schichten hinein, die *Vergegenwärtigungen* von solchen sind, die wir analog in der Wahrnehmung der eigenen Hand, des eigenen Leibes als der „ursprünglichen" Leibeswahrnehmung haben. (Hua XIII: 50)

An apprehension is not a conclusion, otherwise every association would be a conclusion. It is precisely a holding onto, "apprehensive" indeed. That which is seen, that which is perceived here not as raw sensations but as a physical body, is bearer of an "ensouled significance," an ensouled apprehension. And this brings in apprehensive strata that are the *representations* of just such [strata] that we have analogously in the perception of our own hand, of our own living body, as the "primitive" body-perceptions.

---

[38]See Hua XIII: 62-4, discussed by Kern (1973: xxviii).

[39]"Die Einfühlung. Text aus dem Jahre 1910." Hua XIII: 42-55. Edith Stein adopts this example of apprehending the living hand; but her basis for taking it to be alive is *"fusion" in perception*, rather than memory, analogy, and aggregation of sensory fields.

Husserl now accounts for the apprehending of alien living bodies in terms of remembrance, expectation, and empathy. Unlike physical things, whose aspects are available for my perception whenever I please--at least in principle--living bodies have an inner life that is available to consciousness only on very different and restrictive terms. One's memories and expectations, like one's sensory field, are uniquely one's own. A physical thing has no past except insofar as it once was present to someone. However,

Ganz anders verhält es sich mit leiblichem Sein und mit fremden Persönlichkeiten. Sie konstituieren sich bewusstseinsmässig in „*Einfühlungen*", in Akten des *Einverstehens*, wobei das spezifisch Leibliche und Geistige *durch Vergegenwärtigungen* bewusst wird. Das vergegenwärtigte fremde Sinnesfeld ist für mich prinzipiell nicht wahrnehmbar, es ist nicht mein gegenwärtiges, vergangenes oder künftiges Sinnesfeld, nicht mein Erlebnis. Und ebenso das fremde Aktleben. Und es ist nicht wahrnehmbar gewesen und wird nie wahrnehmbar sein. (Hua XIII: 51-52)	It would be entirely otherwise in the case of incarnate being and with alien personalities. As to consciousness, they constitute themselves in "empathy," in acts of inward understanding, whereby that which is specifically somatic and intellectual comes into awareness *through representations*. In principle it is impossible for me to perceive the represented alien sensory field, [for] it is not my present, former, or future sensory field. It is not my live experience. The same goes for the alien act of living. It never was perceivable and never is going to be.

What was called above the ablative or *hier* side of the notion of empathy seems to re-emerge in this connection. Husserl seems to say that a living being feels and understands its own life from within; no one else can. These feelings are reflexive acts of consciousness. They extend in time, taking the form of memories and expectations.[40]

The immediate background for Husserl's *Logos* article, outlined during the Christmas break of 1910-11 and drafted soon afterwards, is the class on "basic problems of phenomenology" that he taught during the winter semester of that year.[41] Among the students in the class was Hedwig Martius, who would become Edith Stein's godmother though they were not yet well acquainted during Stein's student days. The lectures are composed with clarity and uncharacteristic wit. In them Husserl's explorations of time, space, and the human body are succinctly formulated, and empathy is investigated for the first time from the standpoint of

[40]Memory and expectation are mentioned toward the end of the 1909 research paper on *Einfühlung* (Hua XIII: 52). In Hua XIII, the first of the volumes on intersubjectivity, the texts that reveal the development of Husserl's thought on the themes of motivation, empathy, and the body prior to the *Logos* article would include numbers 2, 3, and 4, and supplements IV, IX, and XVII, treated above; and text 6, which we now consider.

[41]Manuscripts are available for October and November of 1910; after that, he says, he lectured ad lib. Apparently this material was never taught again in this form. The lectures and seminars on "nature and culture" seem to have taken its place in the cycle of courses offered by Husserl, and there is some continuity of themes. These lectures also provide a background for *Ideen*, published in 1913 in the *Jahrbuch*.

the phenomenological reduction.

The lectures open with the stipulation that phenomenology is essentially different from any science of spatio-temporal existence such as psychology; they go on to discuss uses of the first-person nominative singular pronoun.

Jeder von uns sagt „ich" und weiss sich so redend als Ich. Als das *findet* er sich *vor*, und er findet sich dabei jederzeit als Zentrum einer *Umgebung*. „Ich", das bedeutet für jeden von uns etwas Verschiedenes, für jeden die ganz bestimmte Person. . . . (Hua XIII: 112)	*Each of us says "I"* and knows himself speaking thus as an i. As that i, he lights upon himself, and in that way finds himself each time as a center of an *environment*. "I." That signifies something different for each of us: for each the entirely particularized person.

The i is not its experiences, its acts, or its characteristics, but is *that which has* the characteristics, *executes* the acts, and *lives* the experiences. Husserl describes the ways in which an i occupies its own fragment of time and space, which include ways of experiencing the appearances of things in space and time. Besides experiencing, the i also knows. The i is a scientific human being.[42]

The i--that is, the experiences of being an i--are *localized* in the living body in a distinctive way. Joy is in the heart, but not like blood; sensitivity is in the skin, but not like cells. An i's live experiences depend upon its own living body's circumstances and operations. Nevertheless, every i comes upon things confronting its own body that it spontaneously regards as alien living bodies, bodies with alien i's pertaining to them.[43] Husserl offers a working definition of two important terms--alien i, and inward awareness--in one stroke:

Jedes Ich . . . sieht die Leiber an als „Träger" von Ichsubjekten, es „sieht" aber nicht die fremden Ich in dem Sinn, wie es sich selbst sieht, erfahrend vorfindet. Es setzt sie in der Weise der „Einfühlung", also auch fremdes Erleben, fremde Charakteranlagen werden „vorgefunden"; sie sind aber nicht in dem Sinn gegebene, gehabte wie eigene. . . . (Hua XIII: 115)	Each i . . . regards the living bodies as "carriers" of i-subjects. It "sees" the alien i, but not in the sense in which it sees itself, meeting up with itself while engaged in experience. It posits them in the manner of "empathy," thus alien experiencings and alien character traits are also "met up with." But they are not given or had in the same sense as one's own.

While the alien i is only quasi-perceivable, its placement in the environment is interchangeable with the placement of any other i, including my own. Each is for itself the center but for others a point located in the environment. To be a center is also to be a *Nullpunkt* or "zero" in a system of spatial coordinates. To be "here" systematically entails "there's" to the left and right, before and behind. To be "now" entails earlier points in time and later ones. Things have backs and fronts,

[42]See Hua XIII: 113-114.

[43]See Hua XIII: 115. Where earlier Husserl had reserved verbs of seeing for the direct seeing of essences, opposing "in-sight" to "in-feeling," here he abandons that distinction. *Einfühlung* is here a "seeing as" or "looking upon as" (*ansehen*). This temporary change of terminology may be an attempt to speak colloquially to students.

which appear to us in a regular way as we move along past them. Husserl proposes it as a governing ideal that:

von zwei normalen Individuen, im Fall sie ihre Orte vertauschen oder vertauscht denken und leiblich in einem ideal-normalen Zustand sind, jedes genau dieselben Erscheinungen in seinem Bewusstsein findet, die früher im Bewusstsein des anderen realisiert gewesen waren. . . . Und jeder von uns würde immer dieselben Erscheinungen gehabt haben, wenn er an derselben Stelle wie der Andere gesehen hätte. . . . (Hua XIII: 117)	of two normal individuals, in case they exchange their places or imagine them exchanged, and if the two are bodily in an ideal-normal condition, then each one finds realized in his own consciousness exactly the same appearances that earlier had been realized in the consciousness of the other. . . . And each of us always would have had the same appearances if he had taken a look from the same place as the other.

But if deviations should occur, Husserl adds, we call it sickness.

At first the articulation of this law may seem to be a renunciation of solipsism. In fact, it is no such thing. This law merely states that any i is capable of access to the same appearances as any other i: that what is *meant by* "i" is the capacity to be appeared-to without contributing anything to the contouring of any appearance. This law has no ontological entailments. It says nothing about whether there *are* any other i's, much less whether their consensus is needed for the certification of the appearances that come on to me. Stated another way, the law merely declares: the world must and does appear to me just as it would appear to anyone else who might stand where I stand; for my own unique individuality is irrelevant to my observations. (In *Ideen* Husserl will develop this principle into the foundation of the physical sciences.) If anything, this reaffirms Husserl's "weak solipsism" even though, to be sure, its application is limited to the world of *physical things*.

Matters stand quite otherwise with empathy, which Husserl treats as an empirical experience whose phenomenological structure can be studied. What, then, is the structure of accusative empathy?

In ihr [Einfühlung] erfährt das einfühlende Ich das Seelenleben, genauer, das Bewusstsein des anderen Ich. Es erfährt es, aber niemand wird sagen, es erlebt es und nimmt es in innern Wahrnehmung, in Lockescher Reflexion, wahr so wie sein eigenes Bewusstsein. Erst recht wird natürlich niemand sagen, es erinnert sich dessen oder erwartet es. (Hua XIII: 187)	In empathy the empathizing i experiences the living soul, or more accurately, the consciousness of another i. It experiences it; but no one would say it lives through it and perceives it in inner perception, in Lockean reflexion, just like its own consciousness. Even more, naturally no one would say it remembers or expects it.

What is experienced of the other is not an imagined picture or an analog. What one is shown, and what one senses, when one looks into the live experiencing of another, always turns out to be frankly counterfeit: the picture *you get* is not the experience *I had*, for it is no longer enlivened by i.[44] The elusiveness of the other's soul stems from its activity, not its content, Husserl says. This can be made apparent through inspection of the double-sidedness of one's own experience of feeling-into another's soul. My act of accusative empathy has a time-location in the flow of my own consciousness just as I do it, *now*. But it is a feeling *of* something occurring in the conscious stream of someone else, at some other *now*. The act of empathic feeling-into must posit that which it captures as another "now"--but that captured "now" isn't the same *now* it used to be at its native location within the conscious stream in which it originated. Husserl offers another law:

> dass prinzipiell ein eingefühltes Datum und das zugehörige einfühlende Erfahren selbst nicht demselben Bewusstseinsstrom, also demselben phänomenologischen Ich, angehören können. Von dem eingefühlten Strom führt kein Kanal in denjenigen Strom, dem das Einfühlen selbst zugehört. Niemals kann ein Datum des einen und anderen Stromes in dem Verhältnis stehen, dass das eine die Umgebung des anderen ist. Die Umgebung! heisst das aber nicht Zeitumgebung, und besagt unser Gesetz nicht, dass eins und das andere nicht *einem* Zeitbewusstsein angehören kann? (Hua XIII: 189)

> that in principle, an empathized datum and the corresponding empathizing experience itself cannot belong to the same stream of consciousness, and therefore cannot belong to the same phenomenological i. No canal runs from the empathized stream back out to the stream to which the act of empathizing belongs. At no time can a datum of one and another stream stand in such a connection that the one is the environment of the other. The environment! But doesn't that mean temporal environment, and doesn't our law stipulate that one and the other cannot appertain to *one* temporal consciousness?

Husserl will not tolerate the objection that empathy itself *digs the canal* inasmuch as its essence is precisely to bring two conscious streams into confluence, that is, to produce simultaneity between i's. He points out that the two now's are invincibly incongruous, for the *now* of the act of empathy is a "self-beheld" (*selbsterschautes*) one, while that of the empathized content is a "represented" (*vergegenwärtigtes*) one.[45] Thus the invariable result of the phenomenological reduction is a doubling of i's. What Husserl terms "the phenomenological i" is actively perceiving, remembering, empathizing others, and executing the phenomenological reduction; yet in doing so it continually posits another i that is beheld, remembered, and so forth. This empathized i is the one that belongs to the

[44]See Hua XIII: 188. The live experiencing underway in another is opaque to me because it both invites and blocks my spontaneous impulse to gain insight into it--that is, to have a live intuition of its i-origination. This is Husserl's contention. However Stein will contend that memories, too, invite me to move from opacity to insight; yet that impulse is *not* blocked in the case of *my own* memories, for as I move to coincide with them I will find that something of them is still alive in me.

[45]In terminology later adopted by Husserl and by Edith Stein, this is the distinction between "originary" and "non-originary" live experience. Here Husserl counts two acts and two different contents (i.e., "data"). Stein will count two acts but only one content, although that content ("datum") will be entertained in two slightly different versions.

living body, says Husserl.[46]

The discovery that Husserl has made is that the structure of one's empathy into the souls of other people is similar to the structure of entertaining a memory. Both are live experiences *of* non-live experiences: in each, a *now* appears to contain a "now" as a ripple in the straightforward flow of the stream of conscious life. Husserl later would identify this discovery as having precipitated his breakthrough into fully transcendental phenomenology. He thought he had solved the riddle of intersubjectivity by extrapolation from the doubling of the *now*/"now" as experienced in memory and expectation.[47] One's inward awareness of other people's experiences would be structured just like one's memories and one's expectations.

While never revoking his earlier affirmation of the absolute incommunicability of the other's life as she lives it, Husserl nevertheless takes certain analogies as axiomatic. In regard to the physical world of nature, the alien i is simply interchangeable with my own: it is just "i yonder" or *Ich im Dort*.[48] In regard to human relationships, the alien i is apperceived in the apprehension of certain physical bodies as living, by virtue of an analogy that can be brought out through reflection. So implicit is the analogy that one does not invoke it by reasoning *from* me *to* you. Rather, having already overlaid "yours" upon "mine" through accusative empathy, one recognizes afterwards that you and I are analogs of each other. In other words, the perception of an alien i as such is no longer asserted to be the result of any analogical inference. Rather, the status of the i "here" and the i "yonder" as analogs of one another is recognized through an act of apperception.[49]

C. The *Logos* Article and the First Book of the *Ideen*

Husserl's impersonal disembodied "phenomenological i" seems to be the one speaking in his manifesto of 1911, his article on "Philosophy as Rigorous Science." It is the voice of logic, claiming to be *norma normans non normata*, portraying itself as unaffected by either nature or history, and offering magnanimously to stabilize and coordinate the opposing scientific pursuits of the natural sciences and the humanities. Yet Husserl like a ventriloquist throws this voice into the particular sciences and makes it speak out of the tacit assumptions that justify their own practices.

In the *Logos* essay, intended as a popular introduction to his work, Husserl

---

[46]See Hua XIII: 190-191.

[47]See Kern (1973: xxxv, xxxix).

[48]See Hua XIII: 266.

[49]These two kinds of analogy--unmediated analogous *apprehension*, and analogy established as the conclusion of an *inference*--are juxtaposed by Kern (1973: xlv), who identifies them in texts dating from 1914-1915. But as we have seen, these themes already have emerged in Husserl's work from before *Ideen*.

Husserl says that each of the two kinds of science would like to assimilate the other, and philosophy along with it.

Den herrschenden Auffassungsgewohnheiten entsprechend neigt eben der Naturwissenschaftler dazu, alles als Natur, der Geisteswissenschaftler als Geist, als historisches Gebilde anzusehen und demgemäß, was so nicht angesehen werden, zu mißdeuten. (Hua XXV: 8)

In accord with each one's dominant habit of interpretation, the natural scientist has the tendency to look upon everything as nature, and the humanistic scientist sees everything as "spirit," as a historical creation; by the same token, both are inclined to falsify the sense of what cannot be seen in their way. (1965: 79)

The aggrandizing ambitions of the sciences are termed naturalism and historicism, respectively. Naturalism wants to find physical causes for everything and exclude teleology. Historicism wants to account for everything in terms of how it fits into the spirit of its age, relegating physical causality to the status of a mere fashionable prejudice. Husserl sees that each of these would amount to allowing only one kind of coherence for the world: either physical causality, or intellectual-cultural motivation. There is no way to resolve the competition given only those two options, inasmuch as neither sort of articulated coherence is capable of demonstrating the invalidity of the other (try as they might). Furthermore, neither can account for *how the scientist recognizes* such a relation as "coherence" when he first discovers it. Each kind of science simply insists that its own kind of coherence is the only conceivable kind.

Husserl's proposal is that the coherence of any articulated items is recognized by virtue of something else, something unlike either those items themselves or what articulates them. It is recognized by virtue of seeing that an essence has been fulfilled: that some *possible* way of fitting together *does in fact obtain* in the case at hand. Something clicks for the scientist. A pattern pops out. However, the *validity of* the coherence registering as "causality" among observed physical events is not itself something that can be physically observed. Nor is that validity a transient cultural formation.

Man sieht leicht, daß der Historizismus konsequent durchgeführt in den extremen skeptischen Subjektivismus übergeht. Die Ideen Wahrheit, Theorie, Wissenschaft würden dann, wie alle Ideen, ihre absolute Gültigkeit verlieren. Eine Idee habe Gültigkeit, bedeutete, sie sei ein faktisches Geistesgebilde, das für geltend gehalten wird und in dieser Faktizität des Geltens das Denken bestimmt. . . . Dann also auch nicht für den Satz vom Widerspruch und alle Logik. . . . (Hua XXV: 43)

It is easy to see that historicism, if consistently carried through, carries over into extreme sceptical subjectivism. The ideas of truth, theory, and science would then, like all ideas, lose their absolute validity. That an idea has validity would mean that it is a factual construction of spirit which is held as valid and which in its contingent validity determines thought. . . . Thus too there would then be no validity to the principle of contradition nor to any logic. . . . (1965: 125)

But both natural scientists and practitioners of the human sciences do subscribe to the ideals of "science." They use them in pursuing the intelligible connections that structure the world. When something "clicks" in scientific discovery, the instantiation of an ideal form of coherence has been recognized. Ultimately, the "click" is what can be shared among scientists, scientifically.

But the twofold character of that sharing is what may lead both physical scientists and humanists astray. Husserl's earlier studies of empathy illuminate his argument here. On one hand, the data studied by physical scientists and humanists, respectively, are always "experiences of other people," albeit in different ways. The physical scientist uses observations available to the *Ich im Dort*, that is, to anyone who might occupy the location in physical space-time that the scientist occupied when he made the observation. And the historian uses representations gleaned from the actual experiences reported by other people and received into the scientist's own consciousness through those reports.[50] Physical observations (for the one) and cultural representations (for the other) are both *eingefühlt* or felt-into in their respective ways. They are captured as *formerly* live "now's." On the other hand, the experience of recognizing their coherence--that "click"--is a *currently* live now. It occurs for the scientist as he considers the observations or representations occupying his consciousness.[51] Why can this alive and self-certifying experience of scientific discovery be shared? Because the forms of logical validity already *are* shared; they inhere in any i. Causality and motivation are not items in the world; much less are they either caused or motivated. They are recognized as appropriate coherences because any i already knows what a coherence is. (Any i *is* such a knowing.)

Hence Husserl can say that psychic entities are windowless monads that commune only through empathy. The inner psychic coherence of any human individual's streaming consciousness has a structure--its motivation--that can be inwardly felt by another individual.[52] The one who understands does so because she can empathize the other's activity of coherence-recognition, thanks to her own intrinsic logic. This recognizing is live or *hier*, and although it cannot be transmitted as such it is available to anyone who takes the trouble to enact it. By comparison, physically existing entities can:

von vielen Subjekten als individuell Identisches erfahren und als intersubjektiv Selbiges beschrieben werden. Dieselben Dinglichkeiten . . . stehen uns allen vor Augen und können von uns allen nach ihrer „Natur" bestimmt werden. . . .

be experienced by many subjects and described as intersubjectively the same. [They] are present to the eyes of all and can be determined by all of us according to their "nature." . . . Each physical thing has its nature (as the totality of what it,

[50]Even practitioners of the natural science of empirical psychology, Husserl points out, traffic in representations imported from other people's experiences, for they receptively feel-into their respondents' descriptions of mental events. See Hua XXV: 306; (Husserl 1965: 97-98).

[51]See Hua XXV: 46 (Husserl 1965: 128-129).

[52]See Hua XXV: 28 (Husserl 1965: 106). Thus every human being's own immanent life *is* an inner window on every other human being. This window is nothing other than the life.

Jedes Ding hat *seine* Natur (als Inbegriff dessen, *was* es ist, *es*: das Identische) dadurch, daß es Einheitspunkt von Kausalitäten innerhalb der einen Allnatur ist. (Hua XXV: 26-27)	the identical, is) by virtue of being the union point of causalities within the one all-nature. (1965: 104)

The physical entity, by nature appearing *as* knit up within a causal web, appears *to* us in a series of partial showings and concealings of its aspects. Those moments of *showing* are also the moments of *looking* by someone moving around the thing. The observer, as a psychic entity, is able to fold back and gather up those looking/showing moments in memory, and render all the showings simultaneously (albeit virtually) present in *knowledge* of the thing as many-sided. Ablative empathy, which enables memory, figures importantly into that synthesizing act of scientific knowledge. Moreover, the possibility of sharing knowledge of physical things among different observers is based in their capacities for empathy as well, for through it they have the sense of how any "thing" must come to be.[53]

Thus the possibility of any science depends, for Husserl, on recognizing series of various kinds: causal series (of physical events), motivated series (of cultural and intellectual events), and serial appearances (of things). Therefore it also depends on the psychic ability to follow, to follow along after or simulate, and to re-follow or replicate one's own former followings. Empathy figures structurally into all of this. However, Husserl has ruled out the possibility of empathy as a *flowing together*, by prohibiting "canals" between streams of consciousness. Science is not communion.

Husserl's *Logos* article attempted to reconcile the physical sciences and the cultural sciences by reining in their excessive claims and illuminating their common reliance upon what was prior to both: the necessary structures of human knowledge and communication. The attempt failed, at least in the short run. Soon after the article appeared, Wilhelm Dilthey wrote to Husserl expressing astonishment at having been portrayed as a historicist, when his life's work had been to establish a generally valid science.[54] The scale of the misunderstanding between these two great theorists of understanding gives one pause. Husserl did not again attempt a general intervention into the course of the development of the sciences until after he retired, some twenty years later.[55]

Meanwhile, Husserl continued to lecture on the relationship between nature and culture, and on the coordination of the sciences whose respective fields of investigation they were. Integral to Husserl's developing thought were his reflections upon the essential structures of empathy. But in his next milestone

---

[53]See Hua XXV: 29-30 (Husserl 1965: 107-108).

[54]See Dilthey's letter of June 29, 1911, in Schuhmann and Schuhmann (1994/6: 43-47).

[55]Manuscripts from 1912 and 1916-17, eventually published as supplements to the Second Book of *Ideas* in 1952, indicate that Husserl continued to work on the problem of establishing a single foundation for the natural sciences and the cultural sciences (*Geisteswissenschaften*).

publication, the *Ideen* of 1913, the structures of empathy receive rather little attention. What is new is what is missing: the term "*Einfühlung*" no longer is used in its ablative sense to indicate my own live enactment of my *every* "originary" conscious move, but only my representations of something that has been "originary" and alive in someone else's consciousness. The term continues to be paired with memory, because both are instances of this doubled or folded sort of consciousness. Unlike memory, which presupposes the rememberer's own existence, this sort of representation does not require the existence of the other.[56] In fact, Husserl characterizes the empathy of another's experience as having, essentially, only imperfect "evidence" in my consciousness, without the possibility of ever coming to the "fulfillment" of evidential actuality as other varieties of imperfect evidence may.[57]

At the end of the first book of *Ideen*, Husserl begins to prepare for the studies on "constitution" that are to be offered in the second. In this connection, he hints that empathy of other people's experiences will be an important consideration in accounting for the unity of "the intersubjectively identical physical thing."

Ihre Konstitution ist bezogen auf eine offene Mehrheit im Verhältnis des „Einverständnisses" stehender Subjekte. Die intersubjektive Welt is das Korrelat der intersubjektiven, d.i. der durch „*Einfühlung*" vermittelten Erfahrung. Somit werden wir verwiesen auf die mannigfaltigen, von den vielen Subjekten schon individuell konstituierten Sinnendingeinheiten; in weiterer Folge auf die entsprechenden, also zu verschiedenen Ichsubjekten und Bewußtseinsströmen gehörigen Wahrnehmungsmannigfaltigkeiten; vor allem aber auf das Neue der Einfühlung und die Frage, wie sie in der „objektiven" Erfahrung eine konstituierende Rolle spielt und jenen getrennten Mannigfaltigkeiten Einheit gibt. (§151; Hua III: 372-3)

[The thing's] constitution is related to an open plurality in relation to subjects "understanding one another." The intersubjective world is the correlate of intersubjective experience, i.e., <experience> mediated by "*empathy*." We are, as a consequence, referred to the multiple unities of things pertaining to the senses which are already individually constituted by the many subjects; in further course we are referred to the corresponding perceptual multiplicities thus belonging to different Ego-subjects and streams of consciousness; above all, however, we are referred to the novel factor of empathy and to the question of how it plays a constitutive role in "Objective" experience and bestows unity on those separated multiplicities. (§151; 1982: 363)

There is a curiously impersonal tone to the re-entry of "*Einfühlung*" into Husserl's formal statement of his project at this point. He needs to provide for *the possibility*

---

[56]In *Ideen I*, the experience of empathizing the feeling that originally was someone else's is cited merely in passing as an illustration by Husserl in §§1, 46, 75. See Hua III: 7-8, 85-86, 140; (Husserl 1982: 6, 100-101, 172). For a brief discussion of Husserl's use of the term *Einfühlung* in *Ideen I*, see Ströker (1993: 130-131).

[57]He promises to discuss this partial evidence in the second book of the *Ideen*, which as it turns out he did not release for publication in his lifetime. See §140, Hua III: 292 (Husserl 1982: 344).

of other people, other knowers. For there to be a natural world, that is, a world that can be scientifically known through its causal coherences, Husserl will need to establish the availability of that world for inspection from multiple viewpoints, and then he will need to establish the possibility that the different views can be coordinated. *Should they occur*, that is. Husserl doesn't need any people in particular, or any people at all. He doesn't need any particular observations, or any observations at all. He must establish only the intended possibility that various knowledges *can* occur and that, if they do, they will be capable of being brought to coherence.

The term "motivation" also has shifted its meaning in *Ideen*. Husserl now increasingly tends to apply it to both kinds of coherences that the sciences study: to causal connections as well as to intellectual, cultural, and biographical ones. (This is a distinct departure from the established usage of Dilthey and Pfänder.) The term "motivation" can even signify the following of a process of thought. For Husserl it now marks the contrast between any discursive, articulated, flowing series, and the "unmotivated" immediacy of direct insight. Thus he will say that physical things appear in a motivated series of continuous adumbrations, and consciousness itself flows along according to its own motivational structure.[58]

D. Nature and Intellect in *Ideen II*

Husserl's *Ideen* had been conceived as a single grand project. It was drafted in one fell swoop in 1912. After its first portion was published in 1913, however, the remainder underwent a series of complicated revisions by many hands. What was published at last in 1952, as Books 2 and 3 of the *Ideen*, is in a curious state indeed. Before the contents are examined, a word about their arrangement is in order. Each volume sorts the material into three groupings: the main text, supplements, and fragments included in the text-critical notes.[59] Marly Biemel is responsible for the final arrangment of the Husserliana edition, but she provides information about the work of two previous editors--Ludwig Landgrebe in 1924-25, and Edith Stein in 1916-17 and 1918--as well as about Husserl's notations on their manuscripts.[60] That information is indispensable. Without it, the philosophical incoherence of *Ideen II* cannot be resolved.

This troubled text pursues two distinct and utterly incompatible strategies to solve the problem that it means to address: that of establishing a single foundation for the natural and cultural sciences. One solution founds both kinds of science in the immediate intuitions available to living animal bodies such as our own. Bodily life is intrinsically empathized in the ablative sense, and therefore

---

[58]The reader should bear in mind that Husserl's vocabulary becomes fluid and irregular. There are exceptions to the generalizations that I offer here.

[59]That Books 2 and 3 are separate volumes at all seems to be owing entirely to the logistics of production. The former is 446 pages in length; the latter is another 171 pages.

[60]See chapters four and six, below, for a discussion of how *Ideen II* was composed and by whom.

empathizable in the accusative sense. As such, bodily life is logically prior to:

- the plurality of living bodies,
- the location of i's in them,
- the individuation of those i's,
- investment of my own i in a body,
- the spatio-temporal structure of the natural world,
- the motivational structure of cultural expressions, and
- the possibility of registering worldly influences (mediately), or personal influences (immediately).

This solution effectively does away with the need to show how mind and matter connect, simply by choosing to begin its analyses at a point prior to the alienation of body and mind. It assumes the mutual permeability of the four levels of human being identified by Scheler: the physical, the sentient, the soul, and the intellect. This solution proposes that the absolute science to found all others will be a cultural science that will show the derivative character of natural sciences.[61]

The other solution founds the sciences in the activity of object-constitution carried out by transcendental subjectivity. Instead of starting with bodily life, this account starts with the constitution of such things as live bodies. "Constitution" is the unification of a series of adumbrated appearances into a thing intended as being what it is on its own.[62] Constitution is logically prior to what is constituted:

- the spatio-temporal world with individual bodies and causal relations,
- my body,
- other bodies like mine,
- ideal essential laws,
- the inner world with temporally regulated streams of experience pooling up into discrete unities, and
- the possibility of other i's who would be constituting all of this just as I do.

---

[61]This is signaled in the titles of the last two chapters of *Ideen II*: "Motivation as Fundamental Law of the Cultural/Intellectual World" and "The Ontological Priority of the Cultural/Intellectual World as Opposed to the Naturalistic One." This solution anticipates contemporary marxian and psychoanalytic hermeneutic programs that attempt to show how natural science is determined by economic or by psychological factors, respectively. Compare, however, Edith Stein's section on "Motivation als Grundgesetzlichkeit des geistigen Lebens," §1 of chapter III of her 1919 essay "Psychische Kausalität."

[62]Husserl's term "constutition" names the phenomenality of the achievement of unity and identity for something. It refers to an i-drenched but nonoptional consolidation of appearances through perception: a process by virtue of which something is there as itself. Constitution presumes the possibility of a cognitive autonomic function that produces consensus about the specific character of something's existence. Something's constitution is owing both to determinations by the thing in question, and to the capacities of intersubjective knowing directed toward that thing or any other. Analysis of how constitution occurs is the overall task of *Ideen II*.

This second solution distinguishes between causality (which operates throughout material nature as a feature of its constitution) and motivation (which operates within the intellect and can be replicated by other intellects). Material things never affect the intellect on their own--that is, insofar as their naturally determined properties go. While worldly items may become stimuli for the intellect or the soul, that can happen only when worth has been added on to them through an additional intentional act after they have already been constituted as natural objects. According to this second solution, the absolute science will be phenomenology, the science of transcendental subjectivity.

These two solutions come to expression in various places in the texts that make up *Ideen II* and *III* in their published form. They can be sorted out as follows.

1. First solution: the priority of bodily life.

The principal articulation of what I have termed "the first solution" is found in the concluding chapters of the first and second major sections of *Ideen II* and in the opening of the third. Let us review these passages individually.

(a) §18/b-e and g-h. "The subjectively conditioned factors of thing-constitution and the constitution of the objective material thing." These lengthy passages comprise most of the chapter[63] concluding the volume's first section, entitled "The Constitution of Material Nature." They are designed to show that constitution is accomplished by a subject who is *necessarily bodily*. The argument is made that constitution presupposes perception, perception presupposes appearances, but appearance is conditioned by the sentience and mobility of the body. For example, things are constituted as material insofar as they appear to the various organs of sense, in ways determined by those organs. Furthermore, material things appear in ordered series of adumbrations, and that ordering is imparted by the locomotion of the eyes, ears, and so forth through space. The testing and correction involved in the completion of the unified sense of any material thing are owing in part to the correlation of impressions from different bodily organs, in part to the possibility of moving around, "trading places," and achieving various viewpoints. Because perception in all these modes is an ingredient in constitution, the body must already be there and functioning *before* constitution can get underway.

But bodily life is not a life of solipsism. Although a lone i endowed with a body might *constitute* objects, it would not go on to *objectivate* what it had constituted *as elements in an objective material world.* Thus an i must intend that there could be other i's, before one can intend that this or that item is "real." To regard something as a real material thing is to regard it as offering to any other observer who might happen upon it the same appearances that it offers to me. The sense "reality" includes the sense "objective determinability." But that sense itself

[63]§18 is a chapter in itself, and runs from page 55 to page 90 in Hua IV (Husserl 1989: 60-95). See chapter four below for an account of how this and the following passages were patched together. Figure 4.2 outlines the differences presented here.

is not reached by consensus, for it presupposes the possibility of consensus. Thus bodily life, the plurality of living bodies, and the possibility of communication among them all logically precede the objectively real material world.

(b) §§43-47. "The constitution of soul reality in empathy." This brief chapter[64] continues the line of thought initiated in §18. It is designed to overturn two common misconceptions: that the soul is something extra injected into the body; and that bodies, first recognized as mere physical things, subsequently are taken to be alive on the basis of physical resemblance to one's own body. Rather, the availability of other souls in other bodies is primordially given immanently within my own bodily capacity to register empathies of the embodied live experiences that others undergo.

This passage argues that the natural world, as both target and context for all the investigative activities of the natural sciences, is constituted through empathy. In my physical surroundings I come upon things that are of the same type as my own live body. I apprehend (*auffassen*) these as living bodies and I feel within them *(einfühlen)* an i-subject having all the contents that pertain to an i-subject as the particular occasion demands. At this point, while my feeling-into is underway, there has *already* been intentionally transferred (*sich überträgt*) to those other bodies the same sort of localization of sensory fields and of intellectual activity that I have with my own body. What is presented to me in my own body goes over into the empathy of the other's body (*geht dann in die Einfühlung über*).[65] Additionally, however, the inward activity of a soul (*seelische Aktinnerlichkeit*) also appertains to the appearance of the other human being. This too is transferred (*übertragene*) to the body whom I see from the body in which I live, and becomes a starting point for *understanding* the being of a soul.[66] That understanding will occur on the basis of the physical movements of the body, which are indications (*Anzeichen*) of the soul's experiences by virtue of its localization in them. When the body moves itself from place to place, the soul moves itself too. This is where one would start to elaborate a systematization or grammar of the "expressions" of the life of a soul.[67] This could be an objective study, because the transfer has *not* been an "introjection" of my soul into the other.[68]

---

[64]Hua IV: 162-172; (Husserl 1989: 170-180).

[65]See §45; Hua IV: 164, 166; (Husserl 1989: 172, 174).

[66]This account contrasts with Husserl's statement, two pages later, that I first recognize individual i-hood in the other, and subsequently *on that basis* recognize it in connection with my own body. See below.

[67]See §45; Hua IV: 166; (Husserl 1989: 174-5). See also §46; Hua IV: 167; (Husserl 1989: 176). One notes here the Schelerian hierarchy of terms: the physical, the psychic, the intellectual, and the soul. This conjunction of terms has not appeared before in Husserl's work.

[68]That is, I would be studying something that was "really there" in the other. Introjection (*Introjektion*) is ruled out at the end of §45. Thus it is puzzling to see the same term crop up again a few pages later, in the subheading for §49/a: "Introjection of

The identification of an i with an alien body is no more problematic than the identification of my i with my own body. There is a more fundamental problem: the individuation of i's and their embodiment at all. Left to myself, I would not invest my own subjectivity in my body.

> Nun würde es mir aber in der Einstellung der „Selbsterfahrung" gar nicht einfallen können, all mein Psychisches, mein Ich, meine Akte, auch meine Erscheinungen mit ihren Empfindungsdaten etc. ernstlich in meinen Leib hineinzustecken, zu „introjizieren". Auch ist wohl keine Rede davon, daß ich in der solipsistischen Selbsterfahrung all mein Subjektives mit meinem wahrnehmungsmäszig gegeben Leib als ein Realität vorfinde. . . Erst mit der Einfühlung und mit der beständigen Richtung der Erfahrungsbetrachtung auf das mit dem fremden Leib appräsentierte und beständig zusammen mit dem Leib objektiv genommene Seelenleben konstituiert sich die abgeschlossene Einheit Mensch, und diese übertrage ich im weiteren auf mich selbst. (§46, Hua IV: 167)

> But now it would not occur to me at all in the attitude of "self-experience" to take all that is psychic in me, my [i], my acts, my appearances as well, with their sense data, etc. and seriously place, i.e.,"introject," all this into my body. It is also out of the question that in solipsistic self-experience I encounter all that is subjective about me, along with my perceptually given Body, as a reality. . . . It is only with empathy and the constant orientation of empirical reflection onto the psychic life which is appresented along with the other's Body and which is continually taken objectively, together with the body, that the closed unity, [human being], is constituted, and I transfer this unity subsequently to myself. (§46, 1989: 175)

In short, I could not *solipsistically* apprehend even my own humanity in its body-soul unity. Rather, I am given to understand that basic incarnational aspect of my own human being through apprehending it first in others. Thus empathizing the embodied soul-lives of others enables me to recognize the ensouled being of my own living body.[69] One might say that humanity is not projected into the other, but retrojected into oneself.[70]

Consciousness becomes essentially anchored in the natural world--becomes "naturalized"--because the soul and the body pursue their experiences in unity and because, by virtue of the unity that they thus fashion, what pertains to soul receives a placement in space and time. But there are not many such unities, only one. The same analogy in which I apprehend the other as a *living* body who can say "here" to designate herself the center of things, is also an apprehension of a living *body* whose location is simultaneously a "there" for me. The mobility of living bodies allows them to trade places and exchange their "here's" and "there's";

---

the soul as presupposition even for the naturalistic orientation" *(Introjektion der Seele als Voraussetzung auch für die naturalistische Einstellung)*. But there the term "introjection" is labeled "misleading" (*mißdeutbaren*). See Hua IV: 176; (Husserl 1989: 186).

[69]The priority of the alien live body as such was already discussed by Husserl in his lectures of 1910; see Hua XIII: 62-64.

[70]If, as was suggested above, Husserl was practicing a "weak solipsism" by counting upon only those perceptions that appear to me with inner evidence, then we see here how such a solipsism overcomes itself.

the text attributes this to empathy as well.[71] Therefore the spatio-temporal arena through which the alien i moves is not a new or alien realm for me. It is my same old realm: the *one objective world*, but now seen to be such. "Otherness" is enforced in it only by this law: two people may perceive the same appearances, but only at different times; and conversely, if two people share a time, they must do so from different places and therefore with access to differing appearances. Nevertheless on those terms we now know how it is that two people can perceive one identical object.[72]

Although my own living body is logically prior to nature, it is an item in nature as well. It is available to be known on the same objective, intersubjective terms as other items are. Similarly, humanity itself can be studied objectively, which is to say scientifically.

Das Objekt Mensch ist also ein transzendentes äußeres Objekt, Objekt einer äußeren Anschauung, und zwar ist es eine zweischichtige Erfahrung: mit äußerer urpräsentierender Wahrnehmung verflochten ist appräsentierende (bzw. in das Äußere introjizierende) Einfühlung, und zwar in einer Apperzeption, die das ganze Seelenleben und Seelensein *realisiert* zu einer Art Erscheinungs-einheit. . . . (§46; Hua IV: 169)	[Human being] as object is thus a transcendent external object, an object of an external intuition; that is, we have here an experience of two strata: interwoven with external primally presenting perception is appresenting (or introjecting into the exterior) empathy, in an apperception, specifically, which *realizes* the entire soul-life and soul-being in a certain sort of unity of appearance. . . . (§46; 1989: 177-178)

Thus it is empathy that leads to the constitution of "intersubjective objectivity" for things in general, and in particular for the thing that is human being. Thanks to empathy, nature is constituted as the field of investigation for the natural sciences, a field in which particular natural things can be reliably investigated.

§47 concludes on an incongruous note. It claims that the foregoing analyses have presupposed an "absolute subject . . . for which common nature, physical as well as animal, constitutes itself" (*das* absolute Subjekt . . . für *das sich die gesamte Natur, die physische wie die animalische konstituiert*). That subject posits nature; therefore it is not going to be one of the subjects within nature, as individual people are. Nature is a unity of appearances, a unity posited of subjects and indeed needing to be imposed in acts of reason. There are presuppositions here, the text concludes, that must be followed into a realm of research that lies altogether beyond the sciences of physical, animal, and human nature. They lead into the realm of subjectivity itself.[73]

---

[71]Reprising the argumentation in Husserl's lectures of 1910, discussed above.

[72]See §46; Hua IV: 168-169; (Husserl 1989: 176-177). This is crucially important for the physical sciences, of course.

[73]See §47; Hua IV: 171-172; (Husserl 1989: 180). This claim seems to accord better with what I call "the second solution." Its placement here may be an editorial attempt to impart balance to the concluding lines of the second of the book's three major sections. In chapter four I will argue that Edith Stein composed §§43-47 of *Ideen II*.

(c) §49. "The personalistic orientation in contrast to the naturalistic." The term "naturalistic" (*naturalistischen*) denotes the propensity of the natural sciences to explain the world and its events by examining the relations of physical causality that obtain among the items in that world. "Personalistic" (*personalistischen*) refers to the human or cultural sciences, which seek to understand what influences persons and the choices that they make. Paradoxically it is the personalistic orientation that is the more "natural," because it takes the human subject to be originally a member of a social world; while the "scientific" naturalistic orientation is abstract and artificial as it considers the physical and sentient strata of human being in isolation from the soul and the intellect. By "person" is meant the i in its social connections with other i's.[74] The personal i is not the pure or transcendental i.[75] Nor is it *merely* an item among the objects of the natural world.

Als Person leben ist sich selbst als Person setzen, sich zu einer „Umwelt" in bewußtseinsmäßigen Verhältnissen finden und in Verhältnisse bringen. (§49/e; Hua IV: 183)	To live as a person is to posit oneself as a person, to find oneself in, and to bring oneself into, conscious relations with a "surrounding world." (§49/e; 1989: 193)

Even the scientist, adopting the naturalistic orientation in order to conduct natural-science investigations, does so as a person. Therefore the personalistic orientation is a prerequisite for the naturalistic orientation. The personalistic orientation is the

Einstellung, in der wir allzeit sind, wenn wir miteinander leben, zueinander	attitude we are always in when we live with one another, talk to one another,

[74]See Hua IV: 175; (Husserl 1989: 184).

[75]Transcendental i is that which conducts the activity of constitution. What I will term "the second solution" wants to place the transcendental i logically prior to the personal i. Accordingly (for the "second solution"), transcendental i will be that with which any pure i constitutes its objects--among which will be its own personal i, or soul, as well as the personal i's of others that are available through empathy. See §27, Hua IV: 110; (Husserl 1989: 117), where: "There are the same number of pure [i's] as there are real [i's], while at the same time these real [i's] are constituted in pure streams of consciousness, are posited by the pure [i's] . . . . Every real [i] belongs, as does the entire real world, to the 'environment', to the 'field of vision', of my--and every--pure [i], as is shown with a priori necessity by a closer study of the intentional constitution of the objective (intersubjective) world. And so . . . every pure [i] that accomplishes the apperception 'I, the human being', has the human [i], the person, as an object in its environment." [Es gibt soviele reine Ich als es reale Ich gibt, während zugleich diese realen Ich in den reinen Bewußtseinsströmen konstituierte, von den reinen Ich gesetzte . . . sind. Jedes reale Ich gehört wie die ganze reale Welt zur „Umgebung", zum „Blickfeld" meines und jedes reinen Ich--wie sich bei näherem Studium der intentionalen Konstitution der objektiven (intersubjektiven) Welt in apriorischer Notwendigkeit herausstellt. Und damit hat . . . jedes reine Ich, welches die Apperzeption „Ich, der Mensch" vollzieht, sich, das Mensch-Ich, die Persönlichkeit zu seinem Umgebungsobjekt.] Nevertheless "the first solution," which we are currently discussing, accords priority to the personal i.

sprechen, einander im Gruße die Hände reichen, in Liebe und Abneigung, in Gesinnung und Tat, in Rede und Gegenrede aufeinander bezogen sind. (§49/e; Hua IV: 183.	shake hands with one another in greeting, or are related to one another in love and aversion, in disposition and action, in discourse and discussion. (§49/e; 1989: 192)

Furthermore, the person is so deeply ensconced in society and in the body that no science can probe any deeper. For the *naturalistic* orientation,

ist . . . alles Erleben leiblich fundiert, also auch der Gesamtbestand des die Welt mit allen ihren Eigenschaften in den Personen intentional Konstit-uierenden. (§49/e; Hua IV: 184)	all lived experience is founded bodily, and hence, in addition, so is the total content of that which, in the persons, intentionally constitutes the world and all its properties. (§49/e; 1989: 193)

This does not rule out the possibility that experience could show itself to be differently founded if considered from another orientation, however. What might that other orientation be? Is there a transcendental perspective, which would investigate the structures of constitution apart from or prior to the body and the interpersonal in-feeling that it facilitates? According to "the first solution" the answer is no! This becomes apparent as the argument continues.

(d) §54-56. "Motivation as the basic lawfulness of intellectual living." It turns out that transcendental constitution is a component of motivation. Constitution does not and cannot occur apart from the activity of perceiving objects *as valued*; that is, as endowed with attractiveness or repulsiveness for me. I who constitute them am vulnerable and receptive to my objects, which therefore are constituted precisely as stimuli in regard to which I may enact some physical, sentient, soulful, or intellectual response. Thus constitution occurs within the personalistic orientation. Subsequently within the constituting of the physical qualities of things (which is accomplished personally and socially), there comes about the possibility of a "naturalistic" scientific orientation toward the things: an orientation toward causal relationships. But the constituting i *is affected* at various levels by the objects that it constitutes. The i is receptive to influences of various kinds from what it constitutes. They register with the i, by means of its body. In fact, it is their bodily registration that has occasioned their constitution as objects at all (inasmuch as the data of perception are the raw material out of which objects are constituted). Yet no object is "merely" physical. It can *matter to* someone in some way.

The mutual recognition among subjects that founds society is a special instance of constitution, and in fact is the one that makes any other constitution possible. This recognition comes from empathy. What is felt-into in this instance is the motivated coherence of other life-streams. There is a registration, within one's own streaming experiences, that similar streams of experience exist. One empathizes that "if-then" is a form--a rational motivational pattern--shared by all persons who understand the natural world and its causal connections. One also empathizes that personal acts, too, have "because" as their formal structure, although this is a "because" of a different kind. These forms of conditionality--"if-then" or "because"---are forms of temporality. The constitution of temporal sequence is a logical presupposition for the recognition of physical causality and personal choice. (For example, a physical cause precedes its effect, while a motive is the telos or end brought into being by a choice.) Even a "lone i" would constitute sequence as the form of its own streaming experiences. However a lone i would not become a practitioner of natural or cultural sciences. The mere notion of sequence is not sufficient to produce the notion of *coincidence*, that is, a plurality of lived-through sequences amenable to comparison and coordination. The possibility of comparison is given only with bodily experience. Bodily life thus is primordially social; otherwise the world would lack objective temporal structure.

(e) §51. "The person in personal associations." There are formal structures that do not incorporate temporal sequence, of course. Nevertheless these forms, too, would be unavailable to the "lone i" for they presuppose a plurality of subjects in communication with one another. The form of reality itself is of this kind.

Mich selbst kann ich „direkt" erfahren, und *nur meine intersubjektive Realitätsform* kann ich prinzipiell nicht erfahren, ich bedarf dazu der Medien der Einfühlung. . . . meine Erlebnisse sind mir direkt gegeben, die Erlebnisse nach ihrem eigenen Gehalt. Aber die Erlebnisse Anderer sind von mir nur mittlebar-einfühlungsmäßig erfahrbar. Dabei ist aber auch jedes meiner Erlebnisse als Bestandstück der „Welt" (der objektiven raum-zeitlichen Realitätssphäre) nicht direkt erfahrbar; die Realitätsform (die der intersubjektiven Objektivität) ist keine imanente Form. (§51; Hua IV: 200)

I can have a "direct" experience of myself, and it is *only my intersubjective form of reality* that I cannot, in principle, experience. For that I need the mediation of empathy. . . . my lived experiences are given to me directly, i.e., the lived experiences in their own content. But others' lived experiences can be experienced by me only mediately, in empathy. And not even each of my lived experiences, as components of the "world" (of the objective spatio-temporal sphere of reality) is experienceable directly, for the form of reality (that of intersubjective objectivity) is not an immanent form. (§51; 1989: 210.)

A lone i would have experiences but would not have them as "real." Thus there is no transcendental constitution of a real world prior to empathy. There isn't even any *priority* prior to empathy; for "priority" requires comparison and validation. Such determinations are necessary for the constitution af anything as real, but they are impossible without empathic communion among i's. Therefore *Einfühlung* is the basis for all sciences. Science is essentially radically personalistic.

2. Second solution: the priority of transcendental constitution.

Intriguingly, the *Ideen* also proposes a contrasting, impersonal account of the foundation of the natural and cultural sciences. That account rests uneasily beside the articulations of the *Einfühlungslehre* in the main text, but it is most effectively stated in the supplements. It declines to found the world upon anything worldly. There must be a science of world-foundation, and it cannot be either naturalistic or personalistic in its orientation.

(a) §53. A late amendment concerning nature and intellect. At a point in the manuscript where the physical, sentient, and ensouled natural world was under discussion, Husserl added the following lines to make an adjustment in the argument:

Diese naturalistisch betrachtete Welt ist doch nicht *die* Welt. Vielmehr: vorgegeben ist die Welt als Alltagswelt und innerhalb ihrer erwächst dem Menschen das theoretische Interesse und die auf die Welt bezogenen Wissenschaften, darunter unter dem Ideal von Wahrheiten an sich Naturwissenschaft. -- Die vorgegebene Welt wird zunächst hinsichtlich *der* Natur erforscht. Dann kommen die Animalien an die Reihe, zunächst die Menschen. Und da ist eben das erste: sie als Ichsubjekte erforschen. Man geht zurück auf ihr Leben. Das ist induktiv -- als individuell faktisches, wie die Person selbst -- stellenmäßig in der Raumwelt nur psychophysisch bestimmbar. Jedenfalls hat man eine Reihe psychophysischer Untersuchungen. Aber das ist eben nicht alles. (§53; Hua IV: 208-209)	This naturalistically considered world is of course not *the* world. Rather, given prior is the world as the everyday world, and within this arise [for human beings] the theoretical interest and the sciences related to the world, among which is natural science under the ideal of truths in themselves. This pregiven world is investigated first with respect to *nature*. Then animalia have their turn, human beings before all others. And this is precisely the first task: to investigate them as [i]-subjects. We are led back to their life. This is--as individual factual life, like the person [her]self--inductively determinable, in terms of its place in the spatial world, only in a psychophysical fashion. In any case, we have here a series of psychophysical investigations. And that is precisely *not everything*. (§53; 1989: 219, emphasis added)

"Psychophysical investigations" would be those starting with the bodily experience of empathy as proposed in what has been termed "the first solution." This passage, however, clearly is looking for something more. It is looking for something already targeted in the earliest portions of the manuscript, drafted in 1912.

(b) §§30 and 34. The soul's reality. Husserl first juxtaposition of the natural and the cultural sciences in the *Ideen* was framed in terms of "something extra" beyond the human soul's dual citizenship in two realms. The soul-i is continually in flux. It undergoes influences from and through the body, and it also exerts influences on the body and through the body. Thus it is in constant touch with the physical world, and also in constant communication with the social community. Human beings find themselves sometimes "bound and constricted," sometimes "unbound and free." They are sometimes "receptive," sometimes "creatively active." But they are always "really related to their thingly and intellectual world" (*auf seine dingliche und geistige Umwelt real bezogen*).[76] Yet in the midst of the constant changes, receptions, responses, and initiatives, there is an enduring character to the person. Even when submerged and passive, the person-i is the same as the soul-i. Or rather, the person-i does not show itself to a science such as empirical psychology whose orientation is toward the psychophysical determinations of the soul by elements of the natural world. That

[76]§34; Hua IV: 141; (Husserl 1989: 149).

orientation discloses only sentience. The person is apprehended by the cultural/intellectual sciences.

Yet apart from the i-realities of person and soul, which indicate substantial unities enduring through changes, there is also "the pure or transcendental i." The transcendental i is not a substantial unity. It neither influences nor is influenced by anything physical or intellectual. Presumably, then, it is not accessible through natural sciences or cultural sciences.[77]

(c) Supplement 1, §§ 1-2 of *Ideen III*. The transcendental i, alone and therefore not individuated, is presupposed as the basis for the constitution of the soul through empathy, according to this passage--originally drafted as part of Book 2 although now appended to Book 3. Here the inversion of priorities between the "first" and "second" solutions is most clearly to be seen. The starting point here is the lone i. The lone i constitutes material things as its objects. Subsequently, it performs optional acts of empathy through which it grasps some of those already-constituted things as having the additional property of sentience: they are animals or human beings. In other words, thingly being arises primordially through constitution, while sentience and soulfulness are added on subsequently through empathy. The plurality of i's does not come into play until the original lone i gets around to empathizing some of its physical objects as ensouled.

In this passage, the term (*Einfühlung*) is paired with a tentative synonym, the neologism *Eindeutung*. This term connotes an explanation that injects significance into something rather than drawing significance out of it: an "implanation," as it were. On this view, the sentience, souls, and intelligence of other human beings would be granted to them by the grace of an implaining lone i. Of themselves, they would have only their physical characteristics.

So hat auch das dem äußeren Ding „*Leib eines Anderen*“ eingedeutete Ich sein nicht wegrückbares Ding „*eigener Leib*“, und dies ist im Sinne der Einfühlung eben dasselbe, das als *Träger* der Einfühlung oder Eindeutung erscheint. Damit is zugleich gegeben, daß die Uumwelt des fremden reinen Ich dieselbe ist wie die meines reinen Ich; und das sagt: *die Dinge* der Umwelt *sind Einheiten höherer Stufe, sich konstituierend auf dem Weg über die Eindeutung*. (Hua V: 109-110)

Thus the [i] interpreted into the external physical thing: "animate organism of another," has its non-removable physical thing, its "own animate organism," and this is, in the sense of empathy, precisely the same thing that appears as carrier of the empathy or interpretation. With this is given simultaneously the fact that the surrounding world of the alien pure [i] is the same one as that of my pure [i]; and this means: the physical things of the surrounding world are unities of a higher level, constituted by way of interpretation. (1980: 95; emphasis deleted)

Sociality and the plurality of i's here are relegated to the status of afterthoughts: secondary creations by a lone i *after* it has surrounded itself with physical objects primordially constituted. In other words, there is to be a stage in constitution where other people are "already" appearing as physical unities but "not yet"

[77]§30; Hua IV: 120; (Husserl 1989: 128).

appearing as sentient intelligent co-constitutors of the natural world.

To be sure, §1 works its way around to acknowledging that bodily life is prerequisite for the lone i's ability to constitute things in the first place. But some important implications are ignored: that the identification of i with body individuates the i, locates it, limits its access to appearances, robs it of its transcendentality, and privileges its access to one soul above all others.

(d) Supplement 1 of *Ideen II*. This passage is another early attempt by Husserl to offer "a step-wise description of the constitution of the mere material thing, of the body, of the ensouled thing or live being, then of the personal subject, etc." (*Versuch einer schrittweisen Beschreibung der Konstitution des bloß* materiellen Dinges, *des Liebes, des* beseelten Dinges *oder des Lebewesens, dann des* persönlichen Subjekts *etc.*).

As before, this attempt also begins with a lone i who is (somehow) embodied amid material things but has not yet considered the implications of that fact. The material environment is constituted first; it is thinkable in terms of the causal relationships obtaining among its members. Next, the appearances of things are seen to depend upon the states and the movements of the body. Nevertheless the i can sort out the core identity of each thing from the multiplicity of appearances that each presents. Recognition of this constancy amid variations allows the lone i to "objectivate" the things and the natural world. This objectivation of the thing is accomplished *without the collusion of other i's.* Husserl writes,

Die Objektivität dieses Dinges ist noch wesentlich bezogen auf das individuell vereinzelte Subjekt und den für es konstituierten Leib. . . . [D]as vorher vereinzelt gedachte Ich [erfaßt] gewisse „seiner“ Objekte als „andere Leiber“ und, in eins mit diesen, *andere Ich*, die aber damit *noch nicht als reale Subjekte konstituiert sind.* (Hua IV: 307).	The objectivity of this thing is still essentially related to the individually solitary subject and to the body constituted for it. . . . [The i] previously thought of as solitary now grasps certain of "its" objects as "other bodies" and in unity with them, other [i's], which, however, are not yet thereby constituted as real subjects. (1989: 321, emphasis deleted)

Apparently overlooking the scarcity of subjects at this point, the argument now asserts that subjects (plural!) can trade places, exchange viewpoints, and thus intersubjectively construct the spatial world in which objects have their objective being. The body is constituted as a field of free will: it is ensouled. The soul is constituted as what seats the i in a body. Now, at last, bodies of other i's are constituted as well. The constituting i is no longer alone. It now inhabits an intersubjective social realm.

Unfortunately, this argument begs the question of how other i's are in-felt in other bodies--and with it, the related question of how an i would identify with a body at all. There is no step in this "stepwise account" where the individuation and plurality of i's are accounted for. That the pre-individuated transcendental i would have access to bodily conditioned appearings as grist for the mill of constitution is simply presupposed.

(e) Supplement 12.2 of *Ideen II*. This logical glitch is repeated in the text that is Husserl's most elegant statement of the transcendental foundation of the sciences.[78] Here, a connection between i and body is both denied and presumed. The soul is said to be inserted (*eingelegte*) into an already constituted body. The intellectual i, on the other hand, "is not something real connected with the body" (*ist nicht ein mit dem Leib verbundenes Reales*).[79] The levels of human being--material, sentient, soul, and intellect--are imagined as discrete and mutually impermeable. This produces the problem of how they could affect one another, and moreover how the i ever gets ahold of the physical appearings out of which it constitutes its worldly objects.[80]

This methodological manifesto, relegated to a "supplement," disagrees with the main text concerning the character of the foundation of the sciences. Here, motivation as the basic lawfulness of intellectual living is *not* psychophysically conditioned. It seems to be "conditioned" only by the absolute subjectivity available to a lone i. Beyond the individual cultural sciences, with their personalistic orientation, there must be a universal absolute *Geisteswissenschaft*: transcendental phenomenology. Husserl asserts that there is:

der fundamentale Unterscheid einer . . . Geisteswissenschaft [fungierenden] als positive Wissenschaft im System der positiven Wissenschaften gegenüber einer absoluten Geisteswissenschaft. Das Interesse an einer Erkenntnis der Welt, der vorgegebenen Realitätenwelt, sie völlig ausgeschaltet, und statt die Welt schlechthin zum absoluten Thema zu machen, mache ich mich und meine kommunikative Subjektivität zum absoluten Thema als diejenige, deren allen gemeinsame Umwelt der Wahrheit diese Welt ist oder als diejenige, die alles, was als seiend gilt, in Geltung setzt, die Subjekt ist für alles, was Objekt--ihr Objekt--ist und als diejenige, die selbst, wenn es ihr paßt, einer Wahrheit as sich nachgehen würde usw. Ich setze mich als Subjekt und nicht als Weltobjekt, wenn ich shon meinen Leib und dann alles, was für mich als weltlich und wie immer

a fundamental distinction between a human science functioning . . . as a positive science in the system of positive sciences versus an absolute human science. I may completely put out of play my interest in a knowledge of the world, of the pregiven world of realities, and instead of making the world pure and simple my absolute theme, *I make myself and my communicative ability my absolute theme* as the subjectivity whose comprehensive surrounding world includes that world and its truth, or as that subjectivity which posits the validity of everything valid as being and which is the subject for everything that is object--its object--and finally as that subjectivity which itself, if it suited its purpose, would pursue a truth in itself, etc. I posit myself as subject and not as world-object already when I posit my body and then everything valid for me as worldly and as being in

---

[78]What now appears as the second part of Supplement 12 of *Ideen II* was written by Husserl in October-November 1916, when Edith Stein had just gone to work for him and was sorting out the manuscript for the 1912 draft and Husserl's revisions from 1913 and 1915. See Hua IV: 340-372; (Husserl 1989: 351-382).

[79]See Hua IV: 349; (Husserl 1989: 360).

[80]As will be seen in the next chapter, Edith Stein's solution to the problem will be to imagine the same four levels as primordially permeable to one another, accepting influences across their borders. She terms this "blending" (*Verschmelzung*).

seiend gilt, als Erfahrenes meiner Erfahrungen etc. setze, als meine Erkenntnishabe, als mein praktisches Objekt, etc. (Hua IV: 353-354)	whatever way, as the experienced of my experiences, etc., as what I possess in knowledge, as my practical object, etc. (1989: 364; emphasis added)

The intellect-i does not consult its body when seeking access to other intellects. Communication is not primarily a psycho-physical affair. Rather, empathy occurs transcendentally among i's. The i over there (*Ich dort*) is the target of empathy. The body of that i is merely "a passageway" (*Durchgang*).[81]

Thus "the second solution" to the problem of establishing a foundation for the sciences--for science as such--relegates bodily life to a derivative status and proposes to investigate a transcendentality imagined to be more primordial.

3. Husserl's choice of "the second solution."

Some of the logical problems with the "second solution" that were identified above are directly addressed in the last long text that Husserl composed for the *Ideen*: Supplement 12.1, with a deleted fragment that now appears in the editor's critical notes. These ruminations on "the intellect and its soul-basis" read as if they were drafted in response to someone's critical review of the treatments of "constitution" in the various manuscripts of the *Ideen*.[82] Husserl is wrestling with a vexing issue: In his account of the transcendental constitution of things, the world, oneself, and other people, just where does empathy fit?[83]

The reflections start by comparing the way in which we encounter the world with the way in which we encounter other subjects.[84] When experiencing people, we come upon their bodies just as we come upon other worldly things. However human being is co-given (*mitgegeben*) as a body-soul unity. We do not live the inner life of the other as we do our own; but we co-live it (*wir leben es mit*). It is "there in the original" although we do not grasp it in the original as we do our own.

When studying the world and its items, what we do is to classify regularities and follow threads of causal dependency. The same tasks are done when we study the soul-like and personal interiority of people and animals; but now there are two ways of doing these tasks: the descriptive way, and the intuitive

---

[81]See Hua IV: 347; (Husserl 1989: 358).

[82]See Hua IV: 418-420 for the beginning of the notes, and Hua IV: 332-340 for the rest; see also Husserl (1989: 344-351). This material was written in January 1917. See chapter four for a discussion of Edith Stein's interactions with Husserl at that time.

[83]Hua IV: 336 (Husserl 1989: 347).

[84]It is significant that, in this passage, Husserl is reflecting as "we" (*wir*) rather than as "one" (*man*). The "editorial we" in effect drags the reader along involuntarily with the writer. This usage denies that the writer is describing singluar experiences, not corroborated experiences. The writer is claiming an intersubjective validation that he has not yet obtained for these experiences. The device of the *wir* thus entails a kind of question-begging in the matter at hand: whether in-feeling of other subjects--intersubjectivity--is the foundation of validity.

way. The soul can be investigated as an "annex" of bodiliness, a part of nature. In that case we study by following real-causal connections between the soul and the body as well as causal coherences within the soul itself. That kind of study aims to identify regularities that operate inexorably, like other natural processes. Freedom is never brought into play. There is no question of subjectivity. Natural science investigates the world as a system of realities, and real being is passive being (*reales Sein ist passives Sein*).[85]

But the more thorough investigation of human being is the intellectual, personal one. A person is part of a personal network and fulfills her own intellectual life within it. Among the personal acts of an i-subject are some that are specifically social. These are not merely empathetic (*einfühlungsmäßig*) in general, intending an anonymous plurality of subjects. Rather, they grasp other subjects as those whose requests one's own acts are meant to fulfill, those whose acts intertwine with and impart sense to one's own. Subjects connect with one another through their acts and their wills. In this way they constitute communities, families, nations, states: personalities of a higher order. A nation (*Volk*) is not a natural reality; it is constituted in the intellectual world (*Geisteswelt*). This world, the social world, is the field of greatest humane interest and the keenest theoretical interest as well. [86]

How, then, is the intellect related to the soul? The intellect is actively productive, while the soul is a repository into which settle the congealing products of intellectual activity. (For example the intellect constitutes, and the soul stores the constituted objects for subsequent retrieval and possible reenactment.) Settling out and retrieval are transformations that occur with law-like regularity that can be studied and understood. All intellectual activity, all i-acts, have a characteristic tendency to find *sames*. They register similarity.[87] Intellectual acts compare, distinguish, think universals, and classify particulars under them. Intellectual acts also determine absolute value and assign relative values in terms of it. Moreover they set ultimate goals and proximate means for attaining the goals. Personal intentionality always originates in such activity and refers back to it.

All of this activity comes under the heading of *validity determination* (*Geltung oder Nichtgeltung*), and therefore all of it pertains to the idea of truth.[88] What is valid is that which can be enacted by the intellect. For example, an absurdity can be rehearsed or represented, but it cannot be brought to original enactment. A memory is certified if I can live through it again; I then affirm that the event remembered did indeed happen. By the same token, a fantasy cannot be so certified. It begins the transition from soul to intellect--that is, from passive notion to active intention--but it is impeded. I simply cannot complete an active

---

[85]See Hua IV: 419.

[86]See Hua IV: 419-420.

[87]There is of course a certain circularity in the assertion that the similarity of X to Y consists in their common tendency to ferret out similarity.

[88]Hua IV: 332-333; (Husserl 1989: 344-345).

intention that the content of my daydream is true.

The transitions from representation to enactment and back again are two of the tendencies whose regularity can be studied. There are others, including association, memory, and synthesis (*Verschmelzung*).[89] Some of the contents that undergo such transitions do not originate with the intellect. That is, they do not stem from the i's initiative. Instincts are of this type. Instincts originate involuntarily in sensibility. Habits somewhat resemble instincts; however habits have their origin in acts. Yet once they are in place within the soul, habits and instincts alike can become *stimuli* for the i. Thus the i has a receptive side. Receptivity to stimuli is the lowest level of i-activity.

Left unmentioned at this point in the manuscript is the question of the status of a particular kind of stimulus: empathy of another person. How is it that I register live experiences in which someone has been active although the someone was not I myself? Such experiences begin in sensibility, perhaps with the sight of a wound or the sound of a cry. How do I register that someone hurts although it is not I who hurt? The content "hurt" tends to follow the regular transition pathways through live body and soul toward the intellect, but it is impeded in a distinctive way. All this is unsaid by Husserl, but it is the missing link to his remark: "Where is empathy to be accommodated?"

The question is not answered, but a pertinent excursus on "impression and reproduction" follows. Impressions are alien while reproductions are native to consciousness, having originated in one's own constituting activity. With any given mental content, there is a tendency to certify whether it is an impression or a reproduction: that is, to test out whether I can reenact or reconstitute it. The constitution of impressions cannot be undone or redone, without undoing the entire constitution of the natural world. The determination "imported or domestic," which distinguishes impressions from reproductions, depends on whether one can fulfill the tendency to reenact its originary content. There are other tendencies; for example, the tendency to give in to an instinct or a habit, and the tendency to reach out and acquire some desired good. There seems to be a kind of dynamic traffic in these tendencies constantly within the i. They are its possessions, its *Habe*. In light of this, Husserl says that the i never acts without first being affected in some way. The i always acts in response to its *Habe*, intentionally.[90]

Motivations, however, are something else entirely. They do not impinge on the i as tendencies do. Tendencies and affections are still a part of nature. Through them, the i has a passive side that is bound to nature. Tendencies are governed by a time-bound causal structure. Even acts have a natural side, for they leave a comet's tail, a residue of memory and habit. Yet the i also has a free side.

---

[89]Hua IV: 334-335; (Husserl 1989: 346). The context does not make clear what Husserl might mean by the term *Verschmeltzung*. As we shall see in the next chapter, Edith Stein uses this term to refer to any influence that crosses from one level of the human being to the adjacent level: from physical organism to sensibility, from sense to soul, or back.

[90]Hua IV: 336-337; (Husserl 1989: 348-349).

It originates its acts. To say that human acts are motivated is to say that they follow laws of reason rather than laws of natural causality. What, then, are laws of reason? Husserl's remarks break off with that question.

The question of rational method, and of the place of empathy within it, framed Edith Stein's initiation into philosophy. During the weeks before he penned these final pages for the *Ideen*, Husserl had been working with Stein on a reformulation of his doctrine of constitution. Shortly before that, she had submitted to him her dissertation, "On the Problem of Empathy," and received her doctorate. Between October 1916 and 1918 Stein edited the manuscripts of *Ideen* that we have been examining but she did not see them before preparing her own dissertation. Yet they provide relevant background for her text. Husserl was already working on the coordination of "nature and culture" in his classes and seminars from 1913 to 1916.[91] Stein was an avid listener; moreover, Husserl's phenomenology was handed down from senior to junior students at the Göttingen Philosophical Society meetings and in other informal contexts. There was ample opportunity for her to know the particulars of his research and to be thinking and talking about his formulations, even before she became his assistant.

In this chapter it has become clear that Husserl employed the term "empathy" *(Einfühlung)* in provocative, diverse, and inconsistent senses. We have seen that the deployment of the term in *Ideen* contrasts sharply with Husserl's earlier usage, and that *Ideen II* offers contradictory solutions to the problem of founding the sciences. The considerable effort that we have expended in sorting through Husserl's pre-Armistice *Einfühlungslehre* affords us some insight (or, if you will, *in-feeling*) into the doctorand Stein's predicament as she prepared her study of this elusive term under Husserl's direction. Her work will be examined in the next chapter.

[91] Husserl offered courses or seminars on "nature and culture" in the summer semester of 1913 (which Edith Stein attended), the winter semester of 1913-14, the summer semester of 1915, and the winter semester of 1915-1916. The *Husserl-Chronik* indicates the topics that were treated.

Chapter Three

Edith Stein's Hermeneutic Theory

The time has come to interpret the interpretation theory of Edith Stein, as presented in the portion of her doctoral dissertation that was printed in 1917 under the title *Zum Problem der Einfühlung*. The author tells us that this dissertation was defended on August 3, 1916. She says that the text originally opened with a chapter offering a historical review of the literature, a chapter that subsequently was left out of the 1917 printing.[1] The loss of that introductory material is unfortunate, for its availability is assumed at many points in the remaining three chapters. Our foregoing survey of the work of Stein's predecessors can perhaps stand in for her own missing historical discussion.

The extant portions of Stein's hermeneutic theory comprise the second, third, and fourth chapters of the original work: "the essence of acts of empathy," "the constitution of the psycho-physical individual," and "empathy as understanding of intellectual persons."[2] The author says that she intends to address a basic problem that will provide the key to understanding all of the various theories of empathy that were current in the literature in her day. This problem is: "die Frage der Einfühlung als Erfahrung von fremden Subjekten und ihrem Erleben" (the question of empathy as encountering of/by alien subjects and their live experiencing.[3]) The writer who is taking on such a task is 24 years of age. As she writes, the First World War is still raging. Many of her friends, including her teacher Adolf Reinach, are at the front. She has recently resumed her studies after several months' interruption for service as a volunteer medical helper in a

---

[1]The 1916 inaugural dissertation was entitled "Das Einfühlungsproblem in seiner historischen Entwicklung und in phänomenologischer Betrachtung." Apparently the manuscript was considered for publication in Husserl's *Jahrbuch*, but not accepted because at that time Husserl himself was planning to bring out something on empathy there. I provide a brief summary of Stein's positions in the first few pages of the next chapter. this chapter presents a detailed commentary.

[2]I will prefer the translation "intellectual" for Stein's term *geistig*, although the latter term connotes the spiritual and the cultural as well. My preference follows Stein's own choice to translate the Latin *intellectualis* in the *De Veritate* of Thomas Aquinas with *geistig*. Latin *intelligentia* is rendered *geistiges Wesen*. See Stein 1934, the second volume of which presents Stein's own Latin-German glossary.

[3]*Von* plus a verbal noun can convey either an objective-genitive sense (how one has met with alien individuals in one's own experience "of them") or a subjective-genitive sense (experiences that have been had "by" the alien individuals themselves). I indicate both prepositions, "of/by," so as to avoid prematurely narrowing the meaning to one or the other alternative. The ambiguity is precisely what is at issue in this work. See Stein 1917: v. The English translation of Stein's dissertation, by her great-niece Waltraut Stein, is not precise enough to be used for philosophical study, but its index is helpful.

convalescent hospital for gravely wounded soldiers.

My presentation of Edith Stein's hermeneutic will be made in the second person. There are several reasons for this choice. most importantly, it will be my thesis in a subsequent chapter that interpreters of Stein too easily draft her into service as spokesperson for their own views. Stein's work has been extraordinarily vulnerable to misappropriation. The mechanism that facilitates such interpretive capture and redeployment of Stein's words is the packaging of them as "Edith Stein," that is, as a reified text available to be mined and hauled away. While no interpretation can remain innocent of appropriating and displacing its target, still I hope to resist this tendency as much as possible by refusing the convention of third-person description. For me, the text is not "it" and the writer is not "she." I will speak to Edith Stein as "you."

My further reason for employing the second-person form of exposition is that I cannot find a comfortable name by which to call Edith Stein when I do try to refer to her in the third person. Her own niece, Suzanne Batzdorff, poetically ponders the problem of address. Batzdorff writes that a Jewish relative does not wish to call this woman like a nun, "Sister," or like a Catholic saint, "Blessed"--but can she still call her "Aunt"?[4] For my part, as a co-religionist I could indeed call Edith Stein both Sister and Blessed; yet in doing so I might compromise or discredit my stance of philosophical interrogation and analysis. Moreover, as a 47-year-old woman I can hardly help calling the 24-year-old author of *Einfühlung* "child" despite my sincere desire to regard her as a senior colleague and teacher. Insofar as I mean to bring this girl into the English-speaking philosophical world for consideration, attempting to gain recognition for her as a citizen of the world's philosophical canon, I might even call her "*my* child, my daughter." Finally, I find it impossible to call her "Stein" -- stone, פסל (Hebrew *pesel*): a statue, an idol, a heart so hard that it cannot hear; or rock, צור (Hebrew *tsur*): a butte, a mountain fortress, a secure stronghold. But neither can I call her "Edith" -- עדות (Yiddish *eydes*): testimony, witness, omen, reminder. Her character and identity remain invincibly elusive. Provisionally, then, I must call her "you."

A. Eidetics of Empathy

Liebe, hochverehrte Frau Dr.: What a ride you took through the philosophical rapids of Göttingen and the confluence of egological theories! In your work, the upwelling transcendentalism of Husserl met the surging personalism of Scheler. Amid these currents you sought to navigate a phenomenology of interpersonal access. Your text is a record of your creative engagement at the front between transcendentalism and personalism. You grasped the dissonance between Husserl and Scheler that had come to expression in the long critical footnote of the

---

[4]See Batzdorff 1990: 105-113, 121-123. Edith Stein took the name "Teresia Benedicta a Cruce" when she entered the Carmelite cloister and she was subsequently called "Schwester Benedicta." In 1987 she was beatified by Pope John Paul II and is now called "Blessed Edith Stein" by Roman Catholics.

Idolenlehre.[5] Scheler there complained that Husserl had gone astray somewhere between the 1901 Appendix to the *Logische Untersuchungen* and the 1911 *Logos* article. Scheler agreed with the 1901 statement, in which Husserl had delineated his own position on the kinds of perception from that of Franz Brentano, his teacher:

Nach Brentano unterscheidet sich die innere von der äußeren Wahrnehmung 1. durch die Evidenz und Untrüglichkeit und 2. durch die wesentlich verschiedenen Phänomene. In der inneren Wahrnehmung erfahren wir ausschließlich die psychischen, in der äußeren die physischen Phänomene. . . . Demgegenüber will es mir scheinen, daß *innere und äußere Wahrnehmung, wofern man diese Termini naturgemäß versteht, von ganz gleichem erkenntnistheoretischen Charakter sind.* Ausführlicher gesprochen: es gibt zwar einen wohlberechtigten Unterschied zwischen *evidenter* und *nicht-evidenter*, untrüglicher und trüglicher Wahrnehmung. Versteht man aber . . . unter *äußerer* Wahrnehmung die Wahrnehmung von physischen Dingen, Eigenschaften, Vorgängen u.s.w., und danach unter *innerer* Wahrnehmung alle übrigen Wahrnehmung: dann koinzidiert diese Einteilung durchaus nicht mit der vorigen. So ist jede Wahrnehmung des Ich, oder jede auf das Ich bezogene Wahrnehmung eines psychischen Zustandes gewiß nicht evident, wenn unter Ich verstanden ist . . . die eigene empirische Persönlichkeit. Auch ist es klar, daß die meisten Wahrnehmungen psychischer Zustände nicht evident sein können, da sie leiblich lokalisirt wahrgenommen werden. (Hua XIX/2: 760-761)

According to Brentano, inner perception distinguishes itself from outer perception: 1. by its evidence and its [infallibility], and 2. by essential differences in phenomena. [According to Brentano,] in inner perception we experience exclusively psychic phenomena, in outer perception physical phenomena. . . . As opposed to this, *inner and outer perception* seem to me [Husserl], *if the terms are naturally interpreted*, to be of an entirely similar epistemological *character*. More explicitly: there is a well-justified distinction between *evident* and *non-evident*, or between infallible and fallible perception. But, if one understands by *outer* perception . . . the perception of physical things, properties, events etc., and [if one] classes all other perceptions as *inner* perceptions, then such a division will not coincide at all with the division [above, i.e., evident versus non-evident]. For not every perception of the [i], nor every perception of a psychic state referred to the [i], is certainly evident, if by the [i] we mean . . . our own empirical personality. It is clear, too, that most perceptions of psychic states cannot be evident, since these are perceived with a bodily location. (1970: 859)

Scheler had taken this 1901 statement of Husserl's to mean that there are two kinds of appearings to consciousness: those of physical objects, which offer aspects in sequence to be unified in the act of perceiving, and those of psychic objects, which also offer themselves sequentially and acquire unification of a different sort in the act of perceiving. In either mode, what was perceived was "in evidence"; nevertheless the act of perceiving remained subject to illusion.[6] But in 1911,[7]

---

[5]See the discussion in chapter one, above.

[6]In the first section of *Idole*, Scheler nicely distinguishes between illusion (*Täuschung*) and error (*Irrtum*). Illusions pertain to appearances, that is, to perceptions

Scheler charges, Husserl shifts his position and asserts that illusion is not possible in connection with the psychic after all, inasmuch as psychic being is not a unity assembled by perception out of sequential appearings. The evidence of the psychic excludes the possibility of illusion, says Husserl in 1911; and Scheler says that *that's* outrageous.

You saw that the key to resolving this conflict was terminological. You took Scheler to task for having failed to notice that he and Husserl meant different things by "inner perception," and moreover that in the 1913 *Ideen* Husserl had in fact substituted the term "reflection" to indicate what he had in mind in 1911 with his claim that in the psychic sphere, appearing *is* being.[8] You saw--although you did not show--the substantial agreement between the positions of Husserl and Scheler on the accessibility of the i. What Husserl now affirmed as given in "reflection," Scheler too affirmed as the i-drenched character of all "action." This insight gave you the first plank for building your own theory.

What you liked about Scheler, and *didn't* find in Husserl, was an account of the four nested levels through which a personal individualized i gradually emerges: the physical, the psychic, the soulful (*seelisch*) and finally the intellectual or spiritual (*geistig*). Scheler usually reserved the terms "active" and "person" to the last of those levels, where the individual's own distinctive contribution to his or her act *made it be* what it was. Clearly, science could not be a personal activity in that sense, for science specifically excludes any role for individual personal peculiarities within its observations and inferences. But you wanted a *scientific account of* interpersonal communication. For that, you had to turn to Husserl and the theory of the "pure i" that he was elaborating. You were convinced of the need for a Husserlian "phenomenological reduction" to bootstrap yourself out of the realm of nature and culture, and into the realm of the pure essences of their appearings. You understood this move as the kind of maneuver at work in Husserl's *Logos* article: an advance toward founding the possibility of the coherence-recognition that occurs in natural and cultural sciences alike.

---

of appearings, while errors pertain to inference and judgment. Illusion occurs when two contradictory appearings are referred to the same thing (e.g., when a partially submerged stick *looks* bent but *feels* straight if I run my hand along it). The illusion is really there; I have done nothing to produce it. By contrast, an error may refer to something nonexistent. It arises in a wrong inference that I have executed, and it subsists in the relation between the state of affairs existing in intuition, and the state of affairs signified in judgment. That much of Scheler's account, you understood. But you seriously misconstrued his further discussion--as we shall see.

[7]In fact, Scheler refers to the 1911 *Logos* article although he incorrectly dates it to 1913.

[8]See Stein ([1917] 1980: 34). Scheler cannot, of course, be faulted for having overlooked a 1913 restatement in his own 1912 article, where the prickly footnote first appeared. (Moreover, it is even possible that Scheler's criticism prompted Husserl's change in terminology.) Nevertheless the *1915 republication* of the footnote, which apparently is what you were reading, does not take Husserl's 1913 reformulation into account.

There were other moves within Husserl's developing phenomenology that you resisted. You paid lip service to his transcendental reduction, but refused its spirit and declined to fantasize the annihilation of the world. You sought a different way to navigate the difficulties entailed by the juxtaposition of "perception" and "interpretation" in the Appendix to his 1901 *Logische Untersuchungen*. There Husserl had proposed that outer perception--perception of mountains, hills, and forests--was really *interpretation*, and therefore inner perception must also be interpretation if it was to claim the title of perception at all.[9] Husserl's 1920 revision of that text exhibits his own transcendental solution: instead of "interpretation" he substitutes "apperception," perhaps hoping thereby to retain the certitude of immediate intuition for our knowledge of other people. An "apperception" would be given along with a perception as its horizon or backside: that is, as something that *could* be brought into the foreground of perception. You specifically refuse that solution, saying instead that the alien individual is "announced" in certain qualities of one's own live experiences but not "apperceived" there.[10] Yet you avoid using the term "interpretation," which leads me to suspect that it had already fallen out of favor in Husserl's circle well before he repudiated it in print in 1920.[11]

Your toughest struggle with Husserl, however, has to do with depicting the relationship between nature and culture so as to allow psychic life to be assigned to one or the other realm. The 1911 *Logos* article states flatly:

Es gibt, eigentlich gesprochen, nur *eine* Natur, die in den Dingerscheinungen erscheinende. Alles, was wir im weitesten Sinne der Psychologie ein psychisches Phänomen nennen, ist, an und für sich betrachtet, eben Phänomen und *nicht* Natur. Ein Phänomen . . . kennt keine realen Teile, keine realen Veränderungen und keine Kausalität: all

there is, properly speaking, only one nature, the one that appears in the appearances of things. Everything that in the broadest sense of psychology we call a psychical phenomenon, when looked at in and for itself, is precisely phenomenon and not nature. A phenomenon . . . knows no real parts, no real changes, and no causality; all these words are here

[9]The German term *Interpretation* connotes musical or theatrical rendition more strongly than does its English counterpart. Husserl was trying to distinguish between what perceiving acts *are* and what they *accomplish*. Their accomplishment is their content as they present it: the interpretation. See 1901: 704-705. In the 1920 revision of this passage, most occurrences of "interpretation" have been changed to "apperception." (Compare Hua XIX/2: 761-762.) The rhetorical effect of this is to downplay the role of the particular individual: to minimize individual creativity as a factor in the comprehension of what appears.

[10]We'll return to this point below in section A.2.

[11]Husserl briefly took up the term *Eindeutung* or "implaination" in manuscripts from 1912; see chapter two, section D.2.c, above. Perhaps as late as 1915, however, Husserl wrote in a research manuscript (Hua XIII: 267): "Fremden Leib kann ich nur erfassen in einer *Interpretation* eines dem meinen ähnlichen Leibkörpers als Leibes und damit als Trägers eines Ich (eines dem meinen ähnlichen)." [I can apprehend an alien live body only in an *interpretation* of a live bodily thing similar to mine, and thereby as bearer of an i (one similar to mine).]

diese Worte im naturwissenschaftlichen Sinne verstanden. Phänomenen eine Natur beimessen, nach ihren realen Bestimmungsstücken, nach ihren kausalen Zussamenhängen forschen -- das ist ein reiner Widersinn. (Hua XXV: 29).	understood in the sense proper to natural science. To attribute a nature to phenomena, to investigate their real component parts, their causal connections --that is pure absurdity. (1965: 106)

But the text of *Ideen II* as we have it--that is, *as you arranged it* during your work as Husserl's assistant--is organized contrary to this principle. It sets the physical and the psychic both on the side of nature, and then contrasts nature with *world*. That is: the first section of *Ideen II* treats "material nature"; the second discusses the pure i, soulful reality, and the living body as "animal nature"; and the third examines the personalistic world and motivation as "the intellectual/cultural world." One wonders how such a Schelerian shrub came to be grafted onto the Husserlian root. In his seminars after 1911, Husserl reworked his approach to "nature and culture" several times, and may have adapted Scheler's insights on his own. Nevertheless, he declined to publish *Ideen II* in the format that you arranged. For your part, you took to heart Scheler's insistence that psychic events do indeed exhibit their own kind of process, partition, flow, and *causality*. You developed this claim and attempted to feed it back into Husserlian phenomenology in your *Habilitationsschrift*, which Husserl would publish in the 1922 *Jahrbuch*.[12]

With these general comments as background, I want to highlight five distinctive features of your eidetics of empathy that distinguish it within the context of pre-Armistice phenomenology.

1. Empathy is led, not projected.

The technical task that you have undertaken in your dissertation is to investigate *das Phänomen des fremden Seelenlebens*, that is, the indubitable appearance to me of living souls that are not my own.[13] You assert that those other souls are given with the same complete immediacy that invests one's awareness of one's own existence. Such awareness is indubitable because it comes on as direct perception, rather than having to be arrived at through a process of inference whose steps could allow error to occur. Further, you have chosen to investigate the essence of such phenomena. Essence-inspection, or eidetics, leaves aside mere *explanatory* questions, such as how the appearances might have their source in actual alien human beings; *practical* questions, such as what the appearances might oblige me to do; and *epistemological* questions, such as whether the appearances justify belief in the existence of other people. You mean simply to ask what these appearances *are*. More exactly, you mean to describe them persuasively enough to evoke in your readers the recognition that your readers, too, have direct access to experience of other people. You contend that immediate

---

[12]*Beiträge zur philosophischen Begründung der Psychologie und der Geisteswissenschaften.* Part One: "Psychische Kausalität."

[13]See Stein ([1917] 1980: 3).

experience of others is an essential component of being an i--any i.

What then is the distinctive character of such experience? The elegance of your proposal lies in its monadic simplicity. Previous authors had modeled the event of communication with *two* non-symmetrical parts; for example, with an aesthete actively interrogating a passive work of art, or with an expressive instinct balanced against an instinct to imitate. Empathy was construed as a "feeling-into," with the projecting feeler and the targeted feeler ontologically distinct from each other and from a third thing, the felt content. This sort of dualistic modeling tended to generate claims that were objectionable to you, such as that the in-feeler must *replace* the other, or that the interpreter must *understand the other better* than the other understood herself or himself. In contrast, you describe empathy as an appearance without any "coming in" or "going out" of personality or information. The feeling registers entirely within one's own consciousness, but it registers there in a way that announces a foreign life. I feel the feeling of another, as such, in that I am aware of *something about* my own feeling that directly presents the other human being. I feel myself led (*geleitet*) in this feeling.[14] My awareness is magnetized and configured to a pattern not of my own design.

To *become aware of* this aspect of my awareness--its having been led--requires a reflective act, you say. You concur here with Husserl's account of the "doubling" of the i as it engages in such experiences as remembering, expecting, pretending, or the imaginative capture of another's perspective through taking the other's place. In the last chapter, this doubled i-experience was indicated with the formula *now*/"now." This serves as a shorthand notation for events such that, in my flowing lifestream, I *now* experience what is a "now" elsewhere in some lifestream--either an alien lifestream, or my own lifestream at some other time. With such events, the other "now" cannot become a live *now* for me because, as Husserl insisted, there are no canals between streams of i-hood. Your monadic model obeys the no-canal rule, because it finds the complete and definitive essence of its target--one's inward awareness of others--entirely within the registrations of appearances occurring within one's own conscious life.

2. What is empathized is neither act nor form, but content.

Receptive accusative empathy of another, then, requires an act of reflection. Lipps had said something similar when he positioned the emergence of individual i's, both mine and the other's, at the conclusion of the original non-differentiated empathy, when "I snap back to myself." You differ with Lipps in that you posit neither a spontaneous loss of self-awareness in ecstatic oneness with the other's experience, nor a deliberate projection of self into an alien body or artifact for the purpose of reaping information. You wish instead to emphasize the similarity between empathy and various other reflective activities, such as remembering and pretending. *As acts*, all of these have essentially identical structures. They simply bring an i into coherence with another i, that is, with an i lying at some temporal or spatial distance.

---

[14]See Stein ([1917] 1980: 10).

In reflective activity, I comprehend the i-drenched character of the acts upon which I am reflecting. My reflection is *memory* if, between the i that is saturating the live experience upon which I am reflecting and the i that I am, I can grasp a flowing sequence of i-to-i face-offs and identity-recognitions in any one of which I can reawaken as an i recognizably my own--without the recognizer's ever fusing with the recognized in perfect identity at any stage.[15] My reflection is *expectation* if I can grasp the possibility of such a sequence between the i that I now am, and the i that would be involved in the fantasized live experience upon which I am reflecting. In any case of reflection whatsoever, i confronts i as *now* confronts "now." Thus, you say, the essential difference of empathy is *not* to be found in the *kind* of act that it is, for--according to its form--*Einfühlung* is just an act of reflection. Moreover, such acts can be performed by any i at all. Capacity for reflection is part of what it means simply to be an i. No personal talent or individual virtue is required.[16]

Empathy of others, then, is an act distinguished by its *content.* Only in regard to content, you say, do individual differences among i's become significant within an eidetics of empathy. I-differentiation appears as an aspect of empathized content. The relevant distinction among i's occurs in two stages. First, I reflect that I have been registering a live experience, an experience in which *someone* lives, and that the i living there is not recognizably identical with my own i. Thus an alien i appears. That much is given immediately. Second, I can go on to discover just who that alien i might be. (But this second step would be a matter for empirical investigation, and therefore lies beyond the scope of your eidetic study.)

Your insistence upon the relevance of the "whose?" within empathic experience places you seriously at odds with Husserl.[17] Your teacher restricted the content of acts of empathy specifically to that which is in principle experienceable by *any i at all*: logic and science. According to Husserl, what can be empathized includes *only* the forms of valid inference, the sciences that they govern, and the empirical observations obtained through such sciences. Therefore, as we saw, Husserl placed empathy at the foundation of his philosophy of science. What you have done is to lay the corrosive little question "why?" next to the ironclad Husserlian insistence that understanding be impersonal. Logic (that is, logic in the

---

[15]Your point here is to preserve the doubleness of the reflecting experience: reflecter and reflected never turn into one.

[16]This is apparent to anyone who has a basic familiarity with phenomenology. Unfortunately it is overlooked by interpreters who believe that you are discussing the "virtue" of empathy, as if it were a character trait that might be strong, weak, or lacking in a given individual, and that moreover were subject to training.

[17]In fact, Husserl earlier had toyed with this very question. Bernet *et al.* (1993: 207) cite an unpublished manuscript from 1912 in which Husserl posed "the ever and a priori question . . . whose *cogitatio*, whose pure consciousness . . . is the pure consciousness that gets thematized in the phenomenological reduction." However in their view Husserl subsequently chose to align the pure i with transcendental subjectivity more primordial than the distinction between i and thou, thus rendering the "whose?" question moot.

sense of the inferential forms whose validity is apparent to "any i at all") cannot be summoned to answer your question or to shoo it away, since the status of that logic itself is what is in question. Your proposal that *the "whose?" matters* goes marching onto the stage of phenomenology, flying a pennant that says "Why not?"[18]

You crossed Husserl because you (like Scheler) tried to adhere to the account of "inner and outer perception" laid out in the 1901 Appendix to the *Logische Untersuchungen*, while Husserl himself did not. "Inner perception" grasps psychic appearances in a way that excludes the possibility of error: it requires no inference and thus no contribution from the one who is entertaining the perception. You took it for a mere tactical adjustment when Husserl substituted the term "reflection" for "inner perception" in the 1913 *Ideen*. Husserl, however, was seeking strategically to wipe out not only the possibility of erroneous inference, but also the fingerprints: the distinctive contribution of a particular "who" to making appearances appear. "Reflection" connotes something more automatic than does "perception." (In a mirror, the reflected image is entirely determined by the original. Supposedly, reflection adds nothing; that is what astronomers presume in regard to the reflecting disks of their telescopes, for example.) This depersonalizing trend is clear in another terminological revision, the 1920 substitution of "apperception" for "interpretation" in regard to either kind of appearance, outer or inner. Husserl had earlier asserted that one hears *einen Leierkasten* (a barrel organ) itself when one registers certain auditory sensations as *Leierkastentöne*. The sensory stimuli are *interpreted* when perceived. The perceiving i is simultaneously an interpreting i.[19] In perceiving, the interpreter-i is contributing of its i-hood to the appearance. But in the 1920 rewrite, the notion of this creative contribution is shut out when "interpreting" is overwritten with "apperceiving." The i now is depicted as having no distinctive contribution to make.[20]

With that stipulation, Husserl nevertheless concurred in the view that the essence of the empathy of alien experience lay in its content. That is, it lay in *the formal aspects of* that content, aspects not owing to the particular "whose" of the content. In other words, he held that the shareable content of experience is *the way it goes*: its coherence, its motivated sequential flow. What can be empathized

---

[18]On the side of "not," it might be argued that in the two-stage distinction of individuals, only the first stage had eidetic significance. That is, the only difference in i's that is relevant to the founding of science is that between my i and i's that are not mine. To be scientific, I need only intend that my claims be claimable by *any* other i at all, not by specific i's. This is Husserl's contention.

[19]See Husserl (1901: 705). Compare *Husserliana* XIX/2: 762.

[20]What I am describing here are developments in the rhetorical construction of Husserl's argumentation. Like any author, he is not completely consistent and he changes his opinions over time. You encountered his work as a repertoire of useable tools. My job here is to illuminate how they variously empowered and impeded you. In this task, I cannot be responsible for making Husserl's life's work make sense.

is what can be followed. Husserl learned this from Lipps, for whom understanding was like dancing through a pattern, so that the bodily capacity to mime was the basic faculty for both intellectual and emotional communication. To Husserl, following a logical inference resembles following a chain of physical causation or a chain of historical events: having followed, quasi-kinaesthetically, I then can reflect and thus grasp all at once the essential form that connected the factual stages as they unfolded sequentially. To empathize another's experience is simply to grasp its logic.

This account appealed to you not even a little bit. From it you salvaged only Husserl's discussions of the living body, which--besides providing the i with a vehicle with which to accomplish all that following--also founds the very spatio-temporal structure of the world itself. Husserl remarked that for this to be so, there must be (the possibility of) a plurality of bodies occupying a plurality of viewpoints. But for him, it didn't matter *whose* bodies or *what kind of* bodies those might be. The Husserlian concepts of nature and the sciences of nature would be served quite adequately given half a dozen Meister-clones promenading up and down the Lorettostraße. The particularities of individual human bodies could be nothing but a distraction to Husserl's phenomenology. By contrast, you wanted to show that personal embodied distinctiveness was not only a value in itself, but phenomenologically indispensable.

3. Empathized content has a quality distinguishing one's own from another's.

In grand phenomenological style, you insist that giving an account of how something has been produced or caused is altogether different from giving an account of what something *is*. Genetic explanations are the province of empirical psychology; but you mean to be doing eidetic analysis proper. Still, as you review the relevant literature you must examine arguments in which the two kinds of accounting have not been scrupulously kept separate. Imitation theories, association theories, and analogical inference theories all have recourse to some productive factor that lies outside the sphere of immediate perception and therefore cannot be phenomenologically grounded, you say. Each fails in its own way to account for our experience of others.

Against Lippsian imitation theory, your criticisms are particularly astute. You take an opposite tack from the usual complaint: that Lipps failed to specify how appropriate targets of empathy are to be selected. Your question arises instead from out of the midst of the experience of empathy, during which the individuality of i's allegedly is dissolved. How, you ask, would one know which body to re-identify with at the conclusion of such an experience? What tells an i whether in this particular case it has been the one driven by the "instinct to express," or the one driven by the "instinct to follow"? When the acrobatic show is over, one of us returns to self-awareness wearing tights, the other trousers; how does each then pick up the thread of on-flowing self-recognitions? Lipps provided no answer, you complain.

In a further criticism of imitation theory, you point out that one's understanding of the gestures and countenance of another is not an instinctively caused spontaneous response at all. It can happen, you say, that someone's face is configured sadly yet you "see" that he is faking. On the other hand, to perceive someone's grief is to perceive more than the bodily expressions of it. Grief is not an outward appearance, nor is it "apperceived" within the perception of tears and moans. Apperception must hold the possibility of progressing toward a direct perception, as when the visible sides of a house appresent its far sides, thus *inviting* one to walk around and take a look. But grief is not like that; it is perceived whole and at once.[21] Therefore Lipps's account of expressive symbolism does not go far enough. The i's meaning overflows the meaning of any outward expression in all directions. As you say, one can understand an unkind remark while at the same time understanding something that in no way has been expressed: that the one who said it now is ashamed and wishes that he had kept quiet.[22] Thus empathy of others cannot be entirely determined by outward appearances and physical perceptions.

You propose, then, that other people's experiences register with us in inner perception. Moreover, the otherly-owned-ness of those experiences registers as a particular quality of the content of the appearances as we perceive them. Lipps went wrong in assuming that the experience of flying was *identical* for the acrobat and the enraptured audience. Your proposal is that the content of that experience, when submitted to reflection, discloses whether the lived experience has been my own or someone else's. Reflection allows the live experience itself to display its quality of being i-drenched, that is, of being something lived-through inwardly. But the experience will give itself thus in one of two ways: "originarily" or "non-originarily," as you say.[23] Reflected contents--such as memories, fantasies, and expectations--all are *non-originary* contents within the *originary* reflective acts that contain them. What is distinctive about receiving another's experience is that, unlike memories and fantasies, this empathized experience appears as never having been my own and never going to be so.

By this criterion of yours, everything that Husserl called *Einfühlung* fails to qualify as such. For logic, sciences, and scientific observations are what they are precisely because they'll be what they are no matter who enacts them. But you say that the impossibility of bringing the empathized experience to originarity must display itself in my reflecting inner perception of it, in order to certify for me that this experience belongs to someone else. Your insistence upon the unique irreplicability of the empathized experience is paradoxically bonded to your

---

[21]See Stein ([1917] 1980: 5). Here Husserl will disagree with you in his 1920 revision of the Appendix to the *Logische Untersuchungen*.

[22]Characteristically, you draw examples from interpersonal situations. Your textual practice increasingly becomes an attempt to simulate bodily contact, culminating in your autobiography. But you do not explicitly apply your hermeneutic theory to the reading of texts, as I do.

[23]The determination "originary or non-originary" applies to inner appearances only. An inner appearance is originary (or not) in relation to the i who makes the determination. By contrast, outer appearances are "caused or not caused" with respect to other outer appearances; and the determination turns out to be the same no matter which i makes it.

insistence--equally adamant--that others' experiences nevertheless do appear immediately within our own stream of live experience. They are differently the same for the other and for me.

It took courage to say this to Husserl. Perhaps you did not fully recognize how thoroughly you had contradicted him. Perhaps you had some fears, and that is why your writing style is so tortured. The stunted definition of *Einfühlung* that you offer is hardly worthy of your arguments:

> Und soll Einfühlung den von uns streng definierten Sinn: Erfahrung von fremden Bewußtsein haben, dann ist nur das nicht-originäre Erlebnis, das ein originäres bekundet, Einfühlung, das originäre aber wie das „angenommene“ nicht. ([1917] 1980: 14)

> And empathy should have the sense rigorously defined by us: a run-in with alien awareness. Furthermore only the non-originary experience that announces an originary experience is [the content of] empathy, not an originary-except-sort-of-received experience.

It would have helped to note that you are singling out here *just one side* of a doubled experience. Obviously, the reflective act of having an empathy is indeed an originary experience in its own right and I feel myself active in it. The non-originary experience appears as *the intentional content of* that originary act, and the non-originarity of the content consists in its having registered with me as something in which another i first lived. It is this registration that you mean to designate with the technical term *Einfühlung*. Amazingly, no inference has come into play in any of this. Scheler therefore would have to admit that it is not subject to error, but only to illusion.

Illusion is the purpose of entertainments such as the circus, the ballet, and the library. But you seem bent upon raining on the parade. Retelling the case of Lipps's acrobat, you seem to imply that you've never forgotten yourself and been swept away in the movement of someone else.

> Ich bin nicht eins mit dem Akrobaten, sondern nur „bei“ ihm, ich führe seine Bewegung nicht wirklich aus, sondern nur -- quasi --, d.h. nicht allein, daß ich die Bewegungen nicht äußerlich ausführe, was ja auch *Lipps* betont, sondern daß auch das, was „innerlich“ den Bewegungen des Leibes korrespondiert -- das Erlebnis des „ich bewege“ --, bei mir kein originäres, sondern ein nicht-originäres ist. Und in diesen nicht-originären Bewegungen fühle ich mich geführt, geleitet von seinen Bewegungen, deren Originarität sich in meinen nicht-originären bekundet und die nur in ihnen für mich da sind. ([1917] 1980: 17)

> I am not one with the acrobat, but only "near" him. I don't really perform his movement, but only quasi. That means not only that I do not outwardly perform the movements (which indeed Lipps also stresses), but also that what "inwardly" corresponds to the movements of the body -- the experience of "I'm moving" -- is with me no originary experience, but a non-originary one. And in these non-originary movements I feel me guided, led by the acrobat's movements, whose originarity announces itself in my non-originary movement. His movements are there for me only in my non-originary movement.

Perception gives me "the same" experience that the actobat is having, but it does not give me *me* as someone having that experience.

4. I's neither overlay nor displace each other while sharing content.

This brings you to a formulation Scheler liked so well that he stole it: "Einfühlung ist also nicht Eins-Fühlung," (empathy is therefore not awareness of being one).[24] There is such a thing as an experience of being at one; but its name is "we" and it is something else than what is under discussion here. Oneness, however, is never perceived between two i's in any reflective act, especially the empathic reception of another's experience such as Lipps proposed in the case of acrobatic performance. Having presented your own arguments against Lipps, you review Scheler's as well.[25] In summary you insist upon the distinction between comprehending (*Erfassen*) a feeling, and taking it upon oneself. The latter, to which you apply Scheler's term "feeling-contagion" (*Gefühlsansteckung*) as well as a term of your own (*Gefühlsübertragung*), is set in contrast to in-feeling (empathy, *Einfühlung)* and with-feeling, also mentioned by Scheler in his 1913 *Sympathiebuch.* To these you add a third contrasting term: one-feeling (*Einsfühlung*), a neologism that you coin in order to characterize what you oppose in previous theories. This list of four options will turn up again in Scheler's 1923 revision of the *Sympathiebuch*, ironically without attribution.

Your doctrine of the irreplacability of the i does not arise from sentimentality or value theory. Rather, it is an implication of the eidetics of empathy that you have rigorously carried out. No reflective act is an *Einsfühlung.* In none of the other reflective acts--remembering, expecting, pretending--does the reflecting i coincide temporally with the reflected i. One i lives *now*, and the other lives at some other "now." But the experience of the alien i presents a special case. My own inner engagement with what someone else is feeling can indeed be simultaneous with his or her act of feeling it. In this case, the temporal barrier to our merging is not there. I and "i" are in sync, so something besides temporality must be keeping them separate. That something is indicated in your eidetics: it is the invincible difference between originarity and non-originarity in the content that is shared.

Fully congruent feeling of oneness is not achieved even in my recollection of my own most intimate and private moments. On the one hand, you say, there is a *tendency toward fulfilling* the overlay (*Deckung*). That is, in remembering I am drawn back into my own past experiences and I seek to relive them. Yet I find that I cannot succeed in becoming identical with the subject of my former acts. (For example, I cannot relinquish my current awareness of the consequences of my former choices in order to experience making "the same" choices again.) In much the same way, I am also drawn along into appropriating the feelings of the other. Yet, you insist, fulfillment of this tendency is not of the essence of reflection. In fact, memory and the empathy of alien experience both display, as an essential feature, the permanent thwarting of this drive short of its fulfillment.[26] Only

---

[24]See Stein ([1917] 1980: 17).

[25]See Stein ([1917] 1980: 24, note 1). See the discussion of Scheler's position above.

[26]In describing an essential drive whose thwarting is equally essential, you have

fantasy is essentially open to the fulfillment of the drive to coincide completely with the reflected i. (I *can* become the i whom I dream, the i whom I hope to be.)

Once again, the contrast between your results and Husserl's is striking. You say that there is no *Deckung* or overlay of subjects. This term, *Deckung*, is used in geometry to indicate what in English we call "congruence" of figures such as triangles. For Husserl, empathy means that subjects do indeed coincide as perfectly as congruent geometric figures, with all sides and angles matching, so to speak. On your view, that perfect overlay of i's would be impossible to achieve and undesirable even to attempt to approximate, because it would annihilate the "difference in sameness" that you require for *Einfühlung*.

5. Empathy requires a new science.

Epistemological difficulties arise from this paradoxical "different sameness" that another human being's experience exhibits to me. There are at least three levels to the problems that you have stirred up in this regard.

First, the status of any claim to knowledge about the other is radically altered. I have no access to the other's experience *as she or he actually has lived it or is living it.* At best, I simulate that experience. You say that while empathy gives me knowledge of the other, this is only an empty and blind knowledge. Metaphorically, the other stands before me but I do not see. My knowledge thus is heavier than a mere conjecture, for it is weighted with something given immanently in my own experience. Yet it is lighter than a perception. Neither way of conceiving of this knowledge seems adequate.[27] Husserl's discontent with the term *Interpretation* casts its shadow across your formulations as well. He yearned for a knowledge automatically produced, tamper-proof, untouched and undetermined by any individual quirks of the knower. You find that that's a doomed quest in the case of knowing people *as people*.

Second, the peculiar doubling of i's in reflective acts raises the impossible quest to a new register. A kind of valence or power gradient obtains between reflector and reflected, knower and known. The knower gravitates irresistibly toward overlaying and becoming one with the known, but invariably encounters an impenetrable barrier. An i essentially withholds its i-hood. It keeps its secret. The instrumentality for knowing the subectivity of others is one's own subjectivity; yet what is ownmost about subjectivity is its ultimate inaccessibility to others. No windows air out the monads; no canals drain the lifestreams. Husserl banished the word *Interpretation* in hopes of making perception more reliable. You suspected that he thereby made it sterile.

Third, you wrote that the usual psychological questions and categories were inappropriate for understanding how one human being experiences the experience of another. Specifically, you resisted the demand that empathy be

---

anticipated a theme that would become commonplace in French existentialist literature a generation later.

[27]See Stein ([1917] 1980: 20).

classified as *either* a perception *or* a representation.[28] These psychological file drawers, as you called them, cannot serve to sort the phenomena that you have brought under investigation. Like an appearance, empathy of another is a live experience indubitably "there" as given within one's own consciousness. But like a representation, it is copied from another. At the disciplinary level, empathy requires a kind of investigation that will respect this phenomenon's own essential being and structures: its dual-sidedness as intramonadic copy.[29]

Before turning to your proposals concerning that new science, I must conclude this discussion of your eidetics with some complaints about your misunderstandings of Scheler.

It was difficult for you to keep track of the convergences and divergences between the phenomenologies of Scheler and Husserl. Scheler was not a careful writer or reader, and Husserl seldom footnoted the texts against which his arguments were framed. Moreover, Scheler's scathing criticism of Husserl evoked in you the loyal desire to defend the Master's position. Yet at the same time, it was Scheler's work that really fired your imagination. You call his theory bold and tempting (*kühn . . . Bestechendes*), and decades later you would recall his engaging manner and big blue eyes.[30] Scheler, not Husserl, was the theorist of personhood who inspired you. In his four-level schematization of egoic function, the activity of the person--*love, bliss, or despair*--was the pinnacle toward which all other functions built: the physical, the psychic, and the soulful. This scheme is based on Husserl's own earlier doctrine of inner and outer perception, which the Master himself was developing along other lines by the time you got to know him in 1913. The similarities still were great, the indebtedness to Lipps was common, the vocabularies were very close, and the distinctions were not always clear.

You recognize a Lippsian theme in Scheler's assertion that we first find ourselves awash in a sea of thoughts and feelings that belong to no one in particular, and that from thence we must undertake to sort out our own from the common pool. Against this view you offer the same objection that you made to Lipps: that there is no way to account for subsequent differentiation into "own" and "alien" thoughts and feelings, if indeed the original condition is utter immersion in "unowned" thoughts and feelings. Fair enough. But you are mistaken when you go on to say that Scheler considers those anonymous opinions to be i-less.[31] Scheler in fact holds that i-hood is given in all appearings and can be brought to consciousness with the act of reflection. *Individual* i's appear with the reflecting *act*; but the old familiar *non*individuated Lippsian busy-i was for Scheler already

---

[28]See Stein ([1917] 1980: 21).

[29]The section in which you frame this case, §4 of chapter 2 of the published version of the dissertation, is entitled "Der Streit zwischen Vorstellungs- und Actualitätsansicht" (the conflict between representation-insight and actuality-insight). Scheler cites this passage in the second edition of his *Sympaghiebuch* (1923: 10; 1954: 13).

[30]See Stein ([1917] 1980: 30; 1986: 259).

[31]See Stein ([1917] 1980: 31-32).

there at work in the physical and the psychic prereflective appearings. Scheler's conception of this is quite consonant with Husserl's conception of the pure i; so it is wrong for you to say that Scheler "recognizes no pure i." You are also incorrect in your views that Scheler equates "i" with "individual soul" and that he excludes any i from experience. (The technical difference between i and person for Scheler is that person cannot become an object of consciousness as i can.)

In contesting the primordiality of non-owned experiences, you make a further statement that, as it stands, is not only incorrect but also inconsistent with the general thrust of your argument toward establishing how it is that I have experiences had by someone else.

[J]edes Erlebnis ist nun einmal wesenhaft Erlebnis eines Ich und auch phänomenal gar nicht von ihm zu trennen. ([1917] 1980: 31)	[I]t so happens that every live experience is essentially a live experience of *an i* and not at all to be separated from him.

The problematic word is the indefinite article "an" (*eines*). "An" presumes individuation, if not individu*aliz*ation. The question here is whether a live experience can be had by the non-individuated executive i, or whether the subjectivity engaged within experience must already be someone's i--*an* i. You presume the latter without having either established it through argument or displayed it phenomenologically.

Nevertheless, there is a kernel of truth within this unfortunate formulation of yours, and it is this: that when I have the experience of someone else, I also in some way *have him or her along with it.* In fact, I may "have" the other even before she reflectively "has" herself. Empathy of another's feelings is ipso facto empathy of some other. Still, your complaint against Scheler would be valid had he actually held that feelings and opinions *must* infect us anonymously from out of the general buzz of everyday opinion, that is, detached from their originators. That is not his position, however. He simply says that this *can and does* happen. And with that empirically founded view, you agree. You agree that when feelings or opinions are caught in that way, they are imbued with the illusion of having originated in the experience of someone--perhaps even in my own experience.

You diagnose two further dimensions of illusion, in connection with the chilling example of race hatred. If someone were raised to hate Jews, you say, she would come by her hatred through empathy. This is an empirical statement about the genesis of the bigotry. The originary appearance of the disvalue of Jews would not have occurred in her own live experience; it would be an *eingefühltes Wertnehmen* (felt-into worth-perception). But if she is phenomenologically inclined, the bigot can discover the non-originariness of her bigotry by reflecting upon it. She is not deceived when, in reflecting, she finds that the bigotry is really there: she really hates Jews, and it is she herself who is doing the hating. Nevertheless, two sorts of illusion may attach to this feeling, you say. She can be deceived about the value of Jews, and she can be deceived about the value of her bigotry (thinking it to be patriotism, perhaps). What she cannot do by reflecting is to recover the originary act of assigning disvalue to Jews, because it is not an act that she ever carried out for herself. She can discover the illusion only if she attempts an originary act of evaluating Jews--presumably, some individual Jews whom she gets to know face to face--and compares the fresh results with her prevailing feeling of hate.

At least, that is the gist of your meaning, although you could not bring yourself to say so explicitly. Your oblique evasion runs like this:

> Über das Fehlen des fundierenden originären Wertnehmens kann ich keine reflektive Klarheit haben, weil ich auf einen nicht vorhanden Akt nicht reflektieren kann. Aber wenn ich einen solchen Akt vollziehe und mir zur Gegebenheit bringe, dann gewinne ich Klarheit und damit die Möglichkeit, durch Vergleich mit diesem Fall die frühere Täuschung zu entlarven. ([1917] 1980: 35)

> I can have no reflective clarity about the failure of the founding originary worth-perception, because I cannot reflect upon an act that is not available. But if I execute such an act and bring it to givenness for myself, then I gain clarity and therby the possibility to unmask the earlier illusion through comparison with this case.

Phenomenological rigor, perhaps, forbids you to require empirical contact with real Jews, Jews existing independently of their physical appearances within consciousness.

Turning to another disturbing example--erotic literature--you manage once again to avoid calling attention to a profoundly significant implication of your *Einfühlungslehre*. Without saying so, you indicate that in order to understand and enjoy cultural artifacts, we must agree to hide from ourselves the fact that the feelings we feel in connection with them are not originary.[32] *To read* is to take on the subjectivity offered by the writer: to ride along on the author's i and vicariously to move through the course of an emotional process, an argument, or a story. *To write* is to display an i as model and guide for such a journey. If one could not feel-into the experience of another human being, one simply could not learn to read or write. Your *Einfühlungslehre* is a theory of literacy.

But back to Scheler. You didn't read him too well. In a single sentence, you manage three different misinterpretations (indicated below with M_1, M_2, and M_3). Fortunately these signal only terminological difficulties not substantive ones, and they are equally owing to Schelerian inconsistency, Husserlian fingerprint phobia, and your own compulsion to iron out differences. You write:

> Wir fragen also zunächst: was ist innere Wahrnehmung? Scheler antwortet darauf: innere Wahrnehmung ist nicht Selbstwahrnehmung (wir können uns selbst [M_1] -- d.h. unsern Leib -- auch äußerlich wahrnehmen), sondern als Aktrichtung [M_2]

> Thus we ask first of all: what is inner perception? Scheler thereupon answers: inner perception is not self-perception (we can also outwardly perceive ourselves--that is, our living body) but, as act-direction, is different from outer

[32]See Stein ([1917] 1980: 35-36). The example that you cite is the schoolboy's falling in love while reading *Romeo and Juliet*. Scheler discussed this example as well. I wish I knew why you turn tongue-tied on the brink of your most astounding discoveries.

von äußerer Wahrnehmung unterschieden; es ist diejenige Art von Akten, in denen uns Seelisches [M_3] zur Gegebenheit kommt. Die Unterscheidung dieser beiden Wahrnehmungsarten soll keine definitionsmäßige sein, die sich auf die Verschiedenheit der in beiden gegebenen Objekte stützt, sondern umgekehrt soll der Unterschied von Physischem und Psychischem nur faßbar sein durch die prinzipiell verschiedene Art, wie sie zur Gegebenheit kommen. ([1917] 1980: 31)

perception. It is that kind of act in which something soulful comes to givenness for us. The distinction between these two sorts of perception should not be, definition-wise, one that stops short at the difference in the two objects given; but on the contrary the distinction of physical and psychic should be conceivable only through the manner, differing in principle, in which they come to givenness.

Here you are clumsily attempting to summarize the standpoint of part two of Scheler's *Idole*, on "Illusion and Inner Perception." But Scheler does *not* say that inner perception is act-directed while outer perception is not. For both are acts, although they are acts directed toward inner appearances or outer appearances, respectively, and toward their respective kinds of givenness. In Scheler's text, we can indicate the misunderstandings as before:

Ist mithin der Unterschied innerer und äußerer Wahrnehmung weder zu leugnen noch weiter zu reduzieren, so ist die Frage, was, abgesehen von dem nur erlebbaren Richtungsunterschied [M_2] der beiden Akte des Wahrnehmens, sie scheidet. Das ist für unseren Zweck vor allem wichtig, daß der Unterschied der beiden Richtungen des Wahrnehmungs *in keiner Hinsicht als relativ auf den Leib* und demgemäß auch auf die Sinnes-funktionen [M_1] und -Organe anzusehen ist. . . . «Äusere Wahrnehmung» als Aktrichtung hat daher mit «Sinnes-wahrnehmung» zunächst gar nichts zu tun . . . . [I]n innerer Wahrnehmung [ist] immer ein Ich überhaupt gegeben, und zwar die *Totalität* eines Ich, auf dessen Hintergrund sich dann dies und jenes abhebt. . . . Der Akt der inneren Wahrnehmung geht dem *Rechte und Können* nach auf *jedes* Erlebnis des Ich [M_3]. . . . (1915: 60, 63-64)

If, nonetheless, the distinction between inner and external perception is indisputable and irreducible, then the question is, What does distinguish them, apart from the difference in the orientation of these two acts of perception? (It is only this difference which we can experience.) For our purposes it is above all important that we not regard the distinction between these two orientations as in any way relative to the body, or, accordingly, to the functions and organs of the senses. . . . Thus "external perception," as the orientation of an act, has nothing at all to do with "sensory perception." . . . [I]n inner perception [an i] is always given, indeed the totality of [an i], as the background against which this or that content stands out in relief. . . . The act of inner perception has the right and the ability to pursue *every* experience of the [i]. . . . (1973: 35-37)

You've put "self" where Scheler has "senses" [M_1]. You've implied that inner perception differs from outer in that inner perception alone is an act [M_2], where Scheler regards both as acts that differ merely in the orientation of their activity. And you've put "soul" where Scheler has "i" [M_3]. For his part, Scheler has been careless in maintaining the distinction among the four ego-functional levels that he laid down in the *Ethik* (according to which "acts" are reserved to "persons" while the physical, the psychic, and the soul involve mere "functions"). You can hardly be blamed for your tendency to write "perception" when Scheler writes "appearance," for Scheler himself cannot keep them straight. At any rate, these terms are so slippery because of the elusiveness of the live experience itself whose dynamics they are trying to pin down.

Besides these inadvertent misinterpretations of Scheler, which are easily cleared up, you make one major and quite deliberate deviation from his phenomenology of perception. Scheler follows the early Husserl in distinguishing two kinds of givenness: physical and psychic. The person, in acting, is for Scheler the one to whom appearances *are given*, the one whose activity perceives them. The person as such does not "appear," and is not "given" for another person. But you argue that other i's are indeed given, with evidence, within one's own monadic lifestream. To illuminate the givenness of other human beings is your distinctive gift to phenomenology.

B. Analysis of the Constitution of Individuals

Human beings, as such, are *given* to our consciousness as more than *objects* of consciousness, in the technical sense of those terms. You have taken over from Scheler the conviction that the person is an actor who never, *as person*, is objectively known. However you insist that the active/creative dimension of other human beings is indeed known, genuinely and reliably, through empathy: that is, through a partnered cogito that does not have the polar structure of subject vis-a-vis object. You mean to describe an intentional consciousness--that is, a "consciousness of"--having an "of" of a special kind. You seek to build that description with the tools of Husserl's transcendental phenomenology, newly introduced in the *Ideen* of 1913; but you do not realize that those tools disintegrate when put to a task for which they were not designed. Specifically, you attempt to conduct an investigation of how the meaning-content "this human being" comes to be constituted in one's consciousness. Constitution-analysis, then, is the task of the middle section of your book.

Unfortunately, you don't tell us what you mean by "constitution." You've simply picked up the term from Husserl. Although he was using it as early as 1909, the 1913 *Ideen* makes constitution "the central point of view of phenomenology" (as Landgrebe said in his analytic subject index to the work). Constitution is a function of consciousness, providing the bridge between the givenness of sequential multiple appearings and the givenness of unified essential form for the thing to which the appearings are referred. This autonomic function achieves the teleological coalescence of multiple aspects or instances into singular identity.[33] Typically, both Husserl and you employ this term reflexively: *sich konstituieren*. (Something "constitutes itself." The grammar indicates that "constituting" is not transitive: it is not something that can be done *by* X *to* Y.)

How does such a curious notion figure into Husserl's work? It serves the interest of establishing phenomenology as the foundation of the sciences, in the following way. The things that will count as science's objects must *be real* and *be*

[33]See Husserl's *Ideen* §86, Hua III: 213; (Husserl 1982: 208).

the same for all possible observers--if science is to be "science" at all. Neither the reality of things nor their qualities can be owing to the arbitrary intervention of some extraneous factor, such as the fancy of the observer. To say that the things "constitute themselves" is to say that they appear without help, clean, untouched, unspoiled. What I have cartooned as Husserl's "fingerprint phobia" is the drive to insure that the knower in no way interferes with the known. It would beg the question if I were to suggest here that Husserl was irrational and neurotic in his desire to insulate the sciences from interference by individual scientists; for Husserl was working on the construction of scientific rationality itself.

If, however, we grant the point that science requires some disciplining of observation and inference, then we can see how Husserl is responding, with this notion of "constitution," to Scheler's charge of "idolatry" and to the twofold danger of error and illusion that Scheler discusses.[34] There is danger of *error* whenever individual thinkers make inferences. Husserl's new transcendental phenomenology is designed to avoid that danger in two ways. It will restrict itself to phenomena that are immediately present to consciousness without any inferences, and it will disengage from everything empirical, including the individual pecularities of this or that thinking human being.[35] But there is also danger of *illusion*, Scheler warns, whenever appearances elicit our perceptive activity. In as many different ways as things appear, they can just as easily misappear or disappear.

Husserl had no good answer for that one. He certainly felt its force, whether in Scheler's accusations or in his own evaluative ruminations, or both. Is there any way to guarantee ourselves against illusion? Descartes had a suggestion: when one is deluded, one's own *existence as deluded* still will not be a delusion. But even though the non-illusory basis of knowledge can be displayed thus, the possibility of error re-emerges as soon as inferences begin to produce consequential claims based upon it. So Husserl sought to disable the means of production. As we observed in the last section, Husserl effectively handcuffed the i in order to eliminate any egoic innovation from the phenomenological realm. He banished the notion of *Interpretation* from perception. He substituted the Kantian term "apperception" as *a denial that the perceiver--qua individual--contributes anything* by completing the "house" that presents only one side at a time, by filling in the meaning behind the marks on a page, or by recognizing the person in the smile and

---

[34]See the 1912 essay, "Über Selbsttäuschungen," that would become Scheler's 1915 "Die Idole der Selbsterkenntnis."

[35]Husserl's transcendental phenomenology is a creative work, and requires of the reader a certain collusion, a certain willingness to suspend disbelief. Literally, Husserl's work is of course full of inferences, as anyone who has ever attempted to work through the text can attest. Artistically, however, it conjures up a "realm" prior to inference where the operations of inference and every other intellective activity can be laid bare for inspection. By the same token, the work is quite obviously the production of a particular man in a particular historical context and language. We as readers are simply asked to extend the courtesy of pretending that this is exactly the work that any rational human being anywhere would produce, given only the questions that Husserl sought to address.

the handshake. (House, meaning, and friend now had to "be appresented" penumbrically in the appearances that presented themselves, and they had to do so irrespective of the particular qualities of whomever they were being appresented to.) Apperception is not a conferral of sense (*Eindeutung*): the individual act of apperception contributes no content to what is apperceived. Variance among versions is excluded.

In short, the i does what it must do; nothing more, nothing less. The i is reduced to following order. Husserl writes:

Das *Problem der Konstitution* besagt dabei klärlich nichts anderes, als daß die geregelten und zur Einheit eines Erscheinenden *notwendig* zusammengehörigen Erscheinungsreihen intuitiv überschaut und theoretisch gefaßt werden können . . . daß sie in ihrer *eidetischen* Eigenheit analysierbar und beschreibbar sind, und daß die *gesetzliche Leistung der Korrelation zwischen dem bestimmten Erscheinenden als Einheit und den bestimmten unendlichen Mannifaltigkeiten der Erscheinungen* voll eingesehen und so aller Rätsel entkleidet werden kann. (Hua III: 371)

Accordingly, the *problem of constitution* clearly signifies nothing else but that the regulated series of appearances *necessarily* belonging together in the unity of what appears can become intuitively surveyed and seized upon theoretically . . . that, in their *eidetic* own peculiarity, they are analyzable and describable; and that the *law-conforming production of perfect correlation between what determinately appears as unity and the determinately infinite multiplicities of appearances* can become fully seen intellectually and thus all enigmas can be removed. (1982: 362)

The i has no discretion in the matter. The "law" of the thing's essence controls how the splintered glimpses of the thing that we get in experience rise up and assemble themselves--automatically, it seems--into the thing as object perceived in what Husserl and Scheler alike call an "outward appearance." What is known of the thing is the essence of the thing: its unity, its binding edges. Constitution is the bridge carrying cognition *forward* from many to one; that is, from process to substance, sequence to essence, appearance to intuition, motion to form. As Husserl depicts it, the forward direction of this crossing cancels out human individual initiative--what I have called the fingerprints. For essential form to be "given," nothing additional is required on the part of the i besides its compliant motion through the series of outer appearings.

Yet . . . what would happen if cognition ran *backwards* across that bridge in the opposite direction? Two things might happen. Husserl himself describes a "reflective" move in which the i brings to consciousness what it has been doing while crossing from the many appearances into the intuited givenness of the essential unity: *nothing special*. (The i has merely been approvingly registering how smoothly the bridge operates; so now the i appears to itself reflectively as nothing but the bridgekeeper.) However, Husserl neglects to describe an obvious alternative: i-hood that flows *from* essential form *to* particular instantiation is creativity. An i who *real*-izes is a creator, a doer. For example (as Schleiermacher saw), in speaking we go from the formal structural rules of the language, toward unique grammatical expressions. An author employs the law-like conventions of a genre as the means of production when writing a literary text. Or in an everyday example (as Pfänder showed), our essential motives give rise to our ingenious and situationally adapted motivated actions.

Constitution and creativity go in opposite directions with respect to essential laws, and in this they are more different than night and day. While both are governed formally, they beckon toward contradictory modes of understanding. To grasp an event of essence-constitution, one simply re-enacts it. Perfectly. But to understand a creative rendition, a second *and variant* creative rendition is required. The interpretation *cannot* be a perfect re-enactment of the original. A creative work becomes itself only through inspiring someone to enact a version of itself. Understanding of a work always is a new production--whether it be an orchestral interpretation, a dance, or the ordinary unique reading that every reader gives to every book. Taking a ride on an artwork, as we said above, entails illusions of various kinds.

Husserl's discussion of constitution mentioned neither Scheler's warnings nor Scheler's terminology.[36] But Husserl does take up an important Schelerian distinction: that between acts and functions. Scheler generally reserved the designation "act" for events imbued with the distinctiveness of the person doing them, to the extent that they would be essentially different if done by somebody else. In contrast, "functions" occurred at the physical, psychic, or soulful level and in principle must be replicable by someone else. Husserl, who wants to handcuff the i, confines phenomenological investigation to "functions" and specifically to the function of constitution.[37] He writes:

Der Gesichtspunkt der Funktion ist der zentrale der Phänomenologie, die von ihm ausstrahlenden Untersuchungen umspannen so zeimlich die ganz phänomenologische Sphäre, und schließlich treten *alle* phänomenologischen Analysen irgendwie in ihren Dienst als Bestandstücke oder Unterstufen. An die Stelle der an den einzelnen Erlebnissen haftenden Analyse und Vergleichung, Deskription und Klassification, tritt die Betrachtung der Einzelheiten unter dem „teleologischen“ Gesichtspunkt ihrer Funktion, „synthetische Einheit“ möglich zu machen. (Hua III: 213)

The point of view of function is the central one for phenomenology; the investigations radiating from it suitably comprise the whole phenomenological sphere, and, finally, *all* phenomenological analyses in some manner or other enter into its service as component parts or preliminary stages. In place of analysis and comparison, description and clarification restricted to single particular mental processes, consideration arises of single particularities from the "teleological" point of view of their function, making possible a "synthetical unity." (1982: 208)

The Husserlian i constitutes *itself* by reflecting upon constitution; that is, by reflecting that it has *been* the on-running activation of that autonomic function, constitution, in which objects have been arising in consciousness to display their essential unities. The things so constituted are said to be "transcendent" to

[36]Scheler wrote of illusions as *Täuschungen*; Husserl discusses *Schein*. See *Ideen* §86, *Husserliana* III: 214; Husserl (1982: 209).

[37]Arguably, Husserl understands even "acts" as functions, since he suppresses the significance of the individuality of the actors.

consciousness, while the functioning i is "transcendental" with regard to them. Something is "transcendent" when it is constituted as *requiring constitution in just that way and no other* by any i at all that would happen upon it. All i's work the same.

What about the body? What are the consequences, for bodies, of Husserl's compulsion to banish individual variance from thought? Let's recall briefly the preceding chapter's discussion. Husserl regarded bodies, too, as interchangeable. One human body was enough to impart organization to the physical world. The live body became the zero-point of a spatial grid, in which all other things made orderly appearances. The live body could move through that grid, and appearances of things would shift in an orderly way. Location and motion of that kind were in principle replicable by any body; they were not owing to the unique sensory apparatus of the particular body in question. Science required the possibility of a second observer but, axiomatically, he would see nothing different from what appeared to the first. An alien body was simply "an i over there" (*Ich im Dort*). Later, Husserl would consider the live body as a sentient being that provided an i with insertion into the world at a specific and mobile place. The *Ideen II*, not released for publication during Husserl's lifetime, suggests that this happens first in regard to the alien i. Having associated the alien i with a live body, I am then enabled to identify with my own body.[38]

While these hints of a "need" for other bodies in general are not well integrated into Husserl's mature transcendental phenomenology, we may surmise that phenomenology of the body was a topic of interest in student discussions and in academic lectures while you were studying in Göttingen.[39] Your dissertation retraces the Husserlian description of body as sentient space-organizer. But where he emphasizes the literal *interchangeability* of bodies, you emphasize their *irreplaceability*. Husserl terms the body the zero or *Nullpunkt* of orientation. You agree, but examine also its distinctive way of appearing. The series of aspects of my body breaks off, and there are parts of it that I cannot see. I need the viewpoint of the other in order to complete my apprehension of my own body. For that, *any* other will do. But you also show that particular individual bodies are necessary for me to know other particular human beings. For Husserl, the possibility of there being others--*Ichen im Dort*--was required for the possibility of science. The possibility of other i's is appresented in the very notion of science. The phenomenon of "the other" for Husserl is simply the sense that nothing restricts the set of i's to one member, or the set of live human bodies to my own. For you, however, others are immediately given. This one and that one appear indubitably in consciousness: not by appresentation, but through the distinctive

[38]You appreciated the significance of this approach to constitution more profoundly than did Husserl himself. See my discussion of your work on the composition of *Ideen II* in the next chapter, below.

[39]For example on July 30, 1913, you likely heard Husserl discuss "apprehension of a body as a living body" in his class on nature and culture, according to Schuhmann (1977: 182).

quality imbuing certain empathized contents of consciousness. The phenomenon of the other, for you, is the sense that a given live experience is mine but did not originate with me.

Husserl suggested that i-hood is constituted prior to any bodily location for the i, and that both i-hood and the alien human body are constituted prior to my recognition of my own i in my own body. (Empirical psyche or soul hardly entered his view at all.) You have followed a different order in designing the chapter that we are about to examine. Starting from the phenomenological pure i, you go to the phenomenological body (if we may use this term), and only then do you turn to the alien embodied i. From Scheler you take the notions of "soul" and of "person" as unique and irreplicable; from Husserl, the description of the live body. Where Scheler's i was the interwoven unity (*Ineinandersetzung*) of live experiences as they flow along in time, and where Husserl's i was constitution's referee appearing reflectively to itself, for you the i will be the creativity and responsibility inhering in any human being and available to body-based empathy by any other human being.

Thus it is misleading for you to indicate that this chapter will study empathy as a constitution problem. No way are you doing a Husserl-style constitution-analysis of the essential unity *"Einfühlung"* here, as if to follow up your eidetics of it from the last chapter. It would have been more accurate for you to say that empathy is *like* constitution, that it substitutes for constitution in certain cases but achieves contrasting results. Constitution, as Husserl displays it, yields objects in their essential unity; but empathy yields the non-objectifiable essential character of a person appearing in the "fingerprints" distinguishing her actions and experiences. In this chapter, your covert accomplishment is to explicate this egoic function, empathy (*Einfühling*), on the model of Husserlian "constitution" but as an alternative principle of essence discernment. The essences that it discerns will be individual human characters, and it will permit them to be understood in a non-objectifying manner. That is, it will permit their creativity and responsibility to be experienced from within, albeit nonoriginarily.

A phenomenology that places empathy alongside constitution, and on an equal footing with it, becomes a technical hybrid. You have not yet seen that this is so, or recognized how radical a departure from Husserl's camp you are going to have to make. You are still hoping that empathy can be an item brought as grist for the mill of Husserl's analytic system. It is maddening to see you play at this charade.

1. The intramonadic i

The middle chapter of your dissertation (as published) begins with an attempt at constitution-analysis in regard to three essential unities--the i, the stream of live experiences, and the soul--in order to see whether any of them will yield the sense of an alien human being.[40] None does so entirely satisfactorily.

---

[40]It seems to me that you are trying to observe Husserl's transcendental reduction in this chapter, just as you observed the eidetic reduction in the last. The transcendental

Your comments on the pure i seem to refer back to an exposition offered in the missing first chapter of the dissertation. In any event, what remains of your text mixes up the "pure i" accessed by transcendental reduction, with the nondifferentiated ecstatic agent of living experience described much earlier by Lipps as prior to differentiation into separate i's. You attribute to Lipps an account of the emergence of individual i's which, though interesting, seems not to appear anywhere in his works: that an i first emerges by withdrawing from "you" and "he."[41] Your point is that individuality is first produced by stepping back from the otherness of other i's; only then does one step forward to confront the other and discover who he or she uniquely is. At the first step there is no difference in qualities among i's but only in modes of givenness; the others differ from me merely in being given otherwise than I am given to myself. The ensuing step, confrontation, brings recognition of the particular distinguishing qualities of individuals.

Next you consider the stream of consciousness, a second way in which the i is constituted. Experiences as we live them are not discrete atoms, but are temporally open to the experiences before and after them in a series whose streaming unity is an i. There are multiple such streams. Once again, these are first distinguished simply in that each is not any other, because each belongs to a discrete subject. Subsequently, you say, the individuated streams are distinguished by their particular contents and also by the placement of each content within each stream in relation to other contents. Yet you state that this is not yet sufficient to establish "selfhood" (*Selbstheit*)--a term more familiar in empirical psychology than in Husserl's phenomenology.

Third, you turn to the soul, the individual unity of the psyche. This is your first major departure from Husserl's previous phenomenology.

Unser einheitlich abgeschlossener Bewußtseinsstrom ist nicht unsere Seele. Sondern in unseren Erlebnissen--so fanden wir schon bei der Betrachtung der inneren Wahrnehmung--gibt sich uns ein ihnen zugrunde Liegendes, das sich und seine beharrlichen Eigenschaften in ihnen bekundet, als ihr identischer „Träger": das ist die substanzielle Seele. ([1917] 1980: 43)

Our unified, sealed off stream of consciousness is not our soul. But in our live experiences--so we found already with the investigation of inner perception --is given to us something lying at their basis, which announces itself and its enduring properties in them, as their identical "carrier": that is the substantial soul.

Now wait just a minute here. You are trying to refer back to work that you never did: an exposition of the constitution of a soul from out of the midst of the stream of consciousness. It is doubtful whether such constitution can be displayed

reduction is an agreement to consider only what is accessible to any consciousness whatsoever. See *Ideen* §33 ([1913] Hua III: 59; 1982: 66).

[41]Theodor Lipps's writings are quite extensive, unindexed, and largely untranslated. I have been unable to find in them a discussion of *Ich* emerging from *Du* and *Er*, but I have not searched exhaustively.

phenomenologically, because it is doubtful whether it happens.[42] You have nowhere shown that a substantial soul is "given" as the basis of streaming experience, as something distinct from that stream itself. What you have done here is simply assign to the term "soul" its usual technical meaning. Let us then stipulate, for the record, that the soul has not been exhibited phenomenologically, but has been more of less pulled out of a hat.

Why this hat trick? Because in your phenomenology, you need soul to be the analog of the physical thing. Soul is to empathy as thing is to constitution. By Husserlian method, soul should appear as the essential unity intuited from out of the many different instances of empathy in which I comprehend an i--whether that of another, or my own. But you do not say this. You say instead that soul is built up out of "categorial elements," which seem to be something like character traits although you indicate that some of them are linked to physical and psychic categories, the latter including "causality and variability."[43] Souls, you say, receive their unique individual structures from the contents of their streams of consciousness, and the converse also is true, for the contents of the stream of consciousness are also dependent upon the soul. Still, this is not yet sufficient to account for the difference among individual human beings, because in principle nothing prevents two streams of consciousness from having the same live experiences in the same order. Husserl must have groaned when he read that, and he certainly winced at the following:

Diese substanzielle Einheit ist „meine“ Seele, wenn die Erlebnisse, in denen sie sich bekundet, „meine“ Erlebnisse sind, Akte, in denen mein reines Ich lebt. ([1917] 1980: 43)	This substantial unity is "my" soul if the live experiences in which it announces itself are "my" live experiences: acts in which my pure i lives.

For Husserl, of course, the i isn't "pure" if it's "mine."[44]

All quibbling aside, you are preparing to make your case for the claim that human individuation and individualization can be secured in no other way than through the body.

2. Monad as live body

The task of showing how we understand other people is only half done when one has established that the plurality of i's is given within any i. It still remains to account for the individual uniqueness of each i--my own no less than any other. How might one go about that? You pursue tactics suggested by Scheler's delineation of different levels of egoic function: physical, psychic, soulful, and personal. Where Scheler worked hard to distinguish these four, you are

---

[42]Husserl's later argument for a substantial soul, constituted analogously to the physical thing, is more detailed than yours but just as weak. See §30 of *Ideen II.*

[43]See Stein ([1917] 1980: 43).

[44]Husserl himself was not always consistent in observing this rule, however. See Ströker (1993: 124-125).

interested instead in their "blending" and reciprocal influence.

You now confess that the discussion of the psychic in the first 44 pages of your book has labored under an artificial abstraction,[45] for what appears as psychic is always the psycho*physical*: the sentient live body (*Leib*). Furthermore, you say, the soul is always in such a sentient body. Thus, phenomena in which soul, psyche, or body appear always blend into one another, so that their fusion (*Verschmelzung*) appears as well. You will display this fusion by describing the distinctive givenness of the live body, arguing thence to the "psychic causality" exhibited in its feelings, expressions, and purposive action.[46]

To begin with, you say, at the physical level we notice that something goes awry in the attempted constitution of one's own body as a thing. Any physical thing will be constituted in perception as the unity of its sequential appearings to outer perception, and each perceptual aspect beckons me to pursue others and so complete the constitution of the thing in its essential unity (according to Husserl's way of describing the matter). But you throw the spotlight upon an anomaly that has escaped Husserl's notice.[47] Unlike all other physical things, my own body disallows access to the appearances I need in order to complete its constitution as a thing. I cannot see every side of it. In its appearance-series there is a gap "der mir mit einer noch größeren Hartnäckigkeit als der Mond seine Rückseit vorenthält" (that denies to me its back side with an even greater stubbornness than the moon).[48] Perhaps I try to resolve this anomaly by switching from the sense of sight to touch, and completing the series by handling what I cannot see. But then another anomaly confronts me. I find that I cannot place this physical thing--my body--at any distance from myself, or go away from it

---

[45]Compare this move with §18f of *Ideen II*, where another artificial abstraction is suddenly removed: "Überlegen wir die Möglichkeit einer *solipsistischen Welt*, die wir bisher annahmen, etwas näher." (Let us consider a little more closely the possibility of a *solipsistic world*, something we have assumed up to now.) In other words, the validity of *Ideen I* and the first 18 numbers of *Ideen II* has rested on the possibility of there being a world without living bodies other than one's own.

[46]You conceive of body, psyche, and soul as more fluid than solid; they run together in creative making, but they are fused at the level of essential being as well. Interestingly, Husserl has suggested that the i is fluid only along the dimension of time, as experiences flow along in the life stream. My comments in this section are based on your phenomenology of the body in §4 of chapter 3, Stein ([1917] 1980: 44-63).

[47]See Supplement 1 of *Ideen II*, Hua IV: 308; (Husserl 1989: 321-322).

[48]See [1917] 1980: 44. The planetary imagery is similar, but the point is entirely different than Husserl's remark in the Seefelder manuscripts to the effect that the center of the earth is not presented but only appresented in perception of the surface. (See Husserl 1973: 1-2, discussed in section A of the previous chapter.) In principle, one could physically relocate oneself so as to complete the series of appearings of materials beneath the earth's crust, or even the backside of the moon (as was done in the late 1960's). One cannot complete the series of the appearings of one's own body by any sort of relocation.

altogether as I can with every other thing. Its appearings never cease.[49] Like Husserl you point out that the body appears doubly: "outwardly" through being seen and touched like other things, and "inwardly" through being a seer, toucher, taster, and so forth--that is, as a sensory field. But you insist that it is *the same* body in the doubled appearings. This body is both less and more than a physical thing: it gives itself with gaps and doublings. It is precisely psychophysical.

Only *one's own* body is given with these anomalies. (In other words, you here establish that I do *not* require others in order to come into my own individuality by withdrawing from them, as you suggested earlier. I can instead begin the journey toward reflective awareness of my i by starting out from the unique givenness of my own body.)

Your next step is to expand the "psycho-" of the psycho-physical into "psycho-soulful," although you do not say as much. Putting a Schelerian spin on the Husserlian notion of "inner perception," you in effect transpose this category of description from the level of the psyche to the level of the soul. Sensory perceptions, you say, are not exhausted in delivering outer appearances of physical things, for precisely in doing so they *are* inner appearances as well. Sensations survive the phenomenological reduction; they are uncancelable components of consciousness. They are live experiences, and so are given just as absolutely as are such acts as judging and willing that Husserl would include among what is available to inner perception (albeit through an act of reflection). In other words, there is something more than empirical facticity to them: they are soulful as well as psychic (employing Scheler's distinction). Yet they differ from the soulful-intellectual acts of inferring, judging, and willing in significant ways. Sensation does not flow out of the pure i, it does not have the form of the *cogito* directed toward an object, and the i does not live-through it. Most importantly, therefore, live experiences of sensation cannot be doubled over in reflection to disclose the i, because they necessarily remain spatially localized and therefore "distanced"

---

[49]Your description of the givenness of the human body follows lines laid down by Husserl in manuscripts from 1905-1910 (discussed in the last chapter). It is likely that you learned this way of describing bodily givenness from Husserl's lectures in 1913 and 1914. But you are the first to *publish* such a description. Furthermore, your version compares favorably with similar material in §§30 and 32 of *Ideen II*, which you subsequently prepared for Husserl although he declined to publish it during his lifetime. In your Foreword to the published version of the dissertation, you say that you did not have access to those manuscripts while working on your own. I take this as your polite invitation to the reader to notice: (a) the coincidences in the two texts, (b) who published first, and (c) your own distinctive departures from Husserl's line. (Those departures are the focus of my own commentary.) At the time you wrote that note, you expected that *Ideen II* soon would appear. But it did not. From 1917 until 1952, then, your phenomenology of the body was the premier *published* source for this approach. Claims of writers such as Maurice Merleau-Ponty to have been given access to Husserl's manuscripts directly, as well as the increasingly numerous unattributed adaptations of Husserlian body-phenomenology, should be re-evaluated in light of this fact. By contemporary standards of scholarly etiquette, at least, one must say that both Husserl and his disciples ought to have acknowledged your ground-breaking work. Why they did not bears further investigation.

from the i.[50]

But this separation of the i from the sensations in its body is a peculiar kind of distance. The live body itself, as "zero-point of orientation," is what organizes space and defines distance, so you are in effect saying that i-hood is "distanced" from *the possibility of* being either close or far at all. Moreover, the zero-point is mobile. I can move it in order to complete series of off-shading appearances of things, in accord with the if-then coordination of locomotion with orderly space.[51]

How a body and a hyper-distant i could then "fuse," how i-hood could imbue the soulful-psycho-physical, is the kernel of the problem. The gist of your solution will be that creative personal initiative instigates the relatively autonomic functions occurring at the other three levels, although in doing so it may be responding to influences first arising from them.

You have succeeded in breaking down the rigid distinctions among the physical, the sentient, and the intellectual. The live body is none of these exclusively, but rather is given as the melting-together of all three. You suggest the term "fusion" (*Verschmelzung*) as the name for the appearance of this coincidence. This term, you say, is borrowed from psychology, where it refers to the co-giving of auxiliary sensory data (as when one perceives the softness of silk along with its sheen in a painted portrait). This blending supposedly is not owing to any associative process.

One can readily agree that by virtue of such fusion, the live human body's physicality presents sentient awareness and intellectual power as well. Unfortunately, your claim that the body also presents *an i* is not yet convincing.

Der Leib ist eben wesensmäßig durch Empfindungen konstituiert, Empfindungen sind reelle Bestandstücke des Bewußtseins und als solche dem Ich zugehörig. Wie sollte also ein Leib möglich sein, der nicht Leib eines Ich wäre? ([1917] 1980: 52)

The live body is constituted essence-wise precisely through sensations. Sensations are solid components of consciousness and as such belong to the i. Thus how should a live body be possible without its being an i's live body?

Rhetorical questions cannot do the work of argument. The difficulty, once again, is the article "an" which presumes individuation, individualization, and the accomplishment of the distinction between "an i" and "the pure i."[52] You haven't

[50]Thus you are trying to accommodate the Schelerian and Husserlian positions. See Stein ([1917] 1980: 46).

[51]Husserl in the *Logische Untersuchungen* had pointed out that the form of consequence or causality, "if/then" is given in the coordination of voluntary movement with the shifting appearances of the environment. As you put it: "Wenn ich mich bewege, so verschiebt sich das Bild meiner Umgebung" (if I move myself, then the picture of my surroundings shifts itself), page 50. Your discussion anticipates the account of the live body in *Ideen II*; but dependence upon *Ideen I* is also detectible in your unreferenced mention of Stumpf on page 47, matching Husserl's note at the end of §86.

[52]The physical experience "I walk down the street" can involve personal creative

yet displayed phenomenologically the relations or distinctions among "person," "an i," and "i." To get on with our reading, we are forced to take this point on faith; perhaps it can be sustained later.

The significance of the "fusion" of egoic levels in the human body becomes apparent in your discussion of what the body *does*: its feelings, expressions, and purposive creative actions. You show how permeable the boundaries are between the psychophysical body, the soul and the i. For example, pain is both in an injured limb and "in me." When I feel weary, all my body's parts and their motions are sluggish. Cheerfulness belongs to the soul and does not localize in the body; nevertheless it affects bodily perceptions and intellectual activities alike. Fusion guarantees the reciprocal influence among the levels of egoic function.[53]

Moreover, a true and bidirectional psychophysical causality obtains between the body and the mind.[54] The activities and condition of the body affect the capabilities of the mind. For example, an overly tired body impedes thinking. By the same token, physical training and practice can cultivate and sharpen one's mental powers of discernment and appreciation. A body-less pure intellect, you say, would experience joy or fear but *not* as the effect of physical or psychic events; nor would its joy or fear precipitate physical or psychic events. For us, by contrast, joy makes the heart pound and fear makes the mind boggle. Moreover, the pounding and the boggling are given precisely as caused effects. They are not the feelings but rather are symptoms (*Begleiterscheinungen*) of the feelings and appear as having been caused by them. Thus *psychic causality*, you say, is immediately given in such experiences.[55]

Altogether different from such symptoms of feelings are *expressions* of

---

decision, but it does not involve pure i as the intellectual/soulful experience "I deduce the conclusion from the premises" does. The distinction between "an i" and "pure i" would later be clarified in *Ideen II*, §§30 and 34. There, the soul or soulful i is equated with the person or personal i--collapsing two categories that you (like Scheler) want to keep conceptually separate yet connected. Opposed to soul and person alike is the pure i, also called the transcendental i. For your part, you either can't or won't conceive of i of any sort apart from the live body.

[53]See Stein ([1917] 1980: 53-54). You express this with the metaphor of fluidity: melting (*Verschmelzung*), influence (*Einfluß*), flowing out (*entquellen*). These lend contour to your depiction of "psychic causality." In German *der Quell* is a spring or fountainhead, and the verb *quellen* means to gush out or originate from, or to swell up. The same word in Yiddish, כולן, (*kwelen*), means to swell up *with joy*. Mothers use this word to describe what pride in their children does to them: it swells their hearts and brings tears of joy. This is a very potent word; surely you heard it at home on happy occasions. But you also say that an oasis spring (*Quell*) can be a mirage ([1917] 1980: 58).

[54]See Stein ([1917] 1980: 54-56). "Mind" is not your term. I use it here because it is a convenient way of referring collectively to the psyche and the soul or intellect, which you contrast with body at this point.

[55]You carefully point out that while you have displayed this phenomenon of psychic causality, you have not worked out an eidetics of it, and so you decline to identify it with the scientific concept of causality that was invoked in the genetic psychology of your day. See Stein ([1917] 1980: 56).

feelings. Symptoms such as a blush, a frown, or a scream occur spontaneously through psychic causality--that is, their having-been-caused is immediately given within the live experience of the one who spontaneously blushes, frowns, or screams.[56] When I reflect upon my symptoms, I notice a *causal* going-out of the somatic live experience from the psychic live experience. (I blushed because I couldn't help it; I can disown the blush as I would a sneeze.) But when I reflect upon my expressions, I find that I lived-through the discharge of the feeling. I rode along on the energy of the feeling into its termination in the expression. I was still alive there inside the expression. (I cannot disown angry words; it was really I who said them. I must ask pardon for them more profoundly than I do for belching.)

Feelings necessarily discharge in some way, you say. They never are sealed off within the psychic level of egoic function. They always come to some expression in the somatic or the intellectual sphere, or both. That is, one lives one's sentiments on through into their expressions. Nevertheless *one chooses* the modes of those expressions. (You will argue below that the one choosing is *a person*, who therefore can be understood in her choices; but that gets ahead of our story.) You adopt from Lipps and Pfänder the term "motivation" to indicate the relation between sentiment and its chosen expression.[57] Expressions are motivated, where symptoms are caused. It is essential to feeling that it motivate *something.*

You discuss five modes of discharge for feelings: (1) motivated action, (2) straightforward physical expression, (3) fantasy substituted for outward expression, (4) expression by selective perception, (5) reflection inhibiting outward expression. The first serves as a model for the others:

Eine Art der Entladung ist uns wohl bekannt: Die Gefühle entlassen aus sich oder motivieren--wie man sagt--Willensakte und Handlungen. Genau dasselbe	One kind of discharge is very well known to us: Sentiments release out of themselves--or *motivate*, as one says--deliberate deeds and actions. Exactly the

[56]See Stein ([1917] 1980: 56-60). You allow that people sometimes scream, frown, or even blush *on purpose*, but that is the exception that proves the rule; for the artifice means to include "spontaneity" within the sense that is deliberately to be communicated.

[57]For the most part, you adhere to the technical sense of the term "motivation" established in Munich phenomenology. You insist that motivation is rational, not causal. But it seems to me that you lapse once or twice into an equivocation and speak of the discharge of feeling as *causing* its expression. This usage would accord with the psychoanalytic conception of "motivation" in the work of Sigmund Freud and his disciples, who proposed to study the non-deliberate expressions of unconscious or subconscious motivations precisely as symptoms. You are utterly opposed to this approach. Yet, having just distinguished somatic symptoms from expressions proper, you turn around and count symptoms *among* the modes of expressions. I see here merely a careless inconsistency in your vocabulary. The underlying distinctions are clear. For an example of your terminological inconsistency, see Stein ([1917] 1980: 59), where you say that when I am happy and catch myself spontaneously stretching my lips in a smile, a "causal connection" is constituted in this "expression." But on the next page, such a caused symptom is (correctly) said not to count as an "expression" at all.

Verhältnis besteht zwischen Gefühl und Ausdruckserscheinung. Dasselbe Gefühl, das einen Willensakt, kann auch eine Ausdruckserscheinung motivieren. Und das Gefühl schreibt seinem Sinne nach vor, welchen Ausdruck und welchen Willensakt es motivieren kann. ([1917] 1980: 57)

same connection obtains between a sentiment and an expression's occurring. The same sentiment that can motivate a deliberate deed can also motivate an occurrence of expression. And according to its sense the sentiment dictates which expression and which deliberate deed it can motivate.

Besides words and deeds that are physically expressed--voluntarily, yet as prescribed by the non-physical sentiment--you mention three further modes of "expression" that are not physically manifest. Or rather, with these modes one chooses to detour the expression away from a straightforward manifestation. For example, you say, when for cultural reasons we forbid ourselves to give any outward indication of our sentiments, we may still express them by producing fantasies of wish fulfillment. In a second example, to be joyful does not mean to be a subject focused upon "joy" as if it were an object separate from oneself. Rather, joy is expressed as a selectivity of perception: one surrounds oneself with what is delightful from memory, fantasy, or the available environment, and one ignores everything else.[58] Your third example is that sentiment may discharge in an act of reflection that turns the sentiment itself into an object. Even this you term an expression.[59]

You complete your argument for the "fusion" of all levels of egoic function with further consideration of purposeful deeds. Purpose, like sentiment, is not a self-contained object of contemplation but rather a dynamic impulse that must discharge itself by taking action.

---

[58]This insight could well be contrasted with Heidegger's discussion of care and concern in his work *Sein und Zeit*, published some eleven years later. You make a nice summary of your doctrine of expression in Stein ([1917] 1980: 59). At the turn of the century, Theodore Lipps had postulated a drive or instinct to express. Subsequent writers formed two camps: those who accepted the involuntary instinctive nature of expression, and those who rejected it outright. Your solution (though you do not identify it as such) is to affirm the necessity of an expressive discharge for every feeling--Lipps's instinctual drive--while at the same time affirming a choice in modes of expression, some of which are compatible with stifling any somatic expression. You accomplish this solution without the cumbersome psychoanalytic apparatus of so-called unconscious and subconscious events and forces.

[59]Perhaps you insist upon making "expression" such an inclusive term because you mean to emphasize that both action and the four expressive modes of discharge *invite understanding*. An expression is anything intelligible by virtue of its motivation. One discerns here the groundwork for a therapeutic practice, comparable with but quite opposed to classical psychoanalysis. Deeds and each of the four modes of expression that you have enumerated would have to brought under therapeutic scrutiny, both in regard to the "sense" determined by the sentiment expressing itself, and in regard to the choice of the person to express in just that mode rather than in another. A formalization of your hermeneutics for more general application in the humanities could also be made along these same lines. You may have had ambitions to do so, but did not have the opportunity to carry them through.

Handeln ist immer Schaffen eines Nichtvorhanden. Dem „fiat!" des Willensentschlusses entspricht das „fieri" des Gewollten und das „facere" des Willensubjekts in der Handlung. ([1917] 1980: 61)

Taking action is always creating something that wasn't available before. To the *fiat* ("let it be done") of the purposeful decision, there corresponds the *fieri* ("it could be done") of the desired outcome and the *facere* ("getting it done") of the purposive agent.

Deciding to do something means having in view a future state of affairs that can come into being through one's actions. We recall that the "realism" of Pfänder and the other Munich phenomenologists was based on the *intention of realizability* that they held to be an implicit content of every act of will. You develop this insight to show that purpose manifests itself not just in choosing a mode of expression for one's sentiments, but in creative engagement with a real world. The body is fused with the will when it takes action, to the extent that when I choose a general purpose the body takes over and takes care of the details. Your example is mountain climbing. I decide to climb, and my hands and feet take it from there. Only when the body gets tired does it put up some resistance. I may then have to produce each step with an act of my will, competing against the contrary somatic inclination to sit down and rest. Following Pfänder, you term these opposing forces *Wollen* and *Streben*. The difference between them is that I am alive in my willing (*Wollen*) even while I am resisting an inclining (*Streben*) that arises within my body. I am the boss of my body when willing wins out over inclining: the inclination persists in me but I do not invest myself in it. Similarly, there are cases when my will struggles to maintain my position as boss of my *soul*; for example, when discouragement and fatigue threaten to dissuade me from completing some intellectual task. In this case, I am inclined (*Streben*) to give up and let my mind go on to more attractive entertainments; but *I* am not living within that inclination. I am living instead within the struggle to maintain my purpose and overcome the inclination that keeps inclining me even as I resist it. Yet such an inclination relentlessly tries to drag the i inside itself: to defeat, seduce, and subvert the i.

In your view, both *Wollen* and *Streben*--i.e., both willing and pre-voluntary inclination--run on psycho-physical causality. I can choose only what I can somehow present to myself as an object. Through willpower alone I cannot conjure up an object such that I intend it as *both* non-existent *and* desirable.[60] Only if I can intend the object as *realizable* can I desire it, for its present non-existence then can be meant as temporary. I am able to make the object. And the intention of realizing something involves the intention to mobilize the body or the mind for productive action. Therefore our lives as purposive beings entail the mutual engagement of body, psyche, and soul. You rephrase the formula cited above to bring out its ontological implications:

Handlung ist immer Schaffen eines Nichtseienden. ([1917] 1980: 62)

Action is always the creation of something that was a not-being.

[60]For it is impossible to want a non-being. One should say, more precisely, that it is impossible to want something that *necessarily* is nonexistent.

Causal chains are involved in creation, you say. They connect bodily actions with physical effects, and they include the psycho-physical influences running through body, psyche, and soul that you described above. But the causal processes are *initiated* by the i, living in an act of decision that is not causally produced.

It seems to me that the i living in the act of decision still is eligible to be considered a "pure i," in Husserl's sense. That is, you are correct to claim that you have displayed the *constitution* of the free i as live body.[61] But it also seems to me that the *creativity of* deciding acts cannot be grasped in constitution analysis. It will need to be grasped instead through your alternative to "constitution," that is, through empathy. The embodied i can be constituted as an individual: it is some body or other, no matter who. But the personal creative i can be known for who she is only with empathy. Before taking the step to the personal i, you offer an account of the constitution of embodied i's that are alien to my own.

3. Plurality of live bodies

The surface rhetoric of your argument makes a turn at this point[62] from talk of "myself" to talk of "others." But this is merely a descriptive device. The underlying technical phenomenological move occurring here is the transition from "pure i" to multiple empirical i's, considered first as the plurality of other individuals and finally as the individual whom I know myself to be through them. Your prior references to "my" experiences, to things that "I know," were meant to invite the reader to notice the direct access to knowledge of i-hood that he or she has and expresses already in the everyday use of first-person grammar. But these references have really been "about" the pure i, not about Edith Stein or any of her readers as individual personalities. The spotlight now shifts to the fact that indeed there are many individuals logged on to i-hood, as it were, and most of them are not myself.[63] It is time for you to review what has been said in general about the constitution of i as live body, and apply it to live bodies as they plurally occur.[64]

---

[61]In Stein ([1917] 1980: 62), you say that your description does not depend upon any particular individual, but holds for the phenomenon of willing as such. This indicates that you mean to be working under stipulations of Husserl's transcendental reduction, although you did not see how this reduction should be more than eidetic.

[62]That is, in the long concluding section of your middle chapter: §5, pages 63-101.

[63]As I have pointed out along the way, I believe that you have not always been careful and rigorous in your discussion of the pure i, or in maintaining the Husserlian reduction that sets aside whatever pertains to individual existence. Too much rigor makes for a deadly dull text, and you have chosen to introduce abstract concepts in concrete terms that sometimes lack precision. But the commentator must elucidate the technical structure of your philosophical argument.

[64]See §5 of Chapter 3, Stein ([1917] 1980: 63-101), for the positions considered below. In that section, you not only apply your previous arguments to the case of individuals, you also engage the earlier literature on various points. This engagement seems to presume that the reader has seen your expositions of your opponents' positions in the first chapter of the dissertation, now lost. Therefore it is impractical to discuss them here, and my commentary is confined to your own positive proposals.

In making this application, you reluctantly face the fact that "constitution" as Husserl proposes it and "empathy" as you propose it are two rival descriptions contesting for the same theoretical territory. The resolution that you will propose is this: empathy (*Einfühlung*) is the source of constitution.[65]

But before we see how that works, we must take a look at how you arrange the evidence. Some terminological reminders are in order. As before, the terms "psyche" and "soul" (*Psyche, Seele*) name two distinct functional levels. The psyche is what distinguishes a live body (*Leib*) from a mere physical body (*Körper*): its sensitivity and responsiveness. The live human body *(Leib)* is sentient matter: a psycho-physical unity. In contrast, the soul is an individual substance bearing qualities or character traits. Its unity is analogous to that of a thing; that is, the soul's unity is *constituted* in the many discernments of its qualities. Nevertheless by virtue of "fusion," the soul is "in" the live body, linked to it by bidirectional psycho-physical causality. The i is not the soul; it is something apart from soul and live body. The i (to be considered eventually as *person*) is the source of purposeful acts that set in motion the chains of psycho-physical causality. Purpose and will (*Willen* and *Wollen*) involve the i, and they necessarily discharge in some sort of expression, just as sentiments do.[66]

But these are Munich words. They derive from the phenomenology of Lipps, Pfänder, and especially Scheler. You load them like sandbags onto the hot-air balloon of Husserl's "transcendental i," in hopes of bringing it to ground on a terrain of concrete individual people. Thus you produce something foreign to Husserl's phenomenology: i's that must be *something more than interchangeable* with one another. You designate the unique embodied i as "the individual."[67]

Wir haben uns in Umrissen wenigstens darüber Rechenschaft gegeben, was unter einem individuellen Ich oder Individuum zu verstehen ist: ein einheitliches Objekt, in dem die Bewußtseinseinheit eines Ichs und ein physischer Körper sich untrennbar zusammenschließen, wobei jedes von ihnen einen neuen Charakter annimmt, der Körper als Leib, das Bewußtsein als Seele des einheitlichen Individuums auftritt.

We have given ourselves, at least in outline, an account concerning what is to be understood by "an individual i" or "an individual:" an integrated object in which the unity of consciousness of an i merges inseparably with a physical body, in the course of which each of them assumes a new character. The physical body surfaces as live body of the integrated individual, and the consciousness surfaces

[65]See chapter four, below, for a comparison of this solution with the alternative that Husserl eventually would adopt.

[66]Thus it comes as a surprise when you announce that you will not consider "purpose" (*Zweck*) because there is no direct live experience of subordinating psycho-physical happenings to it and so it cannot be studied phenomenologically ([1917] 1980: 63, note). Perhaps what you mean to exclude here is consideration of any essential finality or function for human life itself; for the setting and accomplishing of *proximate* goals surely is a commonplace experience within human lives.

[67]*Individuum* in German can mean something like "rascal." The term "person" is your favorite synonym for "individual." But you say elsewhere that a person as such cannot be an object.

Die Einheit dokumentiert sich darin, daß gewisse Vorgänge als zugleich der Seele und dem Leibe angehörig gegeben sind (Empfindungen, Gemeingefühle), ferner in der Kausalverbindung physischer und psychischer Vorgänge und des dadurch vermittelten Kausalverhältnisses zwischen der Seele und der realen Außenwelt. Das psychophysische Individuum als Ganzes ist Glied im Zusammenhang der Natur. Der Leib ist gegenüber dem Körper dadurch characterisiert, daß er Träger von Empfindungsfeldern ist, sich im Nullpunkt der Orientierung der räumlichen Welt befindet, selbst frei beweglich und aus beweglichen Organen aufgebaut, Feld des Ausdrucks der Erlebnisse des ihm zugehörigen Ich und Instrument seines Willens ist. ([1917] 1980: 63)

as his or her soul. The unity certifies itself in this, that certain occurrences are given as belonging equally to the soul and to the live body (sensations, general feelings), and further in the causal binding of physical and psychic occurrences as well as what is mediated through that: the causal relations between the soul and the real outer world. The psycho-physical individual as a whole is a node in the network of nature. In contrast to the physical body, the live body is characterized by this: (1) that it is a bearer of fields of sensation, (2) that it finds itself at the zero-point of orientation of the spatial world, (3) that it is mobile and composed of movable organs, and (4) that it is a field of expression of the live experiences of the i belonging to it and is an instrument of its purposes.

These four constituents of the live body were identified transcendentally by attending to how my *own* is constituted, all the while shutting out whatever differentiates *my* own from *just anybody's* own. These are constituents of the experience of living in a body, no matter whose. Now you extend them to cases where the "whose" does matter, where there is more than one live body. You employ the two phenomena of "blending and tending" in order to make this transition, and you begin with the first level of fusion, that between the physical and the sentient.

The givenness of one's own live body (*Leib*) as such is owing to the blending of materiality and sensitivity in it, made known through the doubling of inner and outer perception simultaneously referred to one thing. Perceiving my hand, I coordinate its sensory field (which I get inwardly) with its shape, size, and weight (as I get them outwardly among those of other things). Moreover the outer perception of my hand tends to complete itself with the inner perception, and vice versa.[68] Thus, you argue, the tendency to completion through inner perception of sensitivity accompanies every outer perception of a hand. Even while I am perceiving outwardly that a hand is *not mine*--perhaps because it lies too far away, or because it is smaller or larger than mine--the possibility that it is *not a hand* is excluded as I recognize that hands are mobile, can trade places, and can come in different sizes, shapes, and colors within one general type. The tendency to

[68]The tendency for understanding to flow from the part to the whole was asserted in various guises throughout the hermeneutical tradition whose heir you are. For Brentano and Husserl, it was owing to the essential unity of the "whole" in question. We noted above that the "molten" or "fluid" character of life is a key metaphor in phenomenology and underlies concepts such as the life-stream or stream of consciousness, the tendency to fulfill memories by reliving them, and the fusion of matter, sensibility, soul, and creative responsibility in the human being.

completion of outer with inner perception, then, gives me a live experience of the hand's sensitivity: not originarily, but *co-originarily*. The alien hand and body are constituted as sensitive through *Empfindungseinfühlung* or *Einempfindung*: receptive empathy of sensations.

Tending, blending, and empathy thus give the alien live body as such. You are right to assert that sensitivity pertains essentially to i-hood, but you are wrong in jumping from there to the assertion that sensitivity must belong to *an* i.

Mit der Konstitution der Empfindungsschicht des fremden Körpers . . . ist dank der wesentlichen Ichzugehörigkeit der Empfindungen auch bereits ein fremdes Ich . . . gegeben. ([1917] 1980: 67)	With the constitution of the level of sensitivity of the foreign physical body . . . an alien i already is given, thanks to the essential i-belongingness of sensations.

(The unsubstantiated hidden premise here is that i-hood entails a plurality of i's--which is what you are supposed to be proving. What your argument needs to say at this point is simply that the alien live body, though of my type, is not mine. It is precisely alien.[69] In your terms, the difference between mine and not-mine would be the difference between originary and non-originary empathy. In general terms, plurality could be introduced among live bodies "from the bottom up," that is, through the obvious plurality of physical bodies. But to guarantee the plurality of live bodies one should also establish it "from the top down," that is, through the plurality of personal agents.) As you correctly point out, a theory of empathy should show how an alien individual is given in the givenness of his or her feelings; it cannot show how feelings are given through an alien individual. Thus the target of your work is not alien feelings but the one who feels them.

Turning to the second constituent of the live body--i.e., its occurrence as the center of organized space--you introduce a compelling challenge to Husserl's treatment of the same phenomenon. The alien center does not appear just as a chance for the overlay of our centers: the possibility of my displacing him and taking his place. It appears as an enduring center in addition to my own center. I have both, simultaneously: one originarily, the other non-originarily or virtually. *To perceive the alien center is not to perceive an opportunity for conquest, but the opposite: to agree to a world with multiple centers.* Nor does the alien center require me to relinquish my own; I simply acquire another center. With that new center comes another world-picture, which is not a mere modification of my prior world-picture. The simultaneity of centers and of world-pictures, plural, is owing to empathy. I am originarily aware of my own through my live engagement with it, but non-originarily aware of the alien through a live engagement that, while *not* my own, is someone's and is available to me by empathy (*Einfühlung*).

In addition you assert that one who has sensations and who centers the

[69]There is no good English equivalent for German and Yiddish *fremd*. The adjective describes what is strange, foreign, alien, other. In the background of the term lies the theologically sanctioned obligation of showing fairness and kindness to strangers who may be living among Jews "because once you yourselves were strangers in Egypt."

spatial world is also one who executes acts, whose perceptions can take the form of the "cogito," and who therefore can reflect and discover i-hood.[70] Like Husserl, you identify the plurality of spatial centers as the basis for the constitution of *one's own* individual bodily being as well as for the constitution of the world *as real*. But unlike Husserl, you have shown that it is empathy that gives rise to the plurality of centers, and therefore is the foundation for the real world and for individual identity alike. Empathy is logically prior to Husserlian constitution.

Given that the alien live body is sentient and is a center of space, we perceive the voluntary character of most of its movements, easily distinguishing these from instances in which the body may fall or be lifted or pushed physically by something else. Thus the third constituent of the living body is seen to occur plurally and not just as a transcendental. The fourth constituent of the live body is that it expresses the sentiments and purposes of (an) i. These, too, have multiple instantiations in the alien live bodies that we meet. The general feelings that diffuse throughout the live body and fill it are expressed--as it were, *adverbally*--in the posture, gait, and attitude of the body, suffusing all its movements. You might have added here a mention of the more obvious example: that besides expressing *sentiments*, the body accomplishes *purposes* through motions that have real effects in the real world--or physically makes attempts at such. But you defer considering how alien purpose is given, in favor of a detailed preliminary consideration of *causality*.

Causality is one of the two forms of "effectiveness" (*Wirksamkeit*); the other is motivation.[71] Both are accessible through empathy. Just as I can follow the physical processes--both outside a live body and within it--that give rise to sensible and emotional responses in someone else, so I can ride along with the flow of an alien i *from* purpose *to* the deliberate physical motions that produce an effect in the real world. At the terminal point of the physical effect, there is causality of the same structure as any other physical event, although at the point of origin the i's purpose was not a mere step in a causal series. What is to be said of the "structure" of causality? At least this: that there is no causing at a distance in time or space, even though causal energy can be stored for later discharge, and even though effects may require a kind of incubation period before they appear. You find this same structure of causality in both the physical and the psychic realms; what accounts for differences between physical and psychic causality is merely owing to the different natures of those realms themselves.[72]

Therefore you take issue with what you take to be Scheler's position: that

---

[70]You refer to Husserl's arguments in *Ideen* §§27, 33-35 (Hua III: 48-49 and 60-62). I have already complained that your move from alien body to multiple i's is much too facile.

[71]See Stein ([1917] 1980: 82, note 3). Your discussion of causality is presented in part (k) of §5 ([1917] 1980: 80-85).

[72]You assert that psychic and physical reality have *different kinds of* structure (*Baus*) on page 85, while on pages 81-82 psychic and psycho-physical effects are given with *the same* structure as that given in the phenomenon of causal relations in physical nature.

psychic causality differs from physical in that psychic effects are produced by events distant in time or space. You argue to the contrary that one's so-called past decisions and future plans are not really distant at all, because the i remains alive in them. A decision or a friendship, even when not in the focus of consciousness, remains in a kind of halo or margin from which it can emerge as a *cogito* at any moment. Thus it remains alive and continues to affect one's expressions and actions.[73]

How, then, do we comprehend these various sorts of effects? We grasp them, you say, according to how they are given; that is, according to whether or not an i flows into them originarily so that its flowing is available for empathy. Earlier you established that symptoms are caused; for example, blushing can be caused physically by exertion, or psychically by shame. The i does not inhabit the blushing--it cannot help blushing--so the blush is not an expression to be empathized. The blush is merely what you now term "an indication" (*Anzeichen*, Husserl's term from the First Logical Investigation). One grasps it, like any causally produced effect, through following the chain of causes backwards.

By contrast, expressions of sentiment or of purpose are given as the discharge of a live experience. That experience, in its discharge, invites empathy. One grasps it from the inside by empathy. That is to say, one *understands* expression. One understands its motivations.

Yet the distinction never occurs so tidily in real life. As you showed, the fusion of physical, sentient, and soulful levels in the human body insures a dynamic traffic of experiences and expressions across the borders among the levels. The *understanding* of expressions, then, presupposes the ability to sort out the "caused" from the "motivated" in what is given in perception. Or rather, the perception of expressions *is* that sorting out. In saying this, I am saying more than you did. Your inspirations seem to have outrun your expository ability in the crucial passage that is the keystone of your hermeneutical theory.[74] But it runs roughly as follows.

The key discriminating factor in interpreting expressions is "how far" they carry the i before they depart from its live experience altogether; in other words, their relative opaqueness to empathy. Most opaque are symptoms. As we just said, symptoms are modifications of the live body that the i had nothing to do with; they were merely caused by physical or psychological processes, and are not transparent to empathy at all. "Signals" come next. They have no significance of themselves. They signify a state of affairs, and do so by virtue of a deliberate decision. (Thus while their construction as well as their placement could be understood as expressions of an i's purpose, nevertheless they deliver their meaning quite apart from such an understanding.)

Next come symbols and words. I can understand them without penetrating back to the live experience that they express. This is because their

---

[73]In fact, Scheler might agree with you, in view of his account of the i as the weaving together (*Ineinandersetzen*) of actions. See chapter one.

[74]See Stein ([1917] 1980: 85-96), which is part (l) of §5.

meanings are ideal objects whose unity the words and symbols constitute. The sound of the word or the perception of the symbol evokes in me the tendency to go over from the expression to its meaning. I can experience this going-over by myself, originarily; I need not empathically share the experience of the speaker, writer, or maker. (Nevertheless, I *can* use words and symbols as vehicles for receptively empathizing the sentiments of one who expressed them. But the necessity of my doing so is not an essential part of their expressivity.) Thus words and symbols permit the expression to come loose from what it expressed and turn it into something different: an object. Through this alienation of meaning, simulation and dissimulation become possibilities.[75].

Finally, the word departs even further from the i than the symbol does. You say that the word has its own living body that grows and develops independently, while symbol remains a modification of another live body. Words deliver a meaning, but in a general one-size-fits-all sort of way. To grasp the verbal meaning is not yet to *understand*. Words whose meaning has been decoded still stand in need of *a live experience of* their expressing of someone's live experience; that is, they need empathy. They need to be particularized and enlivened. By contrast, symbols have no live body of their own but remain modifications of another.[76] As you say, a sad face is the outside of sadness; yet an actor can put on a sad face without being sad.

While words undergo detachment from the live experience that they first expressed, they retain their meaning. It does not become completely arbitrary nor can it be changed at whim. It is not a mere sign. By comparison, as you say, a storm could wipe out the whole system of trail signs in the *Riesengebirge* despite the maintenance efforts of the *Riesengebirgsverein*.

Dergleichen ist beim Wort nicht möglich, sondern es ist immer getragen von einem Bewußtsein (das natürlich nicht das des hic et nunc Redenden ist); es lebt „von Gnaden“ eines Geistes (d.h. nicht kraft seines Schöpfungsaktes, sondern in lebendiger Abhängigkeit von ihm), dessen Träger ein individuelles Subjekt sein kann, aber auch eine Gemeinschaft eventuell wechselnder Subjekte, die durch eine Erlebniskontinuität zu einer Einheit verbunden sind. ([1917] 1980: 91)	Something like that is not possible with the word. Rather, it is always carried by a consciousness (which of course is not that of the one speaking here-and-now). The word lives "by the grace of" an intelligence (that means, not by the power of its act of creation, but in living dependence upon it) whose bearer can be an individual subject, but also a community of possibly succeeding subjects, who are bound into a unity through a continuity of live experience.

[75]It is of course possible to lie with a sign. But in that case, the sign itself does not dissimulate. It still refers faithfully to whatever it is supposed to refer to. The deception consists in the act of placing the sign wrongly. The deception inheres not in the sign itself, but in the deliberate misplacement of it.

[76]Presumably, the symbol could be a modification of a human body, as when a wink changes a scolding into a teasing. Or the symbol could be a modification of a body *of literature*, as when myth becomes allegory. Your term *Wörtleib* begs for elucidation along these lines.

Words are worn more deeply than mountain paths, and they are in the custody of human communities of superlative vigilance. Words are the portals of meaning. At the same time words come in wrappers (*Hülle*) such as speech, handwriting, or print, that do not call attention to themselves until there is some misfit between the wrapper and the "pure type" of the word. In those cases, the odd use or misuse of the word may call attention to the person speaking, who otherwise remains in the shadow of the speech itself.[77] The person may also be foregrounded in speech that relates the speaker to the hearer, such as a question or a request, but the person speaking recedes into insignificance in a sentence such as, "Something's burning!" Here, the meaning comes across clearly no matter who is speaking or what her intentions may be.

Neither person nor community is given in the experience of understanding, which is the empathizing of motivated coherences. You propose the term "motivation" as a general category that includes all instances of non-causal "going forth" (*Hervorgehen*) which in some way involve the streaming live experience of the i: (1) from willing to taking action, (2) from feeling to willing, (3) from experience to any of the various kinds of expression, (4) from one aspect of a thing to the next in a series of its outer appearances, and so forth. Because each of these processes is a live experience, each can also be empathized non-originarily. If there are several i's, then one can empathically grasp what another has lived. And this is what you mean by understanding.

None of these motivated goings-forth is a causal sequence. Thus it is wrong to identify motivation with the causality of the psychic as several writers have done, you say. Causal sequence cannot be "understood" because it cannot be lived-through by empathy. (Yet you also imply that motivation is the mental tendency to follow the form of "if . . . then" wherever it occurs, even in instances of causality.[78] In this derivative sense, may not causal sequences be said to be "understood" as well? Or perhaps you have really made the natural sciences a subdivision of the humanities, and all prediction description.)

Understanding is the empathizng of lived coherence. It is not the "having" of an "object." Intelligible processes are finite. They form wholes. To understand is to fulfill the tendency to go from part to whole, beginning to end. By contrast, causal sequences never begin and never conclude; yet motivated sequences do sometimes break in and manipulate causal sequences.

In practice, understanding does not occur at a single brilliant stroke, but in a campaign of testing and correcting. For example, you say, empathy can clear up the ambiguity of a blush by determining whether it issues from physical

[77]See Stein ([1917] 1980: 93). You have been quite sparing in your use of the term "person." It occurs here as the name for something given in the individual deviation or variation from the general rule. Heidegger will later write about the disappearance of tools within their use until such time as they may break down.

[78]For example, see Stein ([1917] 1980: 96), where you say that one could feel-into a causal connection instead of a motivation-connection. Again, this seems to be a problem of carelessness in vocabulary rather than incoherent thought.

exertion or shame. Empathy also accomplishes the constitution of the individual soul by gradually building from single actions, to habitual qualities, to a firm hypothesis concerning someone's character, subject to subsequent testing. Yet empathy remains susceptible to illusions, especially if one confuses one's own type with humanity in general.

Finally, my own character, my own soul is given to me through iterated empathy. I feel-into the image of me that someone else has constituted through empathizing my expressions. This provides me with something to compare with my own unmediated conceptions of myself, which in principle might be illusions.

C. Analysis of the Empathy of Personal Types

The third of the extant chapters of your dissertation is unlike the other two. A change of pace is demanded of the careful reader. The vocabulary and the issues have shifted. The exposition is no longer conscientiously phenomenological; the argumentation goes by fits and starts. This isn't philosophy. It's a string of examples converging on something left undefined. Through what genre, then, does this text ask to be read?

This chapter is an exuberant manifesto of your convictions concerning interpersonal communication, and it is oriented toward the academic fields of history, psychology, and German studies--the major fields of your university examinations, along with philosophy. Perhaps you originally assembled this material as an independent essay and subsequently reworked it lightly for the dissertation. Whatever its past, this collage has a future in the carefully argued essays of your first attempted *Habilitationsschrift*, which you would publish in Husserl's *Jahrbuch* in 1922.[79] Thus I can save my critical efforts for the more mature formulation. Here I will merely point out what is going on: the breaking and tearing of phenomenology, the hatching of a new model of personality integration, and a flirtation with evil. The young woman who crafted these words was working as a substitute teacher, having recently returned from a field hospital where she cared for dying soldiers. You do not refer directly to those experiences, but they haunt your text.

1. Beyond transcendental phenomenology

In discussing the earlier chapters, I proposed that you were attempting to hybridize Husserlian transcendental phenomenology by displacing the egoic autonomic function of "constitution" with that of "empathy" (*Einfühlung*)--all the while pretending that the latter could easily be made to fit into the Husserlian system. You now abandon that pretense. Husserl's "reductions" are no longer in force: you flit from the eidetic realm, to the realm of nature and history, to the transcendental, and back again, all with nary a glance in the direction of

---

[79]Those essays are entitled "Psychische Kausalität" and "Individuum und Gemeinschaft." See also "Eine Untersuchung über den Staat" in the 1925 *Jahrbuch*.

methodological taboo.[80] It's a crazy quilt.

You can see what I mean if you take a look at your vocabulary. First, there are some terms that retain the specialized meaning that they acquired in Husserl's work. For example, you continue to refer to "constitution" as the way in which seriated appearances are united into the unity of a thing as it is perceived. That unity is a "sense" (*Sinn*) and is held in consciousness as such. The tendency of consciousness to flow toward completion of a sense-unity whenever it embarks upon a series of appearings was among the varieties of "motivation" that you enumerated in the last chapter. These terms--constitution, sense, unity, motivation --continue now to carry meanings that are in accord with Husserl's phenomenology. In contrast, "the pure i" is criticized for its flatness, and in effect it becomes dissolved and diffused within your new tentative modeling of the personality.

Some entirely new vocabulary makes its debut without a proper introduction. From Dilthey come the notions of personal type and of *Geist*, which I will translate as "intellect" or "mind."[81] But the intellect or "intellectual subject," as you describe it, still seems to operate quite like Husserl's transcendental i. For example, it functions to constitute the unity or "sense" of things, it flows along through a stream of live experiences, and it is a kind of arbiter of possibility.[82] Intellect stands over against the natural realm of the live body and the soul, and each side has its own distinctive structuration. There must therefore be:

eine Ontologie des Geistes entsprechend der Ontologie der Natur. Wie die Naturdinge eine Wesensgesetzen unterstehende Struktur haben, die empirischen Raumformen z.B. Realisationen der idealen	an ontology of the intellect corresponding to the ontology of nature. Just as natural things have a structure conforming to essential laws--for example, the empirical spatial forms render realizations of the

[80]Husserl was notorious for never persevering to the end when reading a book, owing both to his poor eyesight and to his impatience. Perhaps you counted on this, and figured you could take liberties in your last chapter.

[81]The alternate term "spirit" carries a religious connotation that is inappropriate. When you compiled the Latin-German glossary for your Aquinas translation, you chose the term *Geist* to render the technical Latin philosophical term *mens* (in English, "mind").

[82]The issue, which I believe you neither resolve nor see clearly, is whether there are two realms to deal with here, or three. To oversimplify: Dilthey discerned two realms, that of the natural sciences and that of the human sciences or *Geisteswissenşchaften*. The coherence of the former was causality, and the coherence of the later was motivation. Moreover, the latter or intellectual realm could account for the former. (The science *of science itself* was history, one of the *Geisteswissenschaften*.) On the contrary, Husserl discerned a third realm, the realm of coherences as lawful connections, whose science was phenomenology. It was phenomenology's job, then, to account for both the coherences of the natural sciences and the coherences of the cultural sciences. (The science of science is neither a natural science nor a human science, but *sui generis* phenomenology). If asked, you would agree with Husserl. But in practice you backslide into the handier Diltheyan position. Thus in what follows, I can content myself with showing how you contrast the natural structures of the live body and the soul, with the structures of the intellect. I need not look for the Husserlian transcendental move, which would make an inquiry into structures of structuration itself. It isn't there.

geometrischen Gebilde darstellen, so gibt es auch eine Wesensstruktur des Geistes und ideale Typen, als deren empirische Realisationen die historischen Persönlichkeiten erscheinen. Wenn Einfühlung das erfahrende Bewußtsein ist, in dem uns fremde Personen zur Gegebenheit kommen, so ist sie zugleich die exemplarische Unterlage für die Gewinnung dieser Idealtypen wie die Naturerfahrung für die eidetische Erkenntnis der Natur. ([1917] 1980: 106)

ideal geometric shapes--so there is also an essential structure of the intellect and ideal types, as empirical realizations of which the historical personalities appear. If empathy is the receptive consciousness in which alien persons come to givenness for us, then it is the exemplary base [or prototype] for the acquisition of this ideal type in the same way that nature-reception is for the eidetic knowledge of nature.

Yet, as you argued earlier, the natural and the intellectual are fused and interactive in the living human body. Now you coin the contrasting terms "structure of live experience" (*Erlebnisstruktur*) and "psycho-physical organization." The latter refers to neurological and chemical events occurring in the sentient body as causal processes governed by essential laws; the former, to the motivated flow of live ex-

FIGURE 3.1: STEIN'S MODEL OF THE HUMAN PERSON

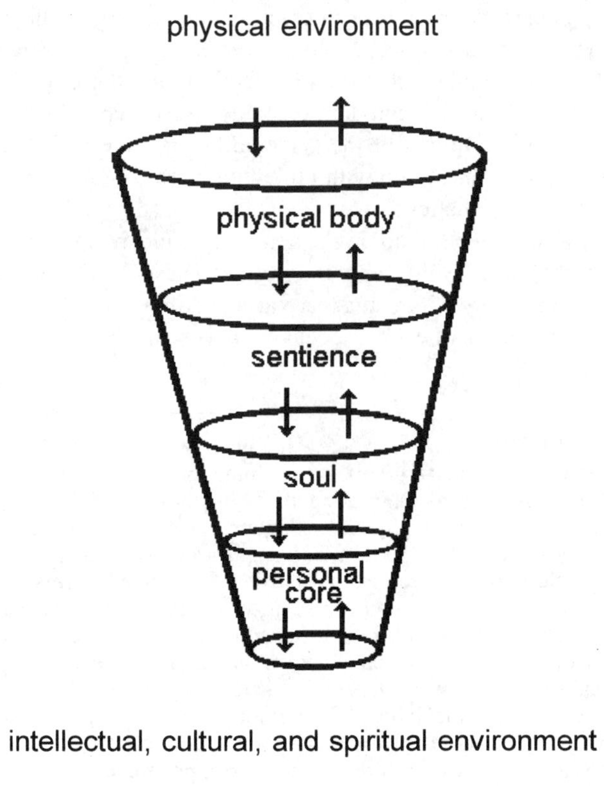

perience governed by laws of reason. You are interested in their interface. If I may say so, you leave transcendental phenomenology behind because it offers no access to the evidence of the mutual limitation of these two structural webs in the conscious organism.

The terms "value" (*Wert*) and "person" now become focal, an indication of your indebtedness to the thought of Pfänder and Scheler. Like those authors, you sketch out a working model of the human person with a "core" and various levels or depths. Had you actually drawn your model it might be funnel-shaped and look something like Figure 3.1. Pfänder had conceptualized the i as a core within a body, as we saw in chapter one. The i's objects could be "located," as it were, within the i's psyche or its body at various "distances" from the core, or they could lie outside of the i altogether. You have picked up this spatialization of an i-core and the notion of directional attraction, but apparently you have forgotten that it comes from Pfänder.[83] From Scheler you take the notion of "levels" and use it to elaborate the core-body model deriving from Pfänder. Scheler's highest or innermost level was that of the person, the one who acts and whose activity cannot be known objectively. The schematizations that I presented in Chapter One are recalled in Figure 3.2, which also summarizes your innovations.

Superimposing these spatializations, you produce a composite map of personality. The advantage of such mapping is that you can represent a hierarchy of levels and indicate their mutual interaction by making them adjacent. The disadvantage is that some concepts do not easily lend themselves to location "within" one level (or depth) rather than another; among these are "the pure i," its autonomic function of constitution, and empathy itself! They become diffuse and ill-defined in your depth model of the personality. Be that as it may, the bulk of your concluding chapter has to do with projecting various questions upon this map and playing out the implications.

The device of depth mapping is helpful in conceiving the relations among the physical body, the psyche, and the soul. These levels are all subject to essential laws of causality. You suggest various "dimensions" for the placement of feelings upon the map. Feelings occur with varying magnitude, depth, reach,

---

[83]See (Stein [1917] 1980: 118), where you quote Pfänder 1911. You take from him the distinction between willing (*Wollen*) and inclining (*Streben*). But you misconstrue his position. You read him as distinguishing them by the "level" in the i at which the attractive *object* arises, when actually his distinction is between an i that places itself into the willing--in your words, takes the form of a *cogito*--and an i that resists a felt attraction that otherwise would pull the core of the i into itself. Pfänder's terms "centrifugal" and "centripetal" connoted *valences* or felt forces; in your version, "central" and "peripheral" connote mere *position*. Thus you are mistaken when you say that these are simply different modes of accomplishing the "act" of inclining (*Strebensaktes*). Rather, for Pfänder the "act" of the i is what makes the difference between willing and inclining. The acting i *accords with* the impulse in willing (*Wollen*); by contrast, the i can be resisting the impulse by *declining* to put itself into the impulse--i.e., declining to accept the impulse into its own core--when the i is merely experiencing an inclining (*Streben*). Naturally, this distinction does not appear in any dictionary; Pfänder was using the terms idiosyncratically to make a point.

duration, and intensity. In case feelings should conflict, a kind of calculus of these dimensions would predict which feeling would win out and come to expression.

FIGURE 3.2: MUNICH MODELS OF INCLINATION

FROM PFÄNDER:

[1] awareness:

Centrifugally, an act of consciousness, directed outward from the core of the i, establishes an object. (The object is "central" if it lies within the live body, or "eccentric" if not.)

[2] desirability:

Centripetally, the object now seems attractive to the i that is aware of it.

[3] desire, prior to consent:

Centrifugally, the i feels tugged toward the attractive object in *Streben*. (Unless and until the i assents and invests itself in *Wollen*, the inclination remains "blind": empty of i.)

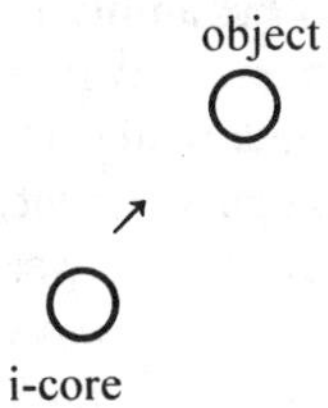

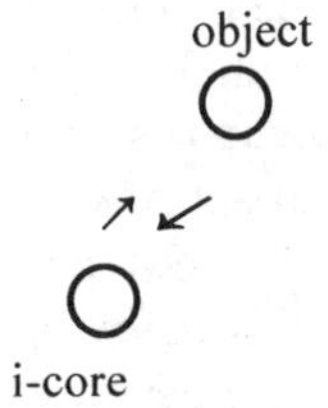

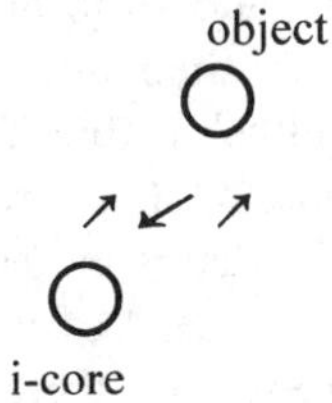

FROM SCHELER:

levels of i-function

free acts, love, by *person*
meaning, thought, logic
sense perceptions
sensations

levels of i-hood

the spiritual/cultural → *geistig* motivation
the soul → *seelisch* motivation
the psychic/vital → *psychisch* causality
the sensible/physical → *sinnlich* causality

EFFECT OF STEIN'S MODIFICATIONS:

- Scheler's "levels" become "depths," as in Pfänder's model of core/periphery.
- Because of "fusion," boundaries between depths (levels) become mere gradations.
- Causality and motivation are described across levels (depths) as well as within them.
- The object-constitutive functions of Husserl's "pure i" are diffused at the depths.
- Value radiates from core. The model is realist, because geared toward realization.
- Different values characterize different persons, and *types* of persons, at their cores.

But you cannot bring this line of thought to coherent and persuasive presentation because of a defect in the model itself. *It doesn't move.* Your static conception of the "psycho-physical organization" neglects its dynamic interface with what you term the "structure of live experience." To represent that *other* structural network, the one conforming to laws of reason, you must (and do) conceive of the i as a stream of live experiences flowing along in a motivated sequence. But the Husserlian modeling of that i is flat, as you complain. In other words, its dimensionality is temporal and consists in forward and reflective movements. We are left with two partially successful schematizatons of subjectivity, neither of which can be brought to completion.

Nevertheless, something useful has resulted from the effort. Your depth model illuminates what has been left out of Husserl's transcendental-phenomenological conception of subjectivity. *The executive i must somehow be seated in a causal matrix.* Reason's laws govern only as nature's laws permit. The coordination of causal sequences with motivated sequences is the job of "the person"; which is to say that you offer the term "person" to name the region of engagement between (a) the ensouled individual organism and (b) the egoic pursuit of logical necessities.[84] On the side of nature, souls have (*are!*) habits, strengths, character traits, and so forth. On the side of intellect, the i has (*is!*) the ability to follow order. At their interface, the person is a distinctive arrangement of *values* at varying "depths" within herself. A value is something meant as both realizable and *to be realized.* Thus every personal act of valuing has two components: the change in being that is envisioned, and the awareness of one's own creative power invested in the "I can" of that change.[85] Creative power (*Schaffenskraft*) always is a source of joy, and therefore a value in itself. The person thus values *herself* within every act of valuing. Creativity, we might say, is the *metavalue* of every person. It is what is loved when I love a person, myself included. Regard for one's accomplishments, good character, or physical beauty would be secondary to regard for creativity. You keep trying to say that love is "deeper" than other values.[86] That's wrong. It's not "deep" at all. It's off the map.

2. Person as limit

"Person" was Scheler's term for the non-objectifiable source of actions at the top of his hierarchy of egoic levels. He insisted that the person cannot be known *as the object* within a consciousness having the usual subject-object structure. Your own major insight was that persons are known to one another instead through empathy, whose structure is the sharing of the sense of a live experience that is "originary" on one side and "non- or co-originary" on the other. In this crazy-quilt concluding chapter, there are lapses in which you describe

---

[84]Your own formulations are not so crisp as these. In fact, on page 121 you explicitly disavow any claim to a doctrine of the person!

[85]See Stein ([1917] 1980: 115).

[86]See Stein ([1917] 1980: 114-115).

objective knowing of persons just as if they were potatoes.[87] Yet you also affirm the liminality or border-status of personhood in some very striking formulations.

The person is not the soul, you insist. The soul is structured by natural laws, the person by the laws of reason. The soul is *constituted* for consciousness like any other natural thing: in perception of the unity of multiple appearances of its qualities.[88] The person, however, is not "constituted" in that technical sense. You say instead that the person "unfolds" or "reveals herself." But how? You approach the question obliquely. The soul, you say, belongs to the natural world and is therefore subject to causal influences that affect its development--up to a point. The soul's "categorial structure" cannot be transgressed. Your attempt to articulate what protects it is hampered by the unresolved problem of indivuation, and ultimately defeated by your rhetoric of spatialization:

Nicht nur, daß die kategoriale Struktur der Seele als Seele erhalten bleiben muß, auch innerhalb ihrer individuellen Gestalt treffen wir auf einen unwandelbaren Kern: die personale Struktur. . . . Die personale Struktur grenzt einen Bereich von Variationsmöglichkeiten ab, innerhalb dessen sich ihre reale Ausprägung „je nach den Umständen“ entwickeln kann. ([1917] 1980: 123)	Not only [do we find] that the categorial structure of the soul must remain intact, but inside of its individual configuration we stumble upon an unchangeable core: the personal structure. . . . The personal structure marks off a domain of possible variation, inside of which its real stamp can develop itself "according to the circumstances."

What goes inside what here? Is the person "inside" the soul? Is one structure a substructure of the other? Generally speaking, the term "structure" in phenomenology stands for an essential lawfulness. Generally, one would say that many different empirical souls all share one *eidetic structure* (i.e., that of substantial unity of qualities), or that every person has the same *structure of live experience* (i.e., that live experiences run together in an intelligible whole[89]). But

---

[87]For example, on page 114 you say that value must be made into an object, through an act of reflection, before it can be valued. The point should have been more carefully framed.

[88]See Stein ([1917] 1980: 116-117, 122).

[89]See Stein ([1917] 1980: 128-129): "Jedes Subjekt, an dem ich einfühlend ein Wertnehmen erfasse, betrachte ich als eine Person, deren Erlebnisse sich zu einem verständlichen Sinnganzen zusammenschließen. Wieviel ich mir von seiner Erlebnisstruktur zu *erfüllender* Anschauung bringen kann, das hängt von meiner eignen ab. Prinzipiell erfüllbar ist alles fremde Erleben, das sich aus meiner eignen Personalstruktur herleiten ließe. . . . Was dagegen meiner eignen Erlebnisstruktur widerstreitet, das kann ich mir nicht zur Erfüllung bringen, ich kann es aber noch in der Weise der Leervorstellung gegeben haben." (Each subject in whom, by empathy, I comprehend a grasp of value, I consider as a person whose live experiences join together into an intelligible whole meaning. How much I can bring to *fulfilling* intuition from his structure of live experience, that depends upon my own. In principle, all alien live experience is fulfillable that allows itself to be derived from my own personal structure. . . . On the contrary, what clashes with my own structure of live experience, that I cannot

here you use the term "structure" equivocally, for you make "personal structure" the principle of individuation rather than the common form. You aren't clear about whether "personal structure" will be a form common to all personal life, or the pattern of valuation unique to each living person.[90]

This conceptual difficulty arises from your insistence upon making a spatial mapping of the personality. Nevertheless, the underlying insight is clear. *Something* preserves individual human beings from being *entirely* determined by the causal chain of events that govern the ensouled psycho-physical organism. There is a kind of semi-permeable membrane, across which influences pass in both directions--from the natural world to the rational, and vice versa. The membrane is alive and active, but not substantial.[91] It is not really a "level" of anything else, whether of the psycho-physical individual or of the i. It is known through the values that it actualizes; or, one should say more precisely, its unique live experiences of creative value-choice can be empathized by others. When you talk about "person," you are talking about this live liminality.

There are intriguing consequences of the fact that persons are proof against alien influences of all kinds. On one hand, persons as such *cannot be educated*. What is trained is merely the character, the soul.[92] On the other hand, persons *cannot be reasoned with*. They can resist the necessities of logic just as easily as they withstand physical and mental influences, although we may call such logic-resisters mentally ill.[93] Therefore personal choices cannot be predicted on the basis of either logic or the material and social environment. Personal choices are not intelligible as either caused or motivated.

How then am I to *understand* other persons, or even my own person? You say that the person "unfolds" or "reveals herself" under certain conditions. Here again, "person" is a limit concept. Persons come in various "types," according to the values that lie deepest in them.[94] Understanding is most complete between persons of like type. Thus differences in values place a limit upon the possibility of interpersonal understanding. Persons are not changed by their environment, but they may "unfold" partially or fully. With persons of similar type, one can empathize in another even something that has not yet unfolded in

bring to fulfillment for myself, but I can still have it given in the manner of empty representation.)

[90]On page 130, you write again of personal structures of different natures.

[91]It is not the soul. In your view, the person is free but the soul is not.

[92]See Stein ([1917] 1980: 123).

[93]See Stein ([1917] 1980: 108). It would be against reason to desire or to attempt something that one also knew to be impossible. You do not make this sufficiently clear, however. You seem to be saying that it's unreasonable to desire something that *someone else* thinks is impossible.

[94]For example, you say that when knowledge is a deeply felt value, the person is of the "scientific" type. Further examples would have been helpful. See ([1917] 1980: 120-121).

herself. Contact with others of one's type, then, can help a person to become fully revealed to herself.

Unlike the person, the character can be educated. Your contrast between the trainable soul and the self-actualizing person permits you to describe what a later generation of "existentialists" would call inauthenticity. You say that someone who received a strict moral education would not know the kind of person he really was until a contrary action burst forth from deep inside him and revealed him to himself. Furthermore, we do not really know love or hate unless we meet someone who evokes it.

Schließlich ist es auch denkbar, daß es zu gar keiner Entfaltung der Persönlichkeit kommt. Wer nicht selbst Werte fühlt, sondern alle Gefühle nur durch Ansteckung von anderen erwirbt, der kann „sich" nicht erleben, keine Persönlichkeit, sondern höchstens ein Trugbild einer solchen werden. ([1917] 1980: 124)

Finally it is also conceivable that no unfolding of the personality comes about at all. If someone does not feel values for himself, but acquires all sentiments only by catching them from others, he cannot experience "himself"; he cannot become a personality, but at most a caricature of one.

Socialization processes and direct educational efforts, then, may very well equip me to live as an impostor, and they can impede or prevent altogether my discovery of the person who I am.

These "existentialist" statements are difficult to reconcile with your assertion of the liminality of the person with regard to environmental and physiological influences. Further development is needed. (Apparently you thought so too, because you would take up these issues again in your *Habilitationsschrift*.)

3. Difference, illusion, irrationality, pathology, evil

In your comments on personal "type," we now come upon the most baffling aspect of this entire work. "Type" is something of a synonym for "personal structure." As I complained above, you have used the latter term equivocally. It has at least four meanings in your text. By "personal structure, " you may mean:

(a) the coherent flow of live experience that is characteristic of *all* human beings and is therefore the basis for understanding through empathy;

(b) the live experience of the eruption of particular valuations, characteristic of all human beings *except* those inauthentic wretches who have come by their values through contagion rather than their own deep feeling;

(c) any of several configurations of felt values that characterize different sorts of human beings;

(d) the distinct configuration of felt values that uniquely characterizes each individual human being.

The term "type" is used in the third and fourth senses. On one hand, you say that I can fully empathize the live experiences that *my type of person* has; that is, I can live along with the *similar* i as it lives through its experiences. On the other hand, with *truly alien* aliens--people not of *my own type*--all I can do is to represent their experiences to myself and watch them from the outside. Objectively.

Frau Doktorin, this is an offensive and dangerous idea. Besides which, it is wrong. There are no aliens so *alien* that we cannot feel inside their live experiences *at all.* We are all of one type.[95] What disables the ability to understand another human being is not differentness, but only the condition stipulated as item (b) above: one's own refusal of depth experiences of valuation, and one's consequent failure to become an authentic person. As you rightly remark (expanding upon an unattributed citation of Dilthey):

„(D)as auffassende Vermögen welches in den Geisteswissenschaften wirkt, ist der ganz Mensch“: nur wer sich selbst als Person, als sinnvolles Ganz erlebt, kann andre Personen verstehen. ([1917] 1980: 129)	"(T)he comprehending capacity that operates in the cultural sciences is the whole human being." Only the one who experiences herself as a person, as a fully significant whole herself, can understand other persons.

There is another sinister aspect to your theory of type. Besides impeding understanding, type is also positioned as its enabler. Citing Dilthey again, you propose that universal knowledge of values allows us to have a priori understanding of all possible types of persons before we even encounter any of the individuals who exemplify the types.[96] This suggests that knowledge of type comes to us without empathy, and with it the ability to assess and classify people.

Your comments about "type," as the factor that variously enables or disables interpersonal understanding, are both spurious and completely unnecessary to your argument. However I do not regard their intrusive presence as accidental, and will consider their significance in the next chapter. Let us note in passing that you select only differences in *valuation* as entailing differences in "personal type" having implications for the possibility of understanding. You are silent about physical, psychological, and cultural differences, such as language, gender, age, and history. Either you see in such differences no impediment to empathy and communication, or you have transposed all the varieties of interpersonal difference into just this difference of values.

However difference may be founded, you have made some disturbing suggestions about how it is *disclosed and assessed.* Empathizing a personal structure at variance with one's own, you say, reveals *what one is not.* I may come to know of values that I have not yet actualized, as well as those that I can never actualize. Thus, understanding another is an opportunity for self-knowledge and self-valuation (*Selbstbewertung*). Therefore it can be an occasion for me to learn

---

[95]Indeed, elsewhere you affirm this, and are ready to admit that even animals are like enough to us that we can empathize their psycho-physical experiences.

[96]See Stein ([1917] 1980: 127).

of my deficiency or my disvalue. As you say, I can learn to assess myself "correctly" (*richtig einschätzen*).[97]

The correction of empathy has been mentioned before. You concluded your previous chapter by noting that I augment my knowledge of myself through empathizing the other's perception of me, and furthermore that repeated acts of empathy on my part will fine-tune what I first grasped of the other's experience. You also noted the possibility of illusion in which I might impute to the other i something that is peculiar to myself and not "typical" of human beings in general.[98] Here the term "typical" marks what is *correct*, while my own special knowledge of myself is what tricks me into the "illusion" of assuming that the other is like me. It seems, then, that the project of empathy unavoidably partakes in the dangerous negotiation of difference and correctness.

Empathy can also serve in the diagnosis and assessment of various forms of mental illness, you suggest. This is owing to the "fusion" of levels in the human being--the physical, the psychic, and the soul--and also to the fact that different sorts of coherence obtain within and among those levels. These coherences are *intelligible* as motivations, that is, as ordered sequences. The causal sequences of psycho-physical events are grasped in their conformity to essential laws. But you also use the term "motivated" in a special sense to label the distinctive intelligibility of the flow from feeling to expression and from purpose to deed, which is owing to the possibility of our living-through that flow along with the human being who originarily experiences it.[99]

Based on this, you have an ingenious observation regarding mental disorders. They can be sorted out according to the *kind of intelligibility* that they allow.[100] This will point to the level or function that is diseased, whether it be some component of the ensouled psycho-physical individual, or the intellect itself. First, the *intellect* is diseased and the human being is irrational when he desires something that is impossible. This is so because a value in its essential structure is something-to-be-realized. (I pointed out above that you have made a mistake here. Technically, in Munich phenomenology a value is essentially *meant as* something realizable and to-be-realized. Its meaning, as intended by consciousness, must include an intention of its possibility. Obviously this can be so even in cases where realization is impossible through factors not known to the one who entertains the desire. But for the sake of hearing out your argument, we can grant provisionally that a case could be found in which expressions or deeds were issuing from a concept that was logically flawed in the way you describe.) The therapist who attempts empathy with regard to the irrational human being will

---

[97]See Stein ([1917] 1980: 130).

[98]This is repeated on page 128.

[99]This summarizes descriptions offered in your previous chapter and recalled on page 107.

[100]The following comments refer to Stein ([1917] 1980: 107-108), but I examine the possibilities in a different order than you do.

be brought up short at some point by her own rational inability to achieve a unified grasp of his purpose and its expression. She finds that she simply cannot reenact his conceptualization. The way in which the therapist's understanding breaks down is the indicator of where the problem lies. This patient is disobeying the laws of reason, even while his soul and psyche may be operating nominally.

Second, there are patients who reason quite correctly but whose souls have been made ill by feelings arising in response to circumstances. These feelings allow the therapist access through empathy. One can receptively live-through what such people have experienced. For example, the depression that is overwhelming the patient can be known from the inside; the therapist feels within herself (but non-originarily) how the patient's feeling has arisen and how it overwhelms him. Such accessibility is the indicator that this illness is an illness of the soul. Third, the therapist may ascertain that the intelligence and the soul are working nominally, and therefore turn to the realm of psycho-physical causality for signs of dysfunction based within the processes of the living body. These processes are not available for empathy, but their causes and effects can be explained.[101]

In view of the course that your own life subsequently would take, it is perhaps not inappropriate to close this commentary on your dissertation by noting its brief treatment of two problematic themes: historical evil, and the deity. On history, you remark that it cannot be understood as a series of caused events, a concatenation of facts. History's intelligibility lies in purposes arising from the intellect. Yet human beings are citizens of the natural world as well as the world of the mind.

Weil der Mensch beiden Reichen angehört, muß die Geschichte der Menschheit beides berücksichtigen. Sie soll die Gestaltungen des Geistes und das geistige Leben verstehen und feststellen, was davon Realität geworden ist. Und sie kann die Naturwissenschaft zu Hilfe rufen, um zu erklären, was nicht und was anders geworden ist, als die Gesetze des Geistes verlangten. ([1917] 1980: 126)

Because human beings belong to both realms, the history of humanity must take both into consideration. It ought to understand the configurations of the mind and the intellectual life, and find out what part of that became reality. And history can call in the natural sciences for help in order to explain what happened otherwise than the laws of the mind required, and what happened in contradiction to them.

These sentences encapsulate a theory of catastrophe. On one hand, you expect the human intellect to be the guiding force of human history and the intelligible component within it. On the other, you expect that events will not always occur *intelligently*. Yet they always will be *intelligible*, for the mindless components of history will be explainable through attribution to causal processes as understood in the natural sciences. In other words, there will be evil, but it will be owing to

---

[101]This application of your *Einfühlungslehre* to the diagnosis of illness would today be challenged on two fronts. A growing body of empirical research suggests that drugs and nutritional variations do affect events that you would have placed squarely within "the soul"; and postmodern and postcolonial critiques undermine the possibility of univocal "rationality." But I believe that your hermeneutic of empathy could not only stand its ground, but resolve some intractable issues as well.

natural events such as floods and earthquakes. There is no room in this theory for evil human purposes, and no way to understand them.

Of the deity, you had little to say. The question of God comes up as a side issue in your discussion of whether understanding requires some medium--if not the live body, then at least texts of some sort. You say that it is an open question whether intellects can communicate directly, and remark that this would be of interest to religious people.[102] Although the deity makes only cameo appearances in your dissertation, it is striking that you don't characterize God as savior, creator, or judge. You cast the deity instead as a potential *communicator*.

[102]See Stein ([1917] 1980: 131-132). Earlier in the text you have made three other passing references to the deity.

Chapter Four

Edith Stein's Hermeneutic Practices

The first three chapters of this work presented expositions and commentary on philosophical texts. The task of the present chapter is quite different. Here we will consider Stein's intellectual *practice* as an interpreter of the writings, characters, and actions of other people. Her life work, which many other scholars have studied, appears in a new light when compared with her own understanding of what it means *to understand.*[1] At some points, it will be helpful to measure Stein's interpretive practices against her own theory, or conversely to test her theory out upon her practices. Disclosure of possible corroboration or contradiction between Stein's theory and practice is an exercise of some intrinsic interest. More significant, however, is the way in which Stein's textual practice--her writing--makes sense against the background of *what she intended to be doing* in the act of interpreting.

Fortunately, we have in her 1916 dissertation an early formal statement of what Stein took to be involved in the activities of understanding and communication. Her earliest professional work, as Husserl's research assistant, required editing and expanding Husserl's texts in line with his evolving intentions. Subsequently, Stein's teaching and her writings were geared to the interpretation of two great textual traditions: German philosophy, and Catholic theology. She worked as a teacher of teachers-in-training and as a translator. These communicative activities required double-sided understanding: of the learners as well as of the traditional texts; of the German vernacular as well as of Newman's English, Thomas's Latin, or Heidegger's jargon. Stein also interpreted *herself* in her unfinished autobiography and in an extensive correspondence. She made a textual artifact of her own life, and in doing so she made more of herself accessible to us than is usual with dead German philosophers.

I will argue that Stein's phenomenology of understanding is complemented by her phenomenology of creativity. The latter is foreshadowed in her dissertation, as we saw in the last chapter, but it is worked out more fully in the essay "Psychische Kausalität," which will be examined briefly below. There, Stein extends her hermeneutic to cover more than the understanding of necessary intellectual movement occurring within other persons' *logical inferences* and the understanding of involuntarily caused *expressions* of people's physical and psychic states. Stein's hermeneutic becomes a poietic, a theory of creativity, when it also covers the noncoerced--yet rationally warranted--*personal choices* occasioned by values posited for, perceived by, and registering with persons who spontaneously give themselves to (or, withhold themselves from) inclinations in regard to those

---

[1]The rich and growing literature on Edith Stein's life and work will be reviewed in the next chapter.

values. Thus the rationality of motivation is also what is understood, but understood within in the realm of free self-determining choices. Moreover (as Stein herself perhaps did not see clearly enough), the understanding of creative choices itself involves creative choice. To in-feel another's choice is to register the other's inclination as *not* having "caused" or required this particular self-investing decision rather than some other response that would appear equally rational. The rationality of another's choice exhibits the other's motivation, but not *in the same way* that physical events exhibit the causal chains that produce them, or logical conclusions display the premises and inferences that support them. The why's of creativity are entirely unlike the why's of physical causality and of rational logic. Every act of interpretation, then, is also creative. The "copy" must differ from the "original." This will be amply illustrated in the next chapter, where we will examine the bewilderng variety of "Edith Steins" that are being created in the expanding literature inspired by her work.[2] Meanwhile, let us recall briefly the main points of Stein's phenomenology of understanding, summarizing the exposition offered in the last chapter.

Stein holds that knowledge about other human beings cannot be completely accounted for through Husserlian analysis of the "transcendental constitution of objects in consciousness." This is so because something elusive is left over after the alien body, psyche, and soul are constituted; or rather, their constitution as objects within consciousness presupposes something extra that cannot be so constituted. Therefore Stein proposes that other human beings are not "constituted" as such but are rather "felt into." Implicitly, Stein proposes that empathy (*Einfühlung*) is not only a technical alternative to constitution, but is prior to it and founds its possibility as well.

What can be understood of other human beings is *that of them that can be empathized.* To empathize (*einfühlen*) is to share a live experience from the inside: to let one's i be led along through the coherent flow of an epidode in which an alien i is living originarily. Where Husserl had allowed that only logical coherence can be perfectly followed, Stein listed at least six varieties of empathizable experiences: (1) Her paradigm was the circus performance in which the spectator inwardly accompanies the acrobat, so that their experiences coincide in everything except that one is originary while the other is led. (2) The deeds of another are "motivated" toward some purpose. Sentiments *must* discharge themselves either in deeds or in some kind of expression, such as (3) straightforward manifestation in words or gestures, (4) fantasy and wishful thinking, (5) selective perception and suppression, (6) reflection, that is, introspective brooding. To understand is to "go with the flow" of the sentiment by living through its discharge in the other human being, but non-originarily. The i flows; that is its essence. Moreover, one can understand that another has *chosen* the modes, or the combination of modes, in which sentiments discharge themselves. One can even understand that another may be dissimulating--and why. Empathized

---

[2]Obviously, I will have to admit that my own portrait of Edith Stein is a creative version of her as well. I will make no claim to have captured "the real" Edith Stein.

understanding of a sentiment--precisely as *someone's* sentiment, as *how someone is*--is something altogether different from objective knowledge of the sentiment. Inner awareness of another human being *as such* is something altogether different from recognizing her living body or her soul or character. Therefore (despite Stein's attempt to harmonize her terminology with Husserl's), an alien human being never is "transcendentally constituted" as an object within consciousness.

Live body and soul, however, are indeed constituted as objects. Objectively, they are given as *fused.* That is, causal processes obtain between physical organs and sentient fields, between psyche and character. Whatever affects one level affects all the levels. From this blending arises the *tendency to complete* one's awareness of the whole psycho-physical-ensouled individual whenever one of those levels is given. Stein's example was the hand. Physical perception of a hand-shaped thing tends to carry through into perception of *a hand* that can make things and gestures, and that can be wounded. "Blending and tending" are twin capacities exhibited in constitution. One might say that they *constitute* constitution for Stein, at least in regard to the constitution of the live human bodies around us.

A second variety of "tending and blending" was left implicit in Stein's dissertation (but would be developed in her subsequent publications, as we shall see). Free choices somehow engage with the intelligent sentient body, for they are expressed and accomplished through it. Thus besides the "blending" of causal influences across the adjacent levels of the physical, the psychic, and the soul, there is a second "blending" of another but possibly analogous order, so that a "person" is disclosed. Persons remain mysterious, even though they are in some unspecified way fused with live intelligent bodies. One person cannot understand another as such, because there is a threshold beyond which one person may not flow-along-with the other. Thus each person is irreplicable. And therefore each human body is indispensable. Here the contrast of Stein's doctrine with Husserl's is very clear. For him, the other's body is a moveable spatial center, because I can displace it with my own body and so enjoy its viewpoint for myself. This is the Husserlian foundation of science. For Stein, I do not displace the other, but rather I consent to live in a world with more than one center--and am the richer for it.

This contrast emerged from Stein's attempt to redesign Husserl's notion of the pure i along lines suggested by Scheler's notion of personal activity. Like Husserl, she held that understanding is a quasi-kinesthetic following of coherences. The i is, essentially, something that can be magnetized and pulled along the live-experiential trails blazed by other living i's. Where Husserl concerned himself with the "highways" of logical necessity, where by the inner lawfulness of i-hood itself *every i goes the same way*, Stein cared for the "byways" rendered valuable by their very difference and irreplicability. She saw that there is a lawfulness to the following of what is non-logical and non-necessary. That lawfulness is the essence of empathy (*Einfühlung*), and for Stein it was to be the basis of the human sciences. She insisted that the experience of another can register immediately within my own experience, not only in cases where we two are logically constrained to infer the same conclusion from premises, but also in cases where the only determination of the content of my experience is owing to the fact that it has originally been yours.

As indicated in the last chapter, Stein's personalization of the Husserlian pure i was not entirely successful and it retains several serious technical flaws. She did not establish how i's are individualized as persons, nor did she account for what I have termed the "second-order fusion" between the streaming life of conscious experience, on one side, and the psycho-physical causal nexus given as the individual human body, on the other. Thus her 1916 dissertation does not achieve the theoretical coherence that one would demand in a mature philosophical work. Yet even a flawed and immature work can present valuable insights and pose provocative problems. We have seen several of these in Stein's work, and we have seen that she already has the means to pursue them further. Four such issues are particularly relevant for our consideration of Stein's hermeneutic practice. Each of the following calls for expansion.

(1) Stein said that sentiments must flow along into a discharge of some kind, and she described various overt and covert modes of expression. But she gave the impression that there is a one-to-one correspondence between feeling and expression. She left no room for the possibility of a cumulative expression of accumulated feelings, or for simultaneous mixed and contradictory feelings. Her atomized account of expression is hardly adequate to the complex intermixture of causes and motivations that characterizes any empirical human life. Moreover, Stein did not take into account the possibility of malevolent reason, and she regarded *mis*understanding as an intrinsically temporary and corrigible condition.[3]

(2) We noted that Stein had an incipient *theory of reading and writing* in her doctrine of the i's intrinsic capacity inwardly to follow another i. To write would be to offer a flowing i-stream as guide and host to accommodate other i's.[4] Stein also remarked, in comparing symptoms, signals, and signs, that *words* have the greatest capacity for alienation from the originary experiences in which they issue as the expressive terminal point. Words, and the system from which they take their meanings, are shared among individual human beings. They can float free apart from the bodies that produce and consume them, so to speak.[5] Stein habitually illustrates understanding with examples drawn from intimate bodily contact. Is her hermeneutic then valid only in the realm of the face-to-face, or can it be transposed beyond body to text?

(3) Stein's doctrine of the person was a mere sketch, but several disturbing aspects of it already are clear. The term "person" names the limit beyond which the "blending and tending" of psycho-physical causality and empathic understanding cannot reach. The person is not the soul, and unlike the

---

[3]See above, chapter 3.C.3, commenting on the possibility of historical evil. See also 3.A.5, on race hatred and pornography.

[4]See above, chapter 3.A.5, where Stein was faulted for not having developed her account of i-following into a coherent theory of literacy.

[5]See above, chapter 3.B.2, on the various modes of expression. See also 3.B.3, on "how far" expressions can depart from one particular individual's expression before losing their immersion in i-hood.

soul cannot be known objectively. The person is impervious to influences from outside herself. She cannot be coerced, but neither can she be educated.[6] At most, one can assist a person in self-discovery by exposing her to a mature person "of the same type" and hoping for some disclosive resonance to occur between them.

(4) The slippery notion of "type" has alarming implications.[7] On one hand, it functions to distinguish what is particular and deviant about myself, in contrast with the typically human. What is human is designated as what is *the same* in, and so "typical" of, every human being. Type thus becomes the principle of understanding between persons. If empathy is attempted on any other basis, it leads to illusion. On the other hand, persons are of *different* types. One lives inauthentically if one has received a character training that was inappropriate to one's native type. Only our own type of person is fully open to us for empathy. Persons "of the same type" naturally understand one another through mutual empathizing of live experiences. For alien types, only objective knowledge is possible.

Our project of understanding Edith Stein is a project of understanding how her textual practices were guided by her hermeneutical theory (and also by her phenomenology of creativity, as we shall see at the conclusion of this study). But our own project is *not* an exercise in Steinian hermeneutics. In a subsequent chapter, we shall have the opportunity to reflect upon what our own interpretive/creative practices have been. For now, suffice it to say that Stein's formal hermeneutic proposal functions as the *means of production* for texts, as the expression of *desire for personal engagement*, and as the *permission for initiative* in textual variation.

A. Classifying Stein's Works

Edith Stein published several books and articles, and she left behind manuscripts of her academic lecture courses, her public lectures, some philosophical studies, and a lively correspondence. How might these best be approached? Oftentimes it is helpful to arrange a philosopher's literary output in chronological order.[8] The milestones of Stein's career are well known, and may be briefly recalled here.[9] Born on Yom Kippur in 1891, she studied with Husserl

---

[6]See above. chapter 3.C.2.

[7]See above, chapter 3.C.3.

[8]The list of references at the end of this chapter presents a chronological list of Stein's writings. A critical edition of her work is underway in the series Edith Stein's Werke, edited under the direction of the Archivum Carmelitanum Edith Stein in Brussels. The volumes are being translated into English under the sponsorship of the Carmelite Fathers as The Collected Works of Edith Stein. Herbstrith (1987 and 1991) and Imhof (1987) offer reliable lists of Stein's publications. However, one should also take into account Stein's contributions to texts that do not bear her name.

[9]The events of Stein's life are reliably recounted in her unfinished autobiography, Stein (1933), and in the narratives of Herbstrith (1987), Imhof (1987), and Koeppel

at Göttingen from the summer semester of 1913 until July 30, 1914, when lectures were cancelled at the outbreak of the First World War. She went home to Breslau and trained for nursing service, but was not immediately needed. So she went back to Göttingen in October 1914 for another semester with Husserl and took her *Staatsexamen* in January 1915. That spring she was called up by the Red Cross and she worked for several months in a soldiers' hospital in Austria. Meanwhile she studied Greek in order to remedy the deficiency of having graduated from a girls' *Gymnasium*. Back home in Breslau, she prepared her thesis on *Einfühlung* and worked as a substitute teacher. Husserl joined the faculty at Freiburg in the spring of 1916, so Stein defended the dissertation and was promoted there that summer.

Husserl then hired Stein, for a tiny monthly stipend, as his private research assistant. She worked for him from October 1916 through February 1918. She brought some order to his mountain of research manuscripts, and she took part in:

- transcribing, expanding, and arranging *Ideen II*;
- transcribing, expanding, and arranging a new introduction to phenomenology;
- revising the Sixth Logical Investigation;
- transcribing, arranging, and revising the manuscripts on time consciousness.
- editing the essays on psychology and epistemology.

She also taught the university proseminar in phenomenology. Stein had her sights set upon *Habilitation*, that is, obtaining a position as *Privatdozentin* on a university philosophy faculty. In Husserl's estimation, however, such a career was impossible for a woman. She left his employ but continued close contact with him. Stein worked now independently in Freiburg and at her mother's home in Breslau, drafting the *Habilitationsschrift* or second thesis required to win appointment as a university teacher. Her major income, as always, was an allowance from her mother, who owned and operated a lumber business in Breslau.

The Göttingen faculty declined in late 1919 even to consider the *Habilitationsschrift* of a woman. Stein then gave private lessons in phenomenology and helped her friend Hans Lipps to write his own thesis, which was accepted for *Habilitation* at Göttingen. Stein was baptized in the Catholic Church on January 1, 1922. Subsequently she took a position as an instructor in German literature and history at a women's teachers college in Speyer, and she lived with the Dominican sisters who ran it. Her translations of Newman and Aquinas were made during the nine years at Speyer, and she also began to lecture and publish essays on women's education.

In 1931 Stein left Speyer to write a second *Habilitationsschrift* but, despite Heidegger's assistance and Husserl's encouragement this time around, that

(1991). Leuven (1983) adds information from memoirs collected at the archive in Brussels. Many other authors have offered biographies of Stein as well, and these will be discussed in the next chapter.

also was rejected both at Freiburg and at Kiel. In 1932, however, she was appointed to teach philosophy at the German Institute for Pedagogical Science in Münster. During the winter of 1932-33 she worked closely with Catholic intellectuals in that city, to develop a pedagogical response to the worsening political situation. But in the spring of 1933 she was removed from her teaching post because of her Jewish ancestry. She began to compile notes for her autobiography at her mother's home in Breslau that summer, applied to the Carmelite cloister in Köln, and was admitted in the fall of 1933.

As a Carmelite nun, Stein wrote the philosophical works of her mature years. She also wrote numerous more popular hagiographical and spiritual essays. Late in 1938, after the violence of *Kristallnacht*, Stein was sent across the border to the Carmelite cloister in Echt for safety. However in August of 1942 she was arrested there and deported back across Germany to Auschwitz, where she was probably gassed and buried in a mass grave. Her story continues with many variant retellings. In 1987 she was declared "blessed" by the pope and now is commemorated in the Catholic liturgical calendar.

One obvious periodization of Stein's literary output suggests itself.[10] The two punctuation points of her baptism (1922) and her internal exile in the cloister (1933) produce three periods:

philosopher, 1916-1922:	→	**translator and popularizer, 1922-1933:**	→	**theologian, 1933-1942:**
dissertation, work for Husserl, *Jahrbuch* essays.		Newman's letters, Aquinas' *De veritate*, talks on womanhood, essays on pedagogy, autobiography.		*Endliches und ewiges Sein*, *Kreuzeswissenschaft*, studies on saints.

These categories fit the textual history fairly well.[11] Moreover, they have the additional advantage of illustrating both the continuity and the gradual shift of Stein's interest toward philosophical theology and spirituality.

Nevertheless, I wish to employ a different classification device that seems better suited to the task of understanding Stein's work on her own terms. I will classify her texts according to the salience of the writer's subjectivity within them. On one hand, Stein is invisible in her work for Husserl and she tries to be so in

[10]For an early proposal of the chronological arrangement of Stein's work, see Gelber (1955: 466). Gelber's chart juxtaposes the events of Stein's life, her intellectual interests, and the major influences upon her development.

[11]With this scheme, there would be a few loose ends, however. For example, Stein would already be into her post-baptismal "popularizer" phase when she published her first comparison of Husserl and Aquinas in the 1929 *Jahrbuch*, an essay which should be characterized as philosophical. Also philosophical were her lecture courses at Münster in 1932-33 on Introduction to Philosophy and Philosophical Anthropology. Moreover her correspondence dances across the chronological and career categories.

her translations. On the other hand, in the autobiography she foregrounds herself explicitly. Somewhere in between lie the studies that she publishes under her own name but which purport to treat various philosophical or religious topics "objectively." The author's "i" is offered in a different way in each of these tactics; thus the texts invite the empathizing reader to "ride along" through their coherent moves on varying terms. This distinction suggests a continuum of classification as follows:

anonymous → →	**← ← autographed → →**	**← ← autobiographical**
Stein tries to be invisible. The work of her hands hides the work of her hands.	*Stein signs her work but presents it as objectively available for anyone else to understand.*	*Stein visibly constructs herself and her "type." She works on the problem of her identity.*
work for Husserl, rewriting Reinach, translations, editing Ingarden, helping Lipps, teaching teachers, possible anonymous challenges to Nazi ideology?	**scientific [philosophy]:** dissertation, *Jahrbuch* essays, book reviews, lectures in Münster. **historical [theology]:** *Endliches und ewiges Sein*, *Kreuzeswissenschaft*, religious essays.	**self portraiture:** autobiography, letters. **other type productions:** essays on women, essays on pedagogy, remarks on Judaism.

Perhaps it needs saying that the writer is still "there" when she is invisible. Effacement and disappearance are textual practices. Stein's creative contributions may be greatest precisely where they are most vehemently denied. (Paradoxically, as we shall see, there may be less of *Stein* where there is most of "Stein," the textually wrought artifact.)

In the rest of this chapter, my intention is to argue that this way of arranging Stein's literary works will enable a very fruitful scholarly appreciation of her contributions to philosophy and to theology. It is beyond the scope of my study, however, actually to *carry out* that interpretive work here. I merely suggest a beginning. The previous chapters have presented thorough commentary, but here I can only indicate the direction that future research should take. My suggestions in this chapter compare favorably, I believe, with the trends evident in contemporary Stein scholarship, which I will survey in the next chapter. It remains now to flesh out the three voices in which Stein's textual practices speak.

B. Anonymous Textual Production

When one writes without signing one's name, the text is "anonymous": unnamed, without authorization. Or rather, as is the case in Stein's earliest professional work, the text may take the name and authority of someone other than its producer. Edith Stein wrote anonymously for Husserl, Reinach, Heidegger, Hans Lipps, Roman Ingarden, John Henry Newman, Thomas Aquinas, and, it may be, eventually also for the Catholic intellectual community of resistance in the Diocese of Münster. As a young philosopher, she regarded this alienation of her intellectual work with wry amusement. "Anonym" was her nickname among her close circle of friends. She customarily signed herself this way in the guest book at Bergzabern when she visited the orchard home of Hedwig Conrad-Martius and Theodor Conrad.[12]

Phenomenologically, we may say that the anonymous writer purports to write as if she were merely reading. (She pretends that) someone else's i has "originarily" lived the experiences that are coming to expression on the paper in the act of writing, while her own i is merely following along "nonoriginarily" when it writes. She flows with and communicates the live experiences, but as if they were not her own. She sides with her readers. The emergent expressions supposedly owe nothing to her, but are entirely determined by the original meaning-enactments of someone else. Husserl's expectation was that Stein would work in this manner.[13] Later, in translating Newman she would try to remain invisible in the text she produced.[14]

Anonymity, as a strategy of writing, does not succeed in its avowed purpose. The writer (like the reader) of course cannot help creating a version of the material in which her own distinctive subjectivity is a decisive ingredient. Nevertheless, the strategy has its uses. Stein employed it to get access to the thought of Husserl for herself and to insure that it would be available for others. She also used it to further the academic careers of her friends. One must ask whether Stein allowed herself to be duped and taken advantage of, to be used and used up for the projects of other people. Her Catholic mentors certainly used her in the cause of reviving Thomism in their defensive campaign against the secular philosophy of the nineteenth and early twentieth centuries. Let us not forget, however, that anonymity was also an effective strategy of political resistance. When creative subjectivity effaces itself, one must investigate the creative, effective, motivated dimensions of that very self-effacement. Why did she hide; what did she accomplish by hiding? This question confronts future Stein research in the following areas.

---

[12]As Sister Maria Amata Neyer told me when I visited the Edith Stein Archiv at Cologne in July 1993.

[13]Exegetes of Husserl's texts subsequently have accused Stein of incompetence or malpractice when they find that important elements in Husserl's published works stem from Stein. See Ingarden (1962) for an example, answered by Ingarden's defense of Stein's creative practices.

[14]See Koeppel (1993: vii), which cites a comment attributed to Stein. Posselt (1952: 14) gives the unacknowledged source of the remark: "A translator must be like a pane of glass, which lets all the light through but is not seen itself." A woman who had been a schoolmate of Stein's in her teens attributed this comment to her young friend.

1. Stein's work for Husserl

Husserl employed Stein as his assistant for seventeen months, from October 1916 through February 1918, and thereafter she continued in close contact with him on several projects.[15] Husserl's own peculiar habits of making philosophy are well known.[16] He thought by writing, and by 1916 he had accumulated a bewildering collection of research manuscripts, lecture notes, and annotations to his earlier publications. Stein was commissioned to bring order into the chaotic manuscripts.[17] She was to translate Husserl's shorthand script into longhand, which then could be typed up by herself or someone else. This task of "working out" Husserl's papers also entailed choosing among manuscripts, placing them in order, expanding sentences that were too abrupt, composing introductions and bridge paragraphs to enhance continuity, separating into paragraphs and sections, arranging and titling the sections, and drafting the overall plan for the publishable form of a work. Husserl was supposed to review and correct what Stein compiled. In fact, she could seldom persuade him to do so. He would rather write something new than edit something old.

Therefore Stein feared that the phenomenological discoveries expressed in Husserl's manuscripts might be lost if not brought to publication. She took on the role of nagging Husserl to get on with the work of publishing, particularly in regard to the second book of the *Ideen* which had been promised in 1913 in the first book. (Hiring Stein to work on *Ideen II* was Husserl's way of avoiding the work that he would have to do to resolve difficulties raised by critics of the 1913 volume.) Stein felt a responsibility to protect Husserl's phenomenology from being lost through his own procrastination and neglect. But she also expected to have a two-way collaboration with Husserl. She hoped he would attend to her substantive suggestions concerning his thought, and would also discuss her own constructive work with her. That did not happen, although she reports in her correspondence that she had many stimulating exchanges with Husserl concerning his own interests. The incident that precipitated Stein's resignation seems to have been a clumsy initiative on Husserl's part to mechanize their collaboration by imposing upon his assistant a written formula for the processing of the manuscripts. She told *den Meister* to follow it himself.

According to Stein's correspondence with Roman Ingarden, the key substantive issue that had Husserl stalled out was the nature of "constitution." This

---

[15]Stein recounts the details of her hiring in a letter of 16 August 1916 to Fritz Kaufman and in her autobiography. (See Stein 1986: 410-11). She recounts her resignation in letters of 19 and 28 February 1918 to Roman Ingarden and 10 March 1918 to Fritz Kaufman.

[16]For example, see Ingarden (1962) and the various editorial introductions to the volumes of Husserliana. Compare Landgrebe's description of his own work on Husserl's manuscripts, as given in the introduction to Husserl's *Erfahrung und Urteil.*

[17]Leuven (1983: 8-9) claims that the Husserl Archives contain 57 significant manuscripts of Husserl's in Stein's handwriting. He says that he counted up 9,669 sheets of her writings for *Ideen II* alone.

problem was tied into the "transcendentalism" introduced in the *Ideen*, which many of Husserl's readers took to be an unfounded rejection of "realism" and a clear departure from the 1901 epistemological statement in the Sixth Logical Investigation.[18] In fact, in 1917 Husserl himself had not yet settled upon a definitive account of constitution. This was a major headache for his assistant. The record of their discussions in her correspondence shows that Husserl could not make up his mind how to formulate the matter. She wrote to Ingarden on 12 January 1917 that she had convinced Husserl to take another look at his doctrine of constitution before proceeding with the publication of the second book of the *Ideen*.[19]

Es war überhaupt ein aufregendes Gespräch. Ich habe sehr abschreckend die Mühsal geschildert, die man bei der Verarbeitung des Rohmaterials der Ideen hat, so daß man sich noch einmal durchaus einverstanden erklärte, mir dieses Vergnügen zu überlassen. Als ich dann erzählte, auf welche Schwierigkeiten ich gestoßen sei, tauchte plötzlich der nicht unberechtigte Gedanke auf, man müßte eigentlich die ganze Lehre von der Konstitution noch einmal überlegen und zu diesem Zwecke den 1. Teil der Ideen wieder ansehen. 2. Tage ist das geschehen, dan war's wieder zu langweilig. Ich habe mir jetzt vorgenommen, unabhängig von den wechselnden Einfällen des lieben Meisters und so schnell es die Nebenaufträge, die ich erhalte, gestatten, das Material, das ich da habe, in eine Form zubringen, die es auch andern zugänglich macht. Wenn ich so weit bin und er sich dann immer noch nicht entschlossen hat, die Arbeit systematisch anzugreifen, dann werde ich auf eigene Faust versuchen, die dunkeln Punkte aufzuklären. (ESW XIV: 28-29)

[I]t was an exciting discussion. I painted so drastic a picture of the trouble involved in processing the raw material of the *Ideen* that, as before, the result was that we totally agreed once again that I should be allowed to have all the pleasure. Then when I told him about the difficulties I had encountered, the not-unjustified thought occurred that one should actually reconsider the entire doctrine of constitution and for that purpose take another look at the first part of the *Ideen*. That was done for two days, then it again became too boring. I am now determined that, independent of the dear Master's sudden and variable fancies and as speedily as other assignments will permit, I will put the material I have into a form in which it will be available to others also. When I have accomplished that, and if by then he is still not resolved to go through it systematically, then *I will attempt on my own to clarify the cloudy points.* (CWES V: 3-4, emphasis added)

---

[18]For example, in late July, 1918, Roman Ingarden sent Husserl a long letter urging him to republish the 1901 version of the Sixth Investigation, accompanied by extensive notations concerning the problems brought to light by the *Ideen*. Ingarden had been kept apprised of Husserl's stuggle with this problem through correspondence with Stein.

[19]These comments come as an addition to a letter begun on 5 January. Four collections of Edith Stein's letters have appeared, and the contents overlap. Hedwig Conrad-Martius published Stein's letters to her in 1960. Volumes 8 and 9 of the series Edith Steins Werke present letters from 1916 to 1934 and 1934 to 1942, respectively. Volume 14 of the Werke offers a fuller selection of Stein's letters to Roman Ingarden, many of which deal with matters of philosophical interest. I have cited the letters by date from the appropriate volume of the Werke. Translations are taken from volume 5 of the Collected Workd of Edith Stein, which came out in 1993 and combines the contents of volumes 8 and 9 of the Werke.

(As we saw in the last chapter, Husserl did in fact draft a new account of constitution in late January 1917; it appeared as Supplement 12.1 of *Ideen II* in 1952.) On 28 January Stein wrote Ingarden that Husserl was still working on constitution:

Er beschäftigt sich jetzt mit Konstitution der Natur (natürlich ohne Berücksichtigung der Ausarbeitung), ich bin indessen mit der Fortführung der Ausarbeitung, zu der ich mich selbst autorisiert habe, ohne auf Widerspruch zu stoßen, bis zur Person vorgedrungen. Natürliche Folge ist, daß man kaum noch miteinander redet. Für mich ist das sehr schmerzlich, denn die Sachen sind sehr kompliziert und das Material, das mir vorliegt, ist höchst unvollständig. Ich arbeite also jetzt ziemlich selbständig, und das ist ja recht erfreulich, aber etwas Gedankenaustausch wäre dabei sehr ersprießlich. Ich könnte Sie also jetzt gut brauchen. . . . (ESW XIV: 32-33)	He is now busying himself with the constitution of nature (of course, without any review of the draft). In the meantime, I have continued working on that draft on my own authority, without running into any opposition about that, and am as far as "Person." The natural consequence of this is that we hardly talk together anymore. For me, this is very painful, for matters are very complicated and the material I have at hand is altogether incomplete. However, I am working pretty independently now, and that is, of course, very enjoyable, but some exchange of ideas could be highly beneficial. So I would have a great deal of use for you now. . . . (CWES V: 6-7)

Stein wrote on 3 February that she felt "co-responsible" (*mitverantwortlich*) for bringing the *Ideen* to publication. She told Ingarden:

Über die Art meiner Arbeit an den Ideen sind Sie sich wohl doch noch nicht ganz klar: vorläufig suche ich nur, aus den Materialien den gesamten Gedankengang (der mir ziemlich klar vor Augen steht, aber nirgends fixiert oder gar durchgeführt ist) in einer einheitlichen Ausarbeitung festzulegen. Das soll die Basis für die Arbeit des Meisters sein, und das möchte ich gern beenden, weil ich glaube, daß er sich durch die Materialien selbst nie hindurch finden und immer in Einzelheiten stecken bleiben würde. Nur für den schlimmsten Fall, daß er gar nicht an die Überarbeitung heranginge, hatte ich es ins Auge gefaßt, es selbst zu tun. Natürlich bin ich dann auf jahrlange Arbeit gefaßt. Aber ich hoffe ja, daß das nicht nötig wird. (ESW XIV: 36)	It seems you still have no clear understanding about the nature of my work on the *Ideen*: I am now seeking to establish from the material at hand a unified draft of the entire thought process (of which I have a pretty clear view even though nothing is fixed or even carried to a conclusion). That is to become the basis of the Master's work, so I would like to finish it because I believe he would never be able to find his way through the material and would forever remain hung up on particulars. Only if [worse] came to worst, and he were never able to rework it himself, had I considered doing it myself. Naturally, that would mean years of work for me. But I do hope that will not become necessary. (CWES V: 8)

She added that she had achieved her own breakthrough to a non-idealistic conception of this constitution.[20] On July 6 of the same year Stein wrote Ingarden

[20]After tracing this correspondence myself, I was pleased to find it reviewed in Müller (1993: 76-82), who makes the point that Stein's discussions with Husserl at this time indicate the strength and substance of her disagreement with him.

FIGURE 4.1: COMPOSITION HISTORY OF THE IDEEN

DATE	CONTENTS OF THE MANUSCRIPTS, AS THEY DEVELOP		
drafted 1912	phenomenology: what it is and what it does +	how phenomenology works: analysis of the constitution of matter, sentience, soul, culture +	how the sciences of all these things arise
1913	↓ Published as *Ideen I.*	↓ Husserl adds new material to the manuscript: how the cultural world is constituted.	↓
1915		↓ Husserl rewrites: expands his treatment of the body-soul connection in what would become §§29, 31, additions to §32, and additions to §34. He deletes nothing.	↓
1916		↓ Edith Stein becomes Husserl's assistant and transcribes his shorthand manuscript. (She had just analyzed the constitution of other people in her dissertation.)	↓
1917		↓ Stein cuts and pastes: compiles §18, recasts §§35-42 [now chapter 3 of section 2]; composes §§43-47 [now chapter 4 of section 2]. She arranges *Ideen II* into three sections: "Constitution of Material Nature" "Constitution of Animate Nature" "Constitution of the Cultural World" (Husserl marks up only sec 3, his 1913 addition.)	↓ Stein sets aside the concluding material on the theory of science, because of space limitations. ↓ (Husserl marks up; this material is unchanged from 1912.)
1924-5		↓ Landgrebe revises: at Husserl's direction, he restores some texts that were deleted by Stein but makes them "supplements."	↓ Landgrebe revises: he restores some of Stein's outtakes from the 1912 draft of *Ideen II* material, making them supplements to *Ideen III*.
		↓ (Husserl marks up both manuscripts, but declines to publish them. He dies in 1938.)	↓
1940's		↓ Editorial work at the Husserl Archive in Leuven. Changes include relegating the opening portion of *Ideen III* to a "supplement" in view of Husserl's criticisms in the margins of the manuscript.	↓
1952		↓ Published as *Ideen II.*	↓ Published as *Ideen III.*

that she had just dug up a parcel of manuscripts on time-consciousness that would have important bearing upon the constitution problem. As we saw in the last chapter, already in her 1916 dissertation Stein had in effect set constitution in opposition to empathy (*Einfühlung*). Thus if the delineation of constitution was not exactly "up for grabs" in 1917, still it was something of a loose canon on Husserl's deck. While he dodged, Stein struggled to get a line around the troublesome notion and secure it within the system of Husserlian phenomenology. That is the immediate background for the particular projects in which Stein took a hand.

(a) Revision of *Ideen I* ? Contrary to a notation by the translator of her letters, Stein did not produce a revision of the first book of the *Ideen*.[21] But she did discuss it with Husserl for two days before he became bored and that particular project was shelved.[22]

(b) Composition of *Ideen II*. Stein worked on the manuscripts for the second book of *Ideen* for five months, from mid-October 1916 through late February 1917. The nucleus of the material was a folio of 84 sheets written by Husserl in 1912 immediately after the composition of the first book. That manuscript had two parts, the second of which would be published posthumously without revision as *Ideen III*. The first part of the 1912 manuscript--treating constitution--had been reworked and expanded in 1915 by Husserl himself, who added material from his lecture courses during the intervening years (which Stein had attended). The first phase of Stein's work was to make a unified longhand copy of Husserl's 1915 shorthand version of the constitution analyses.[23] This was the "nature" half of the work, and it was to lead into a 1913 longhand manuscript by Husserl treating "world." The second phase of Stein's compositional work was to integrate Husserl's more recent thought into that shell. This is schematized in Figure 4.1. Stein's rewrite doubled the size of the manuscript, arranging it into the three major sections that would be retained for its eventual posthumous publication:

> Constitution of Material Nature (now §1 to §18),
> Constitution of Animal Nature (now §19 to §47), and
> Constitution of the Intellectual World (now §48 to §64).

The new texts produced at this time were the following:

---

[21]An editors' note to Stein's letter of 18 January 1917 to Ingarden remarks, ambiguously, that Stein prepared Part I of *Ideen* for publication. Koeppel (1993: 7) adds: "Her work on the manuscript resulted in the second revised edition, published in 1922, although she is not given credit." This seems to be incorrect. All three editions of the first book of the *Ideen* published during Husserl's lifetime were virtually identical. Nevertheless, during the time he worked with Stein Husserl made some significant annotations to some sections in his own copies of the work, including a section mentioning empathy (*Einfühlung*). See Kersten (1982) and Schuhmann (1976).

[22]According to her letter to Ingarden of 5 January 1917.

[23]She also copied out the second part of the 1912 manuscript, to be designated *Ideen III*. For details, see Rojcewicz and Schuwer (1989: xi-xiii), Biemel (1952: xvi-xv11), Kern (1973), and Stein's letters dated from January to August 1917.

§18, on the live body as basis for constituting material things.
(This formed a hinge between sections one and two. To construct it, Stein selected passages from the end of the work and recast them as parts a - e and g - h. In this way she made a frame for part f, which stands in its original place.)

§25 (part), on the polarity of the i in relation to its objects.

§33 (second half), on the concept of reality.

§35-42, on the constitution of soul-reality through the body.
(Here, Stein wrote continuity for the out-takes of Husserl's 1912 draft left over after portions had been moved up into §18.)

§43-47, on the constitution of soul-reality in empathy.
(Apparently Stein herself composed this, the hinge chapter between "nature" and "culture.")

Unspecified expansions in the third section of the book.[24]

Stein then tried repeatedly to get Husserl to examine what she had compiled. She wrote to Ingarden on 20 February 1917 that she believed she had succeded in integrating "*Natur und Geist*" in the manuscript; that is, in making the constitution of nature coherent with the constitution of the intellectual world. She pronounced the work ready for publication--if only Husserl would look it over. No such luck; Husserl put it off. On April 9 she wrote that she had had the chance to read and discuss the manuscript with a friend, and thought that the first two sections could appear as they were in the *Jahrbuch*. In fact, she never managed to bring Husserl around to review the manuscript.

Commentators have assumed that Stein produced her additions to *Ideen II* by consulting Husserl's manuscripts, and that he ultimately approved the work. This assumption of Husserlian authorship for *Ideen II*, with the mere assistance of a scribe, now should be reconsidered. Three pieces of evidence call it into question. (1) According to her correspondence, Stein took matters into her own hands after failing to move Husserl toward producing a coherent account of constitution, because she saw the need for reconciling the 1917 cutting edges of his thought with his earlier formulations in the Sixth Logical Investigation and in the first book of the *Ideen*. In her February 3 letter to Ingarden she wrote that after a discussion with Husserl:

hat sich im Anschluß daran ganz plötzlich bei mir ein Durchbruch vollzogen, wonach	I have experienced a breakthrough. Now I imagine I know pretty well what

[24]This much can be determined from editorial introductions and critical notes to the Husserliana editions, particularly those of Marly Biemel. Biemel dates Stein's second write-through of *Ideen II* "about 1918." In fact, correspondence with Ingarden indicates that the manuscript was completed the previous year. To identify Stein's contributions more exactly, one should examine the manuscripts in the Husserl Archive in Leuven.

ich mir einbilde, so ziemlich zu wissen, was Konstitution ist -- aber unter Bruch mit dem Idealismus! Eine absolut existierende physikalische Natur einerseits, eine Subjektivität bestimmter Struktur anderseits scheinen mir voraudgesetzt, damit sich eine anschauliche Natur konstituieren kann. Ich bin noch nicht dazu gekommen, dem Meister diese Ketzerei zu beichten. (ESW XIV: 36-37)

"constitution" is--but with a break from Idealism. An absolutely existing physical nature on one hand, a distinctly structured subjectivity on the other, seem to me to be prerequisites before an intuiting nature can constitute itself. I have not yet had the chance to confess my heresy to the Master. (CWES V: 8)

If this heresy was recognized as such and confessed, still it was neither repented nor erased. Stein writes on February 20 that after a two-hour debate with her Husserl had agreed to think through his position again, but would not do so immediately.

(2) Stein disagreed sharply with Husserl about matters treated in her dissertation.[25] She mentions in a letter of 20 March 1917 that she has

begonnen mich mit einem der Differenzpunkte zwischen dem Meister und mir (Notwendigkeit eines Leibes für die Einfühlung) etwas näher zu beschaftigen. (ESW XIV: 47)

begun to examine more closely one of the points on which the Master and I differ (the necessity of a body for empathy). (CWES V: 13)

As we have seen, Stein regarded the body as a prerequisite for "trading places" and thus as a prerequisite for constitution. Husserl instead wanted to make constitution a prerequisite for individual embodiment.

(3) Comparison of the main text of *Ideen II* with the passages that Stein excised--now published as "supplements" to *Ideen II* and *III*--indicates that Stein re-engineered the work to support her own theses: that the body and empathy are presuppositions for constitution. She accomplished this both by composing new sections, and by rearranging Husserl's manuscripts to conceal how he had begged the question of *other people*, other i's. Between material nature and culture now the mediating factor isn't *rational logic* (as in the 1911 *Logos* article), but an aspect of nature itself: the psychic, that is, *the sensitivity of the living body.* This approach matches that proposed in work already published by Stein. (In chapter two, above, I termed this "the first solution" to the problem of establishing one foundation for the natural and cultural sciences.)

Curiously, Husserl seems to have accepted Stein's amendment of his project, even though he embraced "the second solution" in a manuscript from January 1917.[26] On the manuscript that Stein produced for him, Husserl left her alterations untouched.[27] He again allowed them to stand (with extensive

---

[25]Indeed, she complained to Ingarden that Husserl never even told her what he thought of it. See her letter of 27 April 1917.

[26]See above, chapter 2.D.

[27]He marked up only the third section, where Stein's changes were minimal; see Biemel (1952: xviii) Stein told Ingarden in her letter of 9 April 1917 that that section still

comments, in places) when he reviewed the work in the mid 1920's after it was copied by Ludwig Landgrebe. At that time, however, Husserl directed that Stein's out-takes from his 1912 draft be restored--but as "supplements" so as to avoid interrupting the argumentation arranged by Stein. Ultimately, however, Husserl declined to release this work for publication. Figure 4.2 summarizes the two solutions that are uneasily juxtaposed in the posthumously published text of *Ideen II*; these were discussed at length in an earlier chapter.

In passages composed in 1912 and 1913, Husserl uses the lone "pure i" as starting point, and imports the plurality of i's as an afterthought--an afterthought that turns out to be quite indispensable. In fact, this same move is made in §18/f of *Ideen II*. But there, the shock of the circular argument is cushioned because Stein has nested that passage into a pastiche of texts torn from lower down in Husserl's draft manuscript, all intimating the priority of the body (§18/b - e and g - h). Woven together, with continuity supplied by Stein, this little nest comprises a case against the lone i. No such case was made in Husserl's 1912 draft; but now this passage concludes the first section--"material nature"--and makes the transition into the second section--animal nature. As one can easily see, this matches the tack taken in Stein's own 1916 doctoral dissertation. But it contradicts the arguments in Husserl's two earlier texts.

Where §18 lists the reasons why constitution cannot be founded in the lone "pure i," §§35-42 make the positive case for starting with the body instead. This chapter in the original draft was the source of the strands that Stein extracted for weaving §18. Here, Stein stitched up the holes left by her extractions and threw in some as-we-saw-earlier's (e.g., §38).

Stein's most substantive contribution to *Ideen II*, I believe, is §§43-47, the chapter on "Constitution of Soul-Reality in Empathy" that concludes the consideration of psychic nature and serves as a bridge from the psychological into the cultural.[28] This hinge chapter asserts that the world of the natural sciences is constituted through empathy. Its argument runs like this: In my physical surroundings I sometimes come upon things that are of the same type as my own live body. Apprehending these as living bodies, I feel within them an i-subject. Yet for that to happen, my empathizing must *already* have transferred to those other bodies the same sort of localization of sensory fields and of intellectual activity that I have with my own body. The i-hood that is presented to me in my own body has already gone over to the other's through in-feeing (§45).

Left to myself, the argument continues, it wouldn't occur to me to invest my subjectivity in my body. I become a live spatial center reciprocally by trading places with other live bodies. Thus the spatial orientations of the world, arising

had to be overhauled.

[28]See above, chapter 2.D.1.b, for a fuller discussion of this passage. As Biemel says in her introduction to *Ideen II* (p. xviii), there is no connected manuscript of Husserl's for §§43-47. By contrast, Biemel's critical notes for *Ideen III* (pp. 115, 119-120, 122-123, and 164) indicate that §18 and §§35-42 of *Ideen II* were composed by cutting and pasting Husserl's 1912 manuscript.

FIGURE 4.2: HOW TO FOUND THE SCIENCES

First Solution

Priority of bodily life.
Bodily life is felt in ablative empathy and is available to accusative empathy.

Empathized bodily life is prerequisite for:

- the plurality of living bodies;
- the location of i's in them;
- the individuation of those i's;
- investment of my i in a body;
- the spatio-temporal structure of the natural world;
- the motivational structure of cultural expressions;
- the possibility of registering worldly or personal influences; and
- constitution of objects.

•

This solution is proposed in Stein's 1916 dissertation and in the following portions of Husserl's *Ideen II* prepared by her in 1917:

§18/b-e, g-h: "The subjectively conditioned factors of thing-constitution and the constitution of the objective material thing."

§§43-47: "The constitution of soul-reality in empathy."

§49: "The personalistic orientation in contrast to the naturalistic."

§51: "The person in personal associations."

§54-56: "Motivation as the basic lawfulness of intellectual living."

Second Solution

Priority of object constitution, as carried out by transcendental subjectivity: a lone i.

Constitution is logically prior to what is constituted, including:

- ideal essential laws;
- the spatio-temporal world with its causal relations;
- my own live body;
- physical things;
- other bodies that resemble mine;
- the inner world, with temporally regulated streams of consciousness pooling up into discrete unities; and
- the possibility of other i's who would constitute all this just as I do.

•

This solution is the one favored by Husserl from 1917 on. It is found in the following passages. Some of these were displaced within or from the main text of *Ideen II* by Stein; others were notes added to her work by Husserl in 1917 or 1924-25:

§18/f (from solipsism to intersubjectivity)

§53 (a late amendment concerning nature and intellect).

§§30 & 34 (distinguishing soul from pure i and transcendental i).

§§1 & 2 of Sup. 1 of *Ideen III* (original introduction to the last section of *Ideen II*).

Sup. 1 of *Ideen II* : "A stepwise description of the constitution of the mere material thing, of the body, of the ensouled thing or live being, then of the personal subject, etc."

Sup. 12.2 of *Ideen II* (a methodological manifesto: the foundation of the sciences).

Sup. 12.1 of *Ideen II* plus fragment in critical notes (explicit rejection of solution #1, penned in Jan 1917).

from the embodied character of subjectivity and required for natural science, cannot be owing to an isolated "pure" subject. Bodies must be plural for spatial order to be established. The mobility of living bodies allows them to exchange their "here's" and "there's." Therefore the spatio-temporal arena through which the alien i moves is not a new or alien realm for me. It is my same old realm: the *one objective world*, but now seen to be such. "Otherness" is enforced in it only by this law: two people may perceive the same appearances, but only at different times; and conversely, if two people share a time, they must do so from different places and therefore with access to differing appearances.

Nevertheless on those terms we now know how it is that two people can perceive one identical object (as they *must* do, for science to be possible). Thus it is empathy (*Einfühlung*) that leads to the constitution of "intersubjective objectivity" for things in general, and in particular for the living thing that is human being. Nature is constituted as the field of investigation for the natural sciences, in which particular natural things can be reliably investigated.

In *Ideen III*, Husserl stops short of attempting to establish the foundation for the cultural sciences, the *Geisteswissenschaften*. Stein sought to do this herself in her (failed) *Habilitationsschrift* of 1919, published in Husserl's *Jahrbuch* in 1922. Her formulations there ought to be compared with those in Husserl 1987--the essays on psychology and epistemology, which she had worked on in 1917--and with Husserl's 1925 lectures on psychology, published in 1962. Stein clearly saw her own work as completing the project of Husserl's *Ideen*: establishing the unified methodological foundation for the two kinds of sciences, natural and cultural. Stein's two-part "Beiträge zur philosophischen Begründung der Psychologie und der Geisteswissenschaften" may be read, in effect, as *Ideen IV*.

(c) A new introduction to phenomenology. While Stein was concluding her work on *Ideen II*, Husserl was preparing his inaugural lecture at Freiburg, which was to be delivered on 3 May 1917. However he got carried away with the task. Instead of a lecture, he produced about 80 pages of notes that would become a pair of essays: "Phänomenologie und Psychologie" and "Phänomenologie und Erkenntnistheorie."[29] (At the last minute Husserl did manage to write the shorter piece that served as the inaugural lecture.[30])

Husserl was apparently quite pleased with this material, and he proposed to publish it in his *Jahrbuch*. But Edith Stein and her friend Erika Gothe persuaded Husserl not to do so. They told him that readers expected to find new research in the *Jahrbuch*, rather than new presentations of established positions.[31] Stein proposed a different use for the essays. Husserl had earlier promised the editor of the journal *Kant-Studien* that he would write a response to articles that had appeared there and elsewhere offering critical evaluations of his own work. Stein therefore proposed to Husserl that he give *Kant-Studien* the essays on

---

[29]These essays were not published until 1987 in Husserliana XXV.

[30]See Nenon and Sepp (1987: xvii).

[31]See Stein's letter to Roman Ingarden of 9 April 1917.

phenomenology and psychology and on phenomenology and epistemology, which together were large enough to appear as a supplement. The direct responses to his critics--Theodor Elsenhans, August Messer, and Heinrich Gustav Steinmann--would then appear as appendices to the main text of the essays.

Husserl agreed to this plan. Stein transcribed the two long essays, working the text of the inaugural lecture into the essay on phenomenology and psychology. Husserl assigned Stein herself to write the appendices in which the critics' objections would be addressed.[32] Between 1918 and 1920, the *Kant-Studien* announced three times that a supplementary volume on "the essence of phenomenology" by Husserl soon would appear. It is not known why the essays ultimately were withheld from publication. Perhaps Husserl was daunted by Stein's and Gothe's arguments that they merely presented old work, not new research.[33]

(d) The Sixth Logical Investigation. When Stein handed over to Husserl her "worked out" version of the two essays, he set about producing clarifications of several difficult issues that were highlighted in them.[34] Meanwhile, he gave Stein the papers in which he had begun to revise the Sixth Logical Investigation. The manuscripts were in a disorderly and unfinished condition. Stein seems not to have made substantive suggestions for bringing them into line with the transcendentalism of the *Ideen*. However she thought that the more finished portions should be published quickly in pieces for the *Jahrbuch* or perhaps in a *Festschrift* for Husserl's 60th birthday. That did not occur.[35]

(e) Teaching the proseminar in phenomenology. In the summer semester of 1917, Edith Stein offered a regular course of instruction to beginnining students in Husserl's own lecture hall.[36] He told her that she would be to him in Freiburg what the *Privatdozent* Reinach had been in Göttingen. Stein's letter of 20 February 1917 to Ingarden indicates that she first began instructing students informally during the preceding semester, and her success then led to a formalized arrangement. But she points out the "touching naïveté" of promising her Reinach's work while withholding any possibility of *Privatdozent* status. Stein would be permitted to teach Husserl's students in Husserl's name, *but never in her own.*

---

[32]See Nenon and Sepp (1987: xvii-xxviii).

[33]The editors of the critical edition of these manuscripts, Nenon and Sepp, suggest this possibility. They add that they find no indication that Husserl was dissatisfied with the quality of Stein's transcription and expansion of the essays. (See 1987: xx). Husserl may also have been distracted by the illness of his mother in the early summer of 1917, the loss of his son in the war, and his own nervous illness in September of the same year.

[34]He produced the manuscripts now appearing as *Beilagen II-V* in the critical edition of the essays. See Nenon and Sepp (1987: xx).

[35]Panzer (1984: lix-lxv) indicates that the extant manuscripts in which an overhaul of the Sixth Investigation is attempted probably date from 1901-06 and 1910-11. The second edition of the Sixth Investigation came out in 1921, largely unchanged, but with a new section appended, on "outer and inner perception: physical and psychic phenomena."

[36]See her letters to Ingarden of 20 February, 31 May, and 7 August 1917.

Twenty-five years later when the director of the Husserl Archive, H. Van Breda, visited Stein in the cloister, she told him that Husserl did not wish to have her at his side as *Dozentin*, that is, as a junior faculty colleague.[37]

(f) The manuscripts on time consciousness. In the context of her persistent arguments with Husserl over constitution, Stein had cited him against himself, calling his attention to the relevance of some of his own earlier but forgotten lectures. On 6 July 1917 she wrote Ingarden that she had just dug up a parcel of manuscripts on time-consciousness that would have important bearing upon the constitution problem. She said that although some of those materials were rather old and on the whole they were unfinished and very disorganized, nevertheless she intended to make them into a publishable monograph. She spent the month of July at that task, and on September 8 could report happily to Ingarden that Husserl himself had been working with her for three days. Stein sorted and sifted the manuscripts, which had been composed and revised over the course of many years by Husserl, and she harmonized their vocabulary.[38] The question of time-consciousness provided Husserl with the new direction in which he would pursue him phenomenology of constitution. Yet Husserl did not immediately proceed with publication of this work either. Much later, Husserl let Martin Heidegger take a look at the manuscript. Heidegger went over it, and in 1928 he published Stein's 1917 draft in the *Jahrbuch*, attaching his own name to it as editor. Accounts differ as to how this came about. Heidegger claimed that Husserl showed him the manuscript in 1926 during spring vacation.[39] But Ingarden says Husserl kept the work locked up until October 1927 when he invited Ingarden himself to undertake the final round of expanding and completing it. (This could not have happened, of course, were the draft already in Heidegger's hands on its way to publication.) Ingarden says that he declined Husserl's invitation, but told Heidegger about the manuscript, and Heidegger then asked

---

[37]See Leuven (1983: 22).

[38]See Brough (1991: xii-xvii) for a reconstruction of how Stein reworked the manuscripts. Her standardization of the vocabulary, designed to produce a unified monograph, has been criticized because it homogenizes Husserl's developing thought, juxtaposes passages from different contexts, and seems to force the author into contradictions in some places. This criticism comes from readers with a historical interest in Husserl, rather than a phenomenological interest in the understanding of time. As Brough explains, Stein intended "to put her selections from the lectures of 1905 into a form compatible with the results Husserl had reached from about 1909 to 1911." More precisely, I believe, she geared the time-consciousness manuscripts to complement *her own work* on constitution in §§35-47 of *Ideen II*. Stein's version of the time manuscripts, as published by Heidegger in 1928, "furnished the controlling text for the understanding of Husserl's phenomenology of time-consciousness for the next forty years," that is, until the publication of a critical edition in the Husserliana series in 1966, as Brough points out. In order to clarify Stein's creative contribution any further, one would have to study the manuscripts in the Husserl Archive in Leuven. Heidegger's own phenomenology of time as represented in his 1925 lectures should also be compared with Stein's arrangement of Husserl's work on time.

[39]See Brough (1991: xiv).

Husserl for it.[40] In any event, Stein's work once again appeared without her name.

(g) Further contributions to Husserl's manuscripts. It is likely that further work by Edith Stein will come to light as Husserl's *Nachlaß* continues to be edited and published, and as the creative process of Husserlian text-production becomes better understood.

2. Stein's work for other philosophers.

The four years between the end of her service as Husserl's assistant in 1918 and the beginning of her service on the pedagogical faculty at Speyer in 1922 were extremely fruitful ones for Edith Stein. Once again, the record indicates that much of her textual production during this period was anonymous.

(a) Reinach's *Nachlaß*. Stein's teacher Adolf Reinach was killed in Flanders late in 1917. Stein began to organize a *Festschrift* in his memory, but by May 1918 it was clear that wartime conditions made that unfeasible.[41] Instead, Stein and the widow Anna Reinach prepared Reinach's own manuscripts for publication. Stein composed his essay on the essence of movement, working from Reinach's lecture notes for a seminar that she herself had attended. In fact, according to a letter to Fritz Kaufmann of 12 January 1917, Stein had gone over this very material with Reinach himself in Göttingen during what was to be his last Christmas furlough, and she had encouraged Reinach to take it with him back to the front for further work. Its final shape, however, was her own doing. On 25 August 1918, Stein wrote to Kaufmann that she had just read her formulation of Reinach's essay on movement to two other of Reinach's students and had gained their approval.[42]

(b) *Festschrift* for Husserl. The first thought of a sixtieth-birthday volume for Husserl seems to have been Stein's. She mentions it in her letter to Ingarden of 27 April 1917, as a possible place for publishing the revised Sixth Logical Investigation. On May 20 of the next year she writes to Kaufmann to ask him for a contribution for either the Husserl or the Reinach volume; in this letter she acknowledges that she instigated the project of the Husserl *Festschrift*. She

---

[40] See Ingarden (1962: 158).

[41] A *Festschrift* also was being planned for Husserl's sixtieth birthday (which would occur in 1919) with the same pool of potential contributors. The decision to change the character of the Reinach volume was made in part because the Husserl volume could not be postponed. Stein's letter to Hedwig Conrad-Martius of 25 May 1918, written as a note upon Conrad-Martius' own letter to her, indicates that Conrad-Martius and Jean Hering had taken over responsibility for editing the Reinach volume. But Stein remained the liaison with Anna Reinach, who controlled her late husband's manuscripts.

[42] While Stein signed her short introduction to the essay, the substantive reconstruction remained anonymous. Compare the critical edition of Reinach's works brought out in 1989 by Karl Schuhmann and Barry Smith, which attempts to disentangle Stein's contributions from Reinach's. This effort disregards the fact that Stein was already working collaboratively with Reinach during their Christmas-furlough visit of 1916-1917, and arguably even as a participant in the pre-war seminar from which the manuscript first emerged.

seems to be responsible for enlisting writers, although Alexander Pfänder now has agreed to take over as editor. Stein committed herself to write an essay on psychic causality for the Husserl volume.[43] Perhaps because of the political turmoil and economic chaos of the times, the *Festschrift* did not appear until 1922, as the fifth volume of the *Jahrbuch*.

(c) Ingarden's early work. Both Roman Ingarden and Edith Stein grew up in what is now Poland, but their native languages were different. Ingarden was a Pole and he neither spoke nor wrote good German when he first came to study with Husserl in Göttingen. He and Stein became close friends, but their friendship passed through some sort of crisis between October 1917 and January 1918 when both were living in Freiburg.[44] In any event, the crisis resolved into a lifelong collegial attachment whose literary legacy is a wide-ranging and philosophically rich correspondence. Stein kept Ingarden informed about developments in Husserl's circle in Freiburg after Ingarden returned to Poland in January 1918. Her accounts of her ongoing debate with Husserl over idealism prompted Ingarden to set other work aside in order to draft the long declaration on the issue that he sent to Husserl in late July of 1918.[45] By Ingarden's own account, this declaration was his first tentative framing of the ontological, metaphysical, and epistemological themes that he would develop throughout his long career of teaching and publishing. While nothing has come to light to indicate that Stein directly collaborated in Ingarden's writing during their time together in Göttingen or Freiburg, he acknowledges her help with his major work *Der literarische Kunstwerk* of 1931.

Frl. Dr. Edith Stein hatte die Güte, die großen Mühen der sprachlichen korrektur des Textes zu übernehmen, und hat mir dadurch einen wertvollen Freundschaftsdienst erwiesen. (1931: xiv)	Dr. Edith Stein was gracious in taking on the great task of editing the text. (1973: lxxvi)

(d) Hans Lipps's *Habilitationsschrift*. A warm friendship developed between Stein and another of her student friends, Hans Lipps, but in this case it did

---

[43]See her letter to Kaufmann of 25 August 1918. She says that she has just outlined the essay on psychic causality. It was to become the first part of the *Habilitationsschrift* that she unsuccessfully presented at Göttingen in 1919 and is discussed below. Eventually both parts were published in the *Jahrbuch* in 1922.

[44]The story that was told to me--in various versions in Freiburg, Tübingen, and Leuven--runs like this. Edith fell in love with Roman but Roman did not reciprocate. Moreover Roman feared to lose Edith's help, advice, and friendship, so he neglected to inform Edith when he married someone else during a trip home to Poland. I have not found textual evidence to corroborate this. However, much is made of the letter that Edith wrote to Roman on Christmas Eve of 1917. She uses the affectionate *Du* form, and she warmly wishes Roman every good thing in life as she sends him her Christmas gift. For further discussion of the biographical significance of this "key document," see Ott (1993: 115-117).

[45]See Ingarden (1918).

not survive a disappointment. The details can be reconstructed as follows.[46] After resigning as Husserl's assistant, Stein wrote her two-part treatise on the philosophical foundations of the sciences that eventually would appear in the 1922 *Jahrbuch*. First, however, she presented it to the philosophy faculty at Göttingen in a bid to habilitate there. By November of 1919 she had received their rejection. Husserl, it seems, had written the required recommendation for her, but in effect his statement had endorsed the longstanding policy of excluding women from university faculties; and in any case he had not explained why he did not sponsor Stein's *Habilitation* himself at Freiburg. Stein's letters from this period make it clear that she had some hard feelings toward the Master over this. Meanwhile back in Freiburg, Lipps had disgraced himself and was sued for paternity by another student. Thus he now had no chance of habilitating with Husserl either.

Stein came up with a plan for both of them to pursue academic careers without Husserl's help.[47] Lipps would habilitate at Göttingen through the sponsorship of Richard Courant, a mathematician who happened to be Stein's cousin and childhood friend. She would help compose Lipps's *Habilitation* thesis. Once he was installed on the faculty, they would work together. Stein would offer private lessons for beginners, and Lipps would offer advanced instruction at the university.

Pursuing this dream of working "hand in hand," Stein labored with Lipps on his *Habilitationsschrift* from the autumn of 1920 through the summer of 1921. She wrote to Ingarden about her plan on 9 October 1920, but asked him not to tell Husserl what she and Lipps were up to. She continued:

Die Habilitationsschrift gedeiht indessen prächtig. Ich bekomme sie in Raten von 5-10 Seiten zugeschickt, und immer bald die 2. Auflage hinterher, in der die von mir geäußerten Wünsche berücksichtigt sind. Lipps hat eine ganz erstaunliche Begabung. Gerade das, was [s]eine Sachen so schwer zugänglich macht, bildet ihren besonderen Reiz: dieser Blick, der ganz direkt und ohne jede Vorbereitung immer auf die tiefsten Schichten geht. Es tut mir selbst immer weh -- ich glaube fast, mehr als ihm --, wenn ich ihn mit der Rücksight auf Darstellung und Publikum plagen muß, was doch aus Zweckmäßigkeitsgründen unbedingt erforderlich ist. (ESWXIV: 130-131)	Meanwhile the *Habilitation* thesis is developing splendidly. I receive it adroitly in installments of 5 or 10 pages, and almost always the second edition afterwards in which the wishes expressed by me have been taken into consideration. Lipps has an altogether astonishing talent. Precisely that which makes his stuff so hard to get at forms its peculiar charm: this glance which always goes to the deepest levels totally directly and without any preliminaries. It always pains me--more than him, I do believe--if I must annoy him with considerations of presentation or readership that are absolutely necessary on grounds of expediency.

The text issuing from this idyllic arrangement was "Die Paradoxen der Mengenlehre" (The Paradoxes of the Doctrine of Numbers). It did indeed win Lipps his

[46] See Herbstrith (1989).

[47] Herbstrith (1989) has reconstructed this episode.

position at Göttingen on 30 July 1921.[48] Thereafter, nothing more is heard of the planned collaboration with Stein. Lipps, who on the side had also been completing his medical training, went to sea as a ship's physician. That was the summer Edith Stein found truth while reading the autobiography of Saint Teresa and decided to become a Catholic. Lipps married a dancer in 1923, but Stein kept his picture on her desk until Hedwig Conrad-Martius, now her godmother, teased her about it.

3. Further ventriloquisms

In the decade after her 1922 baptism, Edith Stein submerged and veiled her own subjectivity in two other kinds of activity that involved the propagation of texts: teaching, and translation. In addition, one must consider evidence that in 1932-33 she participated in a third covert writing practice as well, the anonymously published Catholic challenges to German nationalist ideology.

(a) Teaching teachers. As early as 1912 Stein had been a volunteer teacher of working-class adults for the Humbolt Society in Breslau.[49] In 1916 she worked briefly as a substitute teacher in a girls' school in Breslau, and in 1917 she was teaching Husserl's proseminar at Freiburg. In 1919-20 she held private classes at her mother's home, and as we have seen, she planned to continue that practice in cooperation with Hans Lipps in Göttingen. But her first formal academic appointment, in 1922, was to Saint Magdalena's in Speyer, a teacher-training institute run by sisters of the Dominican order. Her work there may be interpreted as transmitting cultural texts to women who would pass them on to others. Among those texts should be counted both German and religious classics, as well as the "text" of bourgeois Catholic womanhood: that is, the practices of a gender and class system with their apparatus of religious legitimation. After the failure of her second attempt at university *Habilitation* in 1932, Stein made something of a lateral career move to the German Institute for Scientific Pedagogy in Münster. There she hoped to attract university students in addition to those in teacher training.

(b) Translations. In her teens, Stein had received an excellent education in modern languages and in Latin (although not in Greek or Hebrew). At Speyer she lived in the convent, and in the first years of her life as a Catholic she translated John Henry Newman's *The Idea of a University* and his correspondence. That task occupied her from 1922 to 1924, outside of the hours taken up by her teaching duties. It kept her mind off philosophy.[50] From 1925 until 1930, Stein

---

[48]An excerpt appeared in the *Jarhbuch* in 1923.

[49]See Stein (1986: 202).

[50]Among Stein's letters is a request dated 21 February 1926 to the Bishop of Speyer, asking his permission to continue to read works of Hume, Locke, Kant, Spinoza, and Bergson (which were forbidden to Catholics). She assures the bishop: "All of them are writings that I formerly used in the study of modern philosophy. Currently I am principally occupied with the works of St. Thomas. Since it is important for me to attain clarity regarding the connection between Thomistic and modern philosophy, it will scarcely be possible to avoid the occasional use of the above-named writings for

translated a more philosophically challenging work: the *De veritate* of the medieval Dominican theologian-philosopher Thomas Aquinas, which until then Germans had had to read in Latin. She drew up a glossary of correspondences between technical Latin terms and their German equivalents; in effect, she invented the language of German neo-Thomism.[51] The thoughts and words of Aquinas became the words and thoughts of Stein. Perhaps as that merging was underway she occasionally recalled Scheler's observation that it is both easy and dangerous to mistake the convictions of another for one's own. Perhaps she did not.

(c) Talking back to *Der Mythus des 20. Jahrhunderts.* We saw in the first chapter that the philosophy of empathy was effectively extinguished in Germany in the 1930's by the racist ideology of Alfred Rosenberg. His 1930 *Mythus* not only slandered Jews, but also accused the Roman church of numerous deceptions and crimes dating back over many centuries. Despite the prevailing climate of fear, it seems, a group of Catholic scholars in Münster examined the charges made in the *Mythus* and drafted a response documenting Rosenberg's various errors of historical fact. Their identities and their deliberations were kept secret for their own protection, but their report circulated openly. It was reprinted officially by the Archdiocese of Köln and elsewhere, as "Studien zum Mythus des XX. Jahrhunderts."[52] Many copies were seized by the police, but many also were sent to soldiers by their families or were taken abroad. Indications are that the "Studien" went through several expansions and took up seven hefty booklets, over 200 pages in all.[53] Rosenberg was particularly stung by this document, and in

---

comparison." Permission was granted.

[51]Gelber (1952) reviews Stein's remarks on the task of translation. She also gives a sampling of critical responses to Stein's translation of Thomas.

[52]I have been able to examine part 5, a supplement, and an epilog, bound individually as issues of the *Kirchlicher Anzeiger für die Erzdiözese Köln*. Part 5 bears no date, but a note on the first page states that the *Dunkelmänner* of 1935--Rosenberg's answer to the "Studien"--is anticipated but not yet in hand. The supplement is dated December 1934. The epilog is dated autumn 1935, identifies itself as the fourth new edition, and addresses itself to the *Dunkelmänner*, which appeared in April of 1935. It seems plausible that a "first edition" of the "Studien" may have been much slimmer, possibly little more than a sheaf of notes for the Münster diocesan newsletter. Archival research in the neighboring dioceses of Köln and Münster could perhaps clarify the picture.

[53]Various reconstructions have been suggested for the origin of the "Studien." Rosenberg himself (1935: 5) stated that the research had first appeared in the ecclesiastical newsletter of the Diocese of Münster and then was circulated to other dioceses. Chandler (1945: 95) and Nova (1986: 154-157) accept that account. However Cecil (1972: 120-121) claims that the "Studien" were actually composed by Wilhelm Neuss and a "defense staff" of six other young priests in the Archdiocese of Köln under Cardinal Schulte. They were printed in five different cities, but only the courageous Bishop of Münster, Count Galen, dared at first to distribute them. Neuss (1947) tells an insider's story of the distribution of the "Studien." If the text as published in Köln did originate with Schulte's clerical "defense staff" in 1934, then Edith Stein's participation could not have been extensive, since she had just entered the Carmelite cloister in the same city. Nevertheless one must still inquire into her possible involvement in earlier

1935 he published a response entitled *An die Dunkelmänner unserer Zeit.* This work reviles the "Studien" for their very anonymity and discounts their scholarship on grounds that no authors have stepped forward to take responsibility for them. Can Edith Stein have been among those hiding authors? On one hand, her knowledge of church history was hardly detailed enough for her to produce a text with such elaborate footnotes. On the other hand, she was a Catholic intellectual, reasonably well versed in German and Catholic history, who was active in academic circles within the diocese of Münster during the winter of 1932-33. She worked with students and other young intellectuals to whom the *Mythus* was targeted. She had experience in crystallizing the thoughts of an intellectual circle, in organizing cooperative publication projects, and in writing anonymously. There cannot have been too many others like her. What role Stein may have played in the Münster challenge to Rosenberg is a question worthy of further research.

C. Philosophical and Theological Autographs

Signed, overtly authored texts comprise the second category of Stein's literarary production that, according to my proposal, requires critical examination in line with Stein's own hermeneutical philosophy. As before, this is a category constituted by *how* the texts communicate, not *what* they are about. In fact, these texts cover a rather broad variety of topics, with philosophical and religious themes predominating. Of greater interest than their different topics, however (for purposes of my own study, at least), is their common mode of leading the reader through exposition to conclusions. In these texts, this is done by an author who stands forth as a competent guide for the reader's own thinking. She is not writing "about" herself, but she is exhibiting the flow of her own thinking. She offers her own reasoning as a map for the coherent flow of thought. The authorial i blazes the trail, and the reading i follows. The reader is thus afforded a live experience of reaching the conclusions that the author has first reached on his behalf. This is not to say that the reader is coerced or that the conclusions are imposed upon him from without; far from it. While Stein as author of these texts expects that readers will "reach the same conclusion," she expects them to do so precisely by doing the work of actively empathizing her line of reasoning.[54] She invites readers to take her place; or rather, to take her journey through a thinking process. With this set of texts, authorship is an offer to share motivated subjectivity.

This category can be differentiated further into two contrasting but complementary varieties, which can be termed the *scientific* and the *historical.* Stein's scientific texts invite readers into live experience of logical coherence; that is, into *philosophy.* The thought is offered as having originated with the author but

---

deliberations in Münster.

[54]In the next chapter, I will argue that many interpreters today refuse to read Stein in this way. They position her as a dogmatic authority instead of as a persuasively guiding author. In my view, this is to shirk the duty of an adult reader--the duty of *Einfühlung.*

as in principle available on the same terms to anyone else. On the other hand, Stein's historical texts invite readers into live experience of reconstructed personal coherence; that is, into *history* in the Diltheyan sense of biography. Here the experiences are laid open as having originated with someone else, but their recovery or "non-originary" experiencing is, through Stein's version of it, made available for readers to experience non-originarily as well. To complete the critical interpretation of Stein's texts lies beyond the scope of the present study. But a start has been made with the identification of the autographic authorial stance operative in the following works.

1. Philosophical works.

Stein's dissertation on empathy, which was thoroughly examined in the last chapter, is the earliest of her scientific philosophical writings. Subsequently, several themes whose treatment in the dissertation was inconclusive were taken up again in the manuscript planned as the first *Habilitation* thesis (and eventually published in the *Jahrbuch* in 1922), including essays on "psychic causality" and "individual and community"--together labeled "contributions toward philosophical foundation of psychology and the cultural sciences." These were meant to complete the program that Husserl had set forth in the *Ideen*, whose second and third books Stein herself had prepared for publication. Subsequently Stein contributed "an investigation concerning the state" (1925) and a comparative study of Husserl and Aquinas (1929) to the *Jahrbuch* as well. In 1931 she pursued issues first suggested in the 1929 essay as she drafted her second attempt at a *Habilitationsschrift*, "Potency and Act." That manuscript in turn formed the basis for the principal formulation of her mature philosophy, the 1936 manuscript "Finite and Eternal Being," which had a lengthy appendix devoted to a critical review of Heidegger's existential philosophy. In this category also belong several book reviews[55] and the academic lectures that apparently were offered during the winter of 1932-33 when Stein was teaching in Münster: "Introduction to Philosophy" and "Structure of the Human Person." Beginning about 1928, Stein was called upon to address professional societies of women educators, and invitations for public lectures increased after the first volume of her translation of Aquinas appeared in 1931. Some of these lectures were done in a philosophical mode, but Stein found that audiences responded better when she used the historical/biographical mode instead.[56]

2. Historical works.

As a historical author, Stein attempts to share with her readers or hearers the distinctive personal flow of live experience that she has recovered from another life. In effect, she invites others to join her in re-experiencing the coherence of that life. Already in her dissertation, Stein had offered a theory of the disclosive

---

[55]See Stein (1920, 1933a, and 1937).

[56]Indications that a lecture has gone well, or gone badly, sometimes appear in the correspondence. For example see the letter of 7 December 1930.

anecdote as key to unlock historical meaning.[57] The intellectual heritage of that theory goes back to Dilthey, who had argued that the character of an era is expressed in the figure of a genius personality and therefore can best be recovered through biography. Stein's ambiguous theorization of *personal type* as the foundation of understanding became the (rather slippery) basis for her efforts to interpret the Christian tradition. She was not raised and educated as a Catholic.[58] Although her home environment was Jewish, the primary school in which she learned to read and write was imbued with a generally Protestant Christian cultural atmosphere. Stein's access in adulthood to the understanding of Catholic Christianity came through the literary genre of hagiography: "lives of the saints." Saints, accessible in their sympathetic life stories, were one distinctive feature of Catholic Christianity that the Protestant version lacked. For Stein the liturgy, too, was a narrative of the life of the Christ--both in its daily celebration and in the seasonal cycles of the liturgical year. She understood these narratives by grasping their inner motivation, that is, their *intelligible subjective flow*. And this is precisely the sort of understanding that her liturgical and hagiographical texts mean to offer to her audience.

It hardly need be said at this point that Stein's hermeneutic practice is a disciplined, deliberate, quite sophisticated and thoroughly theorized production of texts geared to offer personal access to coherent and intelligible "feeling": feeling felt first and originarily by some significant religious figure, second and nonoriginarily by Stein as author, and third by her readers and hearers. She does not write with blind emotion, nor would she expect anyone to surrender rationality and be swept away by her words. She demands of the reader a disciplined effort to follow the personal motivated path of the religious figure, just as she herself as author is following it.

Among Stein's texts, those that follow the motivated experiences of individual lives, and thus facilitate a vicarious reprisal of those experiences, include lectures and essays collected in volume 11 of Stein's Works (volume 4 of the English edition). The figures examined there include Saints Elizabeth, Theresa of the Child Jesus, and John of the Cross; and some Carmelite heroines. Stein's last, unfinished manuscript, "Science of the Cross," is an extended study of John of the Cross. Stein portrayed Mary the mother of Jesus in notes for her lectures on women, now published in volume 5 of her Works (volume 2 in English). An

---

[57]She said that the most meticulous gathering of facts about the deeds of Frederick the Great from his birth to his death may fail to disclose the mind that transformed Europe, while a chance remark in a short letter can reveal it fully. See Stein (1917: 126). Her use of vignettes in her autobiography is comparable.

[58]Technically, Stein was never catechized. She did not undergo religious formation within the active life of an ecclesial community. Instead, she prepared for baptism largely on her own by studying texts, including especially biographies, by having conversations with her friends and her clerical mentors, and by prayer. After baptism, her Catholicism was formed *intellectually* by thinking the thoughts of Thomas, and *practically* by following the daily schedule of a Dominican convent and visiting a Benedictine monastery for periodic retreats.

underlying theme in all these lives is the *Einfühlung* of the live experience of Jesus of Nazareth and the terms under which it can take place. Stein's 1933 review of Karl Adam's christology is of particular interest in this regard.

There are three major impediments to the success of these texts for contemporary critical readers, even when the texts are approached on their own terms, as I have recommended. First, in style they may seem saccharine and sentimental to today's tastes. Second and more important, they are uninformed by the advances in the academic fields of liturgical studies and historical-Jesus research that had only begun in Stein's day.[59] Third, Stein has not sought corrective responses from the Jewish community to her portrayals of aspects of Judaism, but merely positions Judaism as a foil for the construction of Christianity. These difficulties introduce suspicion about the covert functions of texts that purport to impart the personal experiences of saints. Among those covert functions might be the construction of social realities, and that possibility comes to the fore in the "autobiographical" writings, to which we now turn.

D. Autobiography: Self and Type Under Construction

According to Stein's hermeneutic, understanding occurs when an i is led to live-through the motivated flow experienced by another i. Both leader-i and follower-i must internally register the coherence of this movement as they live-through it. We have seen that Stein the author becomes a "leader of i's" in her signed works; specifically, either a trail-blazing leader, in her philosophical works, or a led leader, in her historical-biographical works. We have also seen that Stein can write *without* authoring, when for some reason she conceals her act of writing and makes herself anonymous. Each of these two modes of textual production--which I have termed the autographical and the anonymous, respectively--invites the readers to *follow along*. The texts succeed, as it were, if readers can and do bring to life within themselves the same experiences of coherence that Stein supposedly has had. For philosophy, this happens when an argument is logically constructed; in that case, it cannot rationally be resisted. For biography, this happens when the reader catches on to the unique irreplicable value and value-choices of the person who has flowed through the events recounted. One cannot take that person's place (as one does in logic, where all "places" are that of the transcendental i), but one *can* savor her unique enactment through empathy of what she has felt.

This presents an intriguing hermeneutical paradox. On one hand, it would seem that biographical narratives cannot but be believed, since they impart live experiences that are in principle unrepeatable. The facticity of occurrences can be checked, of course, but *what they meant* to the person going through them cannot be independently verified. The value of the personally experienced significance of a series of events is something altogether different from the chronological

---

[59]For example, it doesn't occur to Stein to conform her imagination of Mary or of Jesus to the textual and archaeological evidence of their life and times. My own critical work on the conditions of access to Jesus of Nazareth through liturgical and other narrative means has led to quite different results from Stein's, and is reported elsewhere.

accuracy of that series. Access to such valued significance lies not in *replicating* it, but in *receiving* it. On the other hand, as "led leader" the biographer implicitly claims to have climbed aboard someone else's life in order to disclose how his experiences have flowed. Like philosophical writing, biography must exhibit one pattern of coherence while at the same time extinguishing the plausibility of any rival pattern. But unlike philosophy, biography never gets done with that job. The value of the motivation that it asserts cannot be separated from the alien individual, and so cannot be enacted within oneself. The biographer, as led-leader, is no more privileged with respect to certifying the "why" of another life's flow, than is the reader. Other "why's" may suggest themselves, and inexorably do so. That's a permanent hazard. Unlike logic, biography comes undone. So the biographical author has the rhetorical burden of discrediting alternative patternings for the coherence of the feelings she recounts.[60]

The ante goes up in the case of *auto*biography. How could there be grounds for suspicion and for alternative "why's" when the writer is imparting the motivation of a life that has been her very own? Easy. The grounds lie in the fact that writing issues from an act of reflection, not from originary living. Although the *act* of writing is alive, its *contents* are pre-lived. Writing makes "now" out of *now*; or more precisely, it materializes that making. The writer is not still living-through her live experiences when she is writing them up. Moreover, in writing her own life she faces the same twofold rhetorical task that any biographer faces: to offer one coherence while suppressing all possible others. Autobiography favors one "why" (or set of "why's") and suppresses the alternatives.

The "auto" (αὑτη) or self of autobiography is a written product. It is a literary artifact. In this sense it is a "fiction": a made and making thing. The truth of autobiographical fiction lies not in what is said but in how it is said. Holding the "why's" in place is the work that biography must do; and for *auto*biography, this is a desire-driven project of self-defense.[61] To stabilize precisely *this* significance, excluding others, is the act of self-establishment (or self-constitution). Writing accomplishes it by simulating the transcendental i through the individual i. Because the text affords the reader a live experience of coherence, one can point to that inner ride-through as independent corroboration of the particular significance asserted. *It felt right.* It clicked. One is persuaded, by one's very own ability to feel-through the textually stabilized coherence, that it's the *right* one and the *only* one--that is, it's the coherence that would have to register for *any i at all*, given the same concatenation of facts. And the first reader of the

[60]As a preliminary example of alternative coherences to Stein's religious biographical accounts, one might suggest that the religious behaviors of a child prodigy were produced by cultural or economic factors, rather than by divine grace. Further alternatives will be discussed below.

[61]I acknowledge that at this point my discussion jumps to a meta-level, and examines the "why of why's," as it were. I acknowledge that my writing, too, will be a project of stabilizing one reason (or set of reasons) while undermining alternatives--particularly the alternative reasons asserted in Stein's own self-writing and the reasons asserted by othe commentators.

autobiography is, of course, its writer.

Stein creates self through type. When in her autobiographical mode, she does so on behalf of writer and reader alike, in one stroke. As we have seen, there was ambiguity and the shadow of danger in Stein's early theorization of type in her dissertation. There, type was shown to be the principle for marking one's own individuality as deviance, and for restricting the circle of individuals among whom empathic understanding would be possible. But Stein also asserted, incongruously, that all human beings are of "the same type" and therefore are capable of understanding one another. Either way, type was the basis for understanding through empathy.

Stein's textual projects of establishing and communicating her self, then, are pursued by means of the construction of types: recognizable profiles of the *kinds of* human individuals that inhabit the world, and of the *kind of* human being that she and her readers intend to be. The building blocks of this construction include such things as patriotism, elite education, a repertoire of allusions to German history and literature, a preference for certain ways of earning one's livelihood, race-linked somatic characteristics, specification of gender roles, and the degree of facility with German language itself.[62] Edith Stein writes up these realities as she writes her life, both in her unfinished autobiography and in her letters. This writing also effects the propagation of the types into the world beyond her text. They are contagious because they are not new creations, but familiar reinscriptions of social realities that the reader is constrained to recognize. Thus her writing habilitates the reader to understand her, by innoculating him with a booster shot for the type-knowledge that he and she must share prior to the empathic following that reading will be.

1. "Edith Stein" as self-made woman.

The autobiography, entitled "Out of the Life of a Jewish Family," was written in three stages. From April to September 1933 Stein worked on it in Breslau, producing the stories of her grandparents, her parents, and her own childhood. As she says in a foreword dated 21 September 1933, she intended simply to write down the words of her 84-year-old mother but found it necessary to impose somewhat more order than an oral account would have. (These stories were perhaps meant to serve as private notes for a later smoother account that unfortunately never was written.) In the foreword Stein also represents her motivation for writing: a priest had suggested that she tell the truth about what it

---

[62]Sander Gilman has given critical attention to the textual production of distinction between "real" Germans and Jews. He shows how pathologically deformed feet and noses, along with deviant smells and gait, are assigned to Jews. Moreover, he documents the textual disparagement of Jewish speech and writing as unclear, deceptive, and dangerous. Under National Socialism, any literature that was out of favor with the authorities was termed Jewish "asphalt literature." That Yiddish is a German dialect was denied; it was designated not a language but a vice. To speak Yiddish or to use Yiddish expressions when speaking German is *mauscheln*--a verb for which we cannot produce in English an equivalent term or concept. Jews themselves, as participants in German culture, internalized these views in various ways. See Gilman (1986 and 1991).

was like to grow up in a Jewish family. Recently, she says, as caricatures of Jewish life have appeared in fascist tracts and speeches, she has realized that many people might not have had the opportunity to glimpse a different reality. This narrative is (intended to be taken to be) a straightforward account of her experience of Jewish life.

The second and bulkiest portion of the manuscript was written during Stein's novitiate in Carmel, from late 1933 to May 1935. For this portion of the work, her sources would have been her own memories rather than her mother's voice. Finally, a few pages were added after her flight to Echt, in January through April of 1939. The narrative breaks off at the defence of her doctoral dissertation in August 1916, although preceding chapters have already told events that occurred several years later.

This unfinished autobiography of Stein's has proven itself to be factually quite correct. It is a standard, even treasured source used by scholars of Husserl, Ingarden, and other phenomenologists for the purpose of filling in details of professional and personal activities occurring in the circles to which Stein belonged. Yet it is also a fiction: a thing made by constituting events, identities, and the narrative connections among them. The selectivity and creativity of Stein's authorial production must be brought under phenomenological inspection. My proposal that this work should be read *as fiction* must seem jarring to scholars who are accustomed to regard it quite otherwise. But my proposal is founded on two inescapable considerations: first, the formal requirements of Stein's own hermeneutical theory, as discussed above; and second, the coincidence of details between Stein's autobiography and a certain historical novel written at the same time, *Die Geschwister Oppermann* by Lion Feuchtwanger. This novel invokes the genric conventions of the fictionalized documentary, a familiar kind of popular literature at the time.[63]

Feuchtwanger was traveling in the United States when the Nazis came to power early in 1933. He did not go home. He wrote this novel rapidly between April and September--the same months when Stein was composing the first part of her own story. Feuchtwanger's tale was available in several languages by October of 1933, and Feuchtwanger said that he intended the work to enlighten the world about the dangers of National Socialism.[64] Thus like Stein's *Leben*, Feuchtwanger's novel was designed to be a work of self-defense.

It was also strikingly autobiographical in many respects. *Die Geschwister Oppermann*, like Stein's autobiography, offers an insider's view of a Jewish family at the time when Hitler came to power. The main character, Gustav Oppermann,

[63]Gilman (1991: 216-217) says that the genre of the fictionalized documentary was developed at this time by both the fascists and their opponents. He identifies *Die Geschwister Oppermann* as the most successful example of this genre. Feuchtwanger at first had named the protagonists of his docu-drama "Oppenheim," but the publisher changed the name to "Oppermann" because of threats from a non-Jewish family names Oppenheim who were politically powerful.

[64]See Berndt (1972: 141-147).

is usually read as representing Feuchtwanger himself. Gustav is the intellectual brother in a family of businessmen. He fails to resolve the novel's central dilemma, "prudence versus principle" in the face of the Nazi threat. But I would suggest that the character of Gustav's young nephew Berthold is autobiographical as well. Berthold is a *Gymnasium* student faced with a new teacher, a Nationalist, who sets him up to make a speech on an ancient Germanic tribal leader and then accuses the young man of slandering the German nation. Berthold faces an impossible choice: either he retracts what he knows to be the truth, or his school and his family will be destroyed. Refusing both alternatives, Berthold poisons himself and dies. Uncle Gustav, facing an analogous dilemma, is equally suicidal in that he first prudently flees the country and the life he has made; afterwards, acting now on principle, he returns with another identity but is caught and sent into forced labor anyway.

Before considering the nature of the connection between Stein's work and Feuchtwanger's, a listing of their coincidences is in order. On the one hand, the "factual" details in Stein's story attest that Feuchtwanger has succeeded in simulating a true-to-life family situation in his fiction. On the other hand, the "factual elements" conjoined through Stein's *authorial selection* of them correspond to just those details that a skillful novelist has invented. This indicates at least that she is constructing her narrative through the same sort of authorial subjectivity as that with which Feuchtwanger constructed his. Edith Stein uses the textual character "Edith Stein" for self-constitution and for self-defense, just as Lion Feuchtwanger uses the textual characters of "Gustav" and "Berthold." The following coincidences between Feuchtwanger's novel and Stein's autobiography demand attention:

The Oppermann family has a prospering furniture business.

The Stein family has a prospering lumber yard.

The Oppermann furniture store was founded by a patriarchal grandfather, whose portrait looks down from the wall upon business and family affairs.

A matriarch runs the Stein lumberyard, founded by her dead husband.

The Oppermanns' story examines the characters, exploits, tribulations, and achievements of many uncles, aunts, cousins, and friends.

The Steins' story examines the characters, exploits, tribulations, and achievements of many uncles, aunts, cousins, and friends.

The novel flaunts but gently satirizes the luxury, hospitality, and bourgeois customs of the Opperman home.

Bourgeois customs, nice manners, and practices of hospitality are proudly exhibited throughout the autobiographical narrative. The Stein home is comfortable and well-run. Character assassination is accomplished by recounting lapses in manners.

Working-class viewpoints are represented by the sympathetic characters of the school porter, a furniture salesman, and his wife. However, a non-Jewish cabinetmaker is the business rival and unworthy enemy.

While landladies and farm wives are always kindly, the narrative incessantly disparages cabinetmakers, their wives, and Poles.

Gustav is writing a treatise on Lessing, demonstrating his gentility.

Edith writes a thesis on Lessing, demonstrating her refined taste.

The novel offers detailed descriptions of student life in the *Gymnasium* and of pedagogical practices there.

The autobiography presents detailed memoirs of experiences in primary and secondary school, and at university.

Young Berthold learns that the essence of excellence in culture is German history and literature. However, German tribal origins can be compared with Hebrew origins.

Edith concentrates on German history and German literature in her studies. But she has no knowledge of parallel Jewish history.

Students must, above all else, learn to speak and write good German. Education revolves aroud this imperative. Satirically, the fascist press, a nationalist teacher, and Hitler himself are all sketched as deficient in German diction.

Schoolgirl Edith reads German classics precociously. She writes excellent German essays and loves to study German political history. She loves foreign languages as well.

Young Berthold diligently prepares an oral report, but the presentation is sabotaged by his nationalist teacher. His being tested and failing at school is the critical turning point of the novel.

Oral examinations are the milestones in Edith's life. The story elaborates on the preparations for exams, how they are conducted, and the aftermath.

Honor among schoolboys is sacred. School friendships are of major significance.

Schoolgirls form intimate friendships that last. School friends must help one another.

Boys vie to excell one another at patriotism.

Edith is aflame with patriotism at the outbreak of World War One. Her brief service in a field hospital takes up a disproportionately great part of the autobiography.

A commendation for patriotic service is displayed prominently on the wall of the Oppermanns' furniture store. One brother died in the war.

Edith mentions Jews who served in the World War whenever possible. Many, like Husserl, lost sons.

The professional difficulties of a woman writer are recognized. She survives as Gustav's mistress for a while. Nevertheless she is portrayed as shallow and disloyal.

Working-class women in night school find an advocate and protector in Edith, who also takes an active part in the feminist movement at the university.

Gustav has a Zionist niece who emigrates to Israel.

Edith has a Zionist niece who emigrates to Israel.

"Eastern Jews" are portrayed as quaint and sometimes clever. But their customs are troubling anachronisms for the Oppermanns.

"Eastern Jews" are backward, compulsive, and hard to understand for Edith. Religiously, she blames them for handing out "stone, not bread."

Western Jews are just like other Germans.

Western Jews are even nicer than other Germans.

Feuchtwanger writes from voluntary exile in 1933.

Edith enters the cloister in 1933 and continues to write.

There are suicides for both of the autobiographical characters: Berthold's rejection of the forced choice between prudence and principle, and Gustav's vacillation between acting prudently and acting on principle.

Edith cuts off the life of her family through her vow of celibacy and fantasizes her own death in imitatio Christi.

Besides these particular details, one remarks that both works portray characters through intricate descriptions of disclosive vignettes taken from middle-class urban life. It is particularly striking that both works focus attention upon educational institutions and practices. In this connection, Stein's attitude seems purely nostalgic, while Feuchtwanger appears to be more pointedly aware of the schools' function in the maintenance or change of social institutions.

The similarities between the texts demand some explanation. There are at least three possibilities: pure coincidence, plagiarism, or the sharing of socially constructed and maintained social "types." Let us consider each possibility.

(a) Is it an accident that the two works are so similar? No.

(b) Did Stein plagiarize her autobiography? Maybe. The possibility cannot be ruled out. Feuchtwanger was a popular writer, Stein loved literature, and the literature of resistance would have been particularly appealing to someone who had lost her job because of Nazi repression--someone who was cooling her heels at her mother's home for the last few weeks before leaving all worldly literature forever outside the cloister gate. Stein would just barely have had time to read *Die Geschwister Oppermann* before entering Carmel on 14 October 1933. That she would take such a book along into the cloister is unlikely, although not impossible. But she very well may have taken vivid memories of Feuchtwanger's story about a merchant family resembling her own. When, in late 1933 and 1934, she again took up her pencil to write the stories of her own student days, Feuchtwanger's narratives could still have hovered in the background of her consciousness. Stein's was a mind keenly trained for philosophy; but in those first cloistered months it was also a mind being re-engineered to disciplines of meditation, silence, solitude, and physical mortification. She may have been homesick and missed her friends. At the very least, having read such a well crafted narrative as Feuchtwanger's may simply have given Stein a model of *how one ought to write* when one's life depends upon it.

This possibility requires further research. (But we should also research whether it is possible to write one's life entirely *without* plagiarising.)

(c) Do Stein and the novelist share "types"? Yes, this seems most likely. At the factual level, Feuchtwanger's composite portrait has captured circumstances that were common enough among German Jewish bourgeois families. Stein just happened to come from a "typical" economic background, in that sense. But the novel also makes sophisticated hermeneutical use of "type" in a way that the autobiography does not--at least at first glance. Both works are operating on a principle that Stein enunciated in her dissertation: that understanding happens on the basis of shared type. Moreover, both texts are staking their defensive strategy upon the claim that westernised assimilated Jews are "just like" regular old bourgeois Germans.

But there the similarlty ends. Feuchtwanger's novel goes on to make the point that middle-class Germans and Jews *alike* are complacent, blind, and therefore vulnerable to the violent repression being unleashed by Nazis in 1933. It projects, but does not resolve, the paralyzing dilemma of the impossible choice between prudent compliance and principled resistance. The novel ends on a despairing note: to become aware of the danger is not to become able to act. By contrast, Stein's autobiography veers off in another direction. Its fine-grained portraiture of bourgeois life and customs is geared to demonstrate to Germans that Jews are "just like" them. The work is funded by the naive and pathetic hope that Germans will not hurt Jews if only Germans can be brought to see that Jews are people. Stein, ever unable to believe that truly evil choices are possible, works on showing that her Jews are of *the same type* as her Germans.

It is painfully apparent in the autobiography's construction projects, of course, that not all Jews are "her kind" of Jew, and not all Germans are "her kind" of German. Stein must show herself--that is, *textually design herself*--to be *the right kind* of German if she is to win the understanding of the readers whom she hopes to attract and convince. Therefore she will display her education in German political history (and she will be ignorant of Jewish history, naturally). She will rhapsodize about German literature from Ulfias to Schopenhauer (finding Polish and Yiddish accents uncouth; don't we all?). She will parade the doctors and lawyers among her cousins (with a knowing grimace at the reader over the rascally behavior of tradesmen and cabinetmakers).

I must confess, right here in the text and not in a footnote, that my own empathic reception of Edith Stein stops dead in its tracks when Stein employs such ethnic and class disparagement to construct her own bourgeois identity. Her text tries to infect me with that type-knowledge, but I'm congenitally immune. My father spoke Polish until the public schools of Baltimore taught him what a stupid language it was. My mother's immigrant great-grandfather was a German cabinetmaker. In my own childhood, the very word "cabinetmaker" was entwined with stories of hard work and good sense brought here as our people's *only* legacy from the land of the Kaiser. My own "type" obviously is quite different from Stein's. When her text tries to lead my reading i along the contours of these type constructions, I jump the track and sparks fly.

Yet this acute failure of empathy through type is not the failure of understanding altogether. Perhaps the very dissonance that I experience when reading Stein, produced because her meticulous construction of her type so thoroughly backfires in my case, is what brings to light the constructed character of type. Where the construction runs smoothly, it does not exhibit itself so clearly.

2. Stein's collaboration in the production and maintenance of "types."

While Feuchtwanger's novel ends in despair, Stein's autobiography does not come to an ending at all. Naturally, one cannot write the conclusion of one's own life; even a suicide note isn't the final word, because the hand that wrote it can still tear it up. But in Stein's case, the writing project was broken off in favor of other writing, and before she could return to it she was arrested, deported, and gassed. Nevertheless, Stein's standpoint as a writer in 1933-35 is a hopeful one. Like Feuchtwanger, Stein exiles herself from "Germany" and she means to transmit her life story back from beyond--beyond the cloister grate, that is. She writes as if she has an answer, a solution, to what she terms "the Jewish Question" in her 1933 foreword.

What is that solution? Stein's solution is like a coin tossed into the air that lands sometimes on one side, sometimes on the other. On one side it says, "Jews are people, they speak good German and play violins, so don't hurt them." On the other side it says, "Catholicism is truth, so I wanna be a Catholic Jew."[65] One fears that neither Judaism, nor Catholicism, nor humanity itself, is adequately conceived in such a conjunction. Stein's solution to Germany's broader problems is equally entangled with type, particularly with types that establish the social realities of ethnicity, gender, and class. Her pedagogical theory is her autobiographical writing practice writ large. She theorizes a literacy of type, requiring the replication of the gender, ethnic, and class systems that have

[65] I use the metaphor of the tossed coin to indicate that Stein has no single univocal position on Judaism. Many examples can be found of her friendship and respect for individual Jews. For example, after baptism she continued to accompany her aged mother to synagogue. Yet it remains true that she found aspects of Jewish life and culture distasteful, and gave many offhand indications of that attitude in her autobiography. Moreover, she subscribed fully to the theological position that Jewish religion has been fulfilled and superceded in Jesus of Nazareth. I believe that she did this through ignorance more than arrogance or racial prejudice.

materially and emotionally supported her.

The critical question that should guide future readings of Stein's theology is this: To what extent does the religious solution lift the discussion beyond the hermeneutical difficulties of type-bound understanding; or conversely, to what extent is Catholicism merely a dimension of the interwebbed this-worldly maintenance of the types? Stein can be forgiven for reinscribing types in the tentative formulations of her life-writing philosophy, as long as the general thrust of the thought ultimately carries it beyond types. Otherwise, her religion remains part of the problem, not the solution. To determine which, one must balance the understanding achieved through empathy with a more suspicious, even cynical, kind of understanding. A beginning for this kind of critical reading should be made along the following avenues.

(a) Portraiture of women and gender relations. The lectures and essays "on woman" collected in the fifth volume of Edith Stein's works (second volume in the English series) make positive proposals about women's education and preparation for professions. These proposals can be taken at face value. But between the lines, as it were, the covert construction of gendered realities is under way.[66] As Husserl might say, the meant-object "woman" is being constituted out of a series of glimpses. Therefore, by rights, Stein should counterclaim that no person can become identical with such an object. Moreover, Stein earlier argued that a person cannot be educated. What then is occurring in these instructions for the education of women?

Further disclosive dissonance can be evoked between Stein's theorization of women, and her portraiture of individual women in her autobiography and in her letters--especially her mother, but also such figures as Malvine Husserl, the Master's wife. Stein portrays men as well in her life and in her letters. She tells disparaging tales about the two men who disappointed her--Ingarden and Lipps--and about another man who seems to have loved her honorably, Eduard Metis, an Orthodox Jew. These anecdotes should be allowed to tell us something about Stein's collusion in the cultural work of gender construction.

(b) Portraiture of ethnicity. To enhance the normativity of German language and culture, the "other" is both produced and disparaged. This project is in full swing in Stein's autobiography and in her letters. As we have seen, Stein disliked *Yiddishkeit* as well as all things Polish. This bias needs to be read politically in the context of her family origins in Breslau, where the peasant majority were Slavic Catholics and the urban minority were German-speaking bourgeoisie--Jewish and otherwise. The privileged economic position of the Steins is not unrelated to the cultural politics of ethnic differentiation and disparagement. By the same token, Stein admires Austrian and Suabian accents. Why should that be?

(c) Work, class, and the body. Stein's theoretical remarks on the

[66]For a contemporary effort at gender-construction from a writer sympathetic to National Socialism, see Bergmann (1932). Stein's arguments about women's place in society should be understood against this background.

irreplicability of the body in her dissertation should be read alongside her various narrative treatments of the bodily realities of labor. She hates cabinetmakers, but heaps praise upon a well-bred bride who ever so tastefully designs the cabinets that are to be built *(by whom?)* for her husband's study. Stein's memoirs of service in the soldiers' hospital are particularly vivid, and include accounts of patients' suffering and of her own harrassment by a physician. Such compassion seems to be lacking, however, in her remarks about working men and women, and especially in her being ashamed of her own mother's work-worn hands and clothing. To be sure, in telling of this afterwards, Stein is ashamed of her shame for she acknowledges that those working hands kept her in style as she studied phenomenology. Her philosophy of the body ultimately is engulfed in her theology of the eucharistic real presence of Jesus.

Thus it is not a coherent i, but quite a fractured i that beckons the reader of the autobiographical, self-creative writings of Stein. And it is not a seamless web of type constructions that the those writings project, but many ribbons leading readers of various types in various directions. In the next chapter, we will follow some of those ribbons as we survey the variety of interpretations to which Edith Stein lends herself.

Chapter Five
Interpretations of Edith Stein

Having discussed Edith Stein's interpretive practices in the preceding chapter, we come now to a consideration of how she herself has undergone interpretation. It was suggested above that Stein's writing put into practice her distinctive theory of empathy. The authoring i, laying out a path of thinking, offers itself as the guide for readers to follow--even the vehicle for readers to ride--along that path. We saw that authoring subjectivity stands forth in Stein's works in three contrasting modes, which were designated the anonymous, the autographic, and the autobiographic strategies of writing.

Thus one might well expect to discern multiple strategies of *reading* among Stein's interpreters. Indeed, there is an intriguing breadth and variety in the growing corpus of commentaries on the life and works of Edith Stein. This variety is owing in part to the differing interests of those who have studied her. Perhaps the most striking impression one retains from dipping into this literature is that Stein elicits affection and an uncanny sense of rapport.[1] This is apparent not only in texts, but in conversations with scholars who have undertaken serious investigation of Stein's philosophy. I feel it myself, even as I attempt to conduct disciplined and dispassionate research.

Whence comes this rapport? Is it owing to Stein's intellect or her sanctity, or does it perhaps arise through pity over the injustices of her life and the horror of her death? Without denying that such may be contributing factors, I suggest that rapport with Edith Stein is engineered through the skillful use of "type" as a hermeneutic device. Type is the means of production of rapport. We saw in the preceding chapter that Stein's autobiography--as a self-conscious work of self-defense against the racist policies of National Socialism--deliberately constructed her "self" as a cultured middle-class German Jew, adamantly portraying "the right kind of Jews" to be "the right kind of Germans." Stein derived the notion of "type" from Dilthey, and in her dissertation she accorded to it a key function in human communication. Like Dilthey, she believed that the individual "genius" served as the paradigm of his age and revealed its identity. She also believed that understanding and learning could occur only between persons of "the same type." We had occasion to scold her for this doctrine in chapter three. "Type" is a pernicious notion when it usurps the category of humanity and demotes people of "other types" to subhuman status. What was needed in 1930's Germany, and what is still needed today, is a phenomenology of the humanity *of strangers*.

The sharing of type between Stein and her readers was something that she

---

[1]I have found only one exception: Moossen (1987), whose slanders of Stein are calculated to undercut the fulsome praises that Stein was receiving in Germany in the year of her beatification by the Catholic Church.

counted on and for which she took trouble to provide textual support--not only in the autobiography and the correspondence, but in her phenomenological and inspirational works as well. Anecdotes about everyday experiences that "we all share" are a common compositional device of Stein's. If her writing summons the reader to follow a thought-path, then these anecdotes and the other devices of type-construction serve as convenient short-cuts. The reader who is inwardly enacting the cognitive chains offered by Stein's text can conserve some effort there. The stereotypes come pre-fabricated. They are presented as being *already* held in common between writer and reader. They receive reinforcement, but they don't have to be built from scratch. Pre-fabricated concurrences between Stein and her commentators comprise various facets of social identity; for example, elite academic status, bourgeois femininity, Catholicism, patriotism.[2]

Another trend discernable in the secondary literature on Edith Stein, a trend not unrelated to infectious rapport through type, is the drift toward biography. The events of Stein's life are quite often regarded as the interpretive key to her philosophy. Her personality is hermeneutically superimposed upon her scholarly work. This can foster two unacceptable assumptions: first, that when one has understood the woman one has *ipso facto* grasped her philosophy; and second, that the phenomenological evidence pertaining to her philosophy was available only to Stein herself--whose formulations must therefore remain authoritative and beyond question. Either of those assumptions, if left unchallenged, is enough to disable any attempt at critical appraisal of Stein's work. Rapport may also make criticism seem disloyal or unworthy, while the sharing of social stereotypes between author and readers renders "type" very difficult to bring under critical attention at all.[3] These preliminary observations begin to account for the peculiarly sanguine character of the literature of Stein Studies--if we may so name a field opened some forty years ago by the ground-breaking scholarship of Mary Catharine Baseheart.[4]

Before commencing our survey of what people have made of Edith Stein, let's reorient ourselves to what is meant by literacy. Writing and reading are practices. Moreover, they are both *interpretive*--although this has been a hard-won

---

[2]Yet Stein's own remarks earlier in her dissertation had made a potent critical case *against* hermeneutic use of "type" by either writers or readers. She says that the child of racist parents and the consumer of erotic literature alike are infected with ideas that did not originate in their own experience. They did not enact first-hand the warranted value-assignments that they now entertain. See chapter three, above.

[3]A helpful antidote to the seductive influence of type-contagion is the backtalk of those who do *not* share Stein's type constructions: for example, unassimilated Jews, or descendants of cabinetmakers like myself.

[4]Baseheart's 1960 doctoral dissertation at the University of Notre Dame was the first independent critical study of Stein's philosophy. It was preceded by: Verbillion (1960) [an essay citing only the selections of Stein's works translated by Graef (1956)]; book reviews in scholarly journals; and critical introductions to the series Edith Steins Werke, which began in 1950. I have collected titles of 20 dissertations and academic theses on Stein. Copies of most of them may be found in the Stein Archive of the Carmelite Sisters at Cologne, and some have been published subsequently as books.

point in phenomenology, and Husserl did his best to resist it. We saw in chapter two that the term "interpretation" was largely expurgated from the Sixth Logical Investigation when Husserl revised it for reissue in 1922. After his turn to transcendentalism, Husserl wanted to deny that one's particular subjectivity innovates or contributes anything distinctive to the constitution of the objects of consciousness. As pure transcendental ego, subjectivity was indeed responsible for the arising of a unified object out of many appearances; yet Husserl insisted that *any subject at all* would accomplish the constitution of an identical object, given identical appearances. Is reading, then, something like Husserlian constitution? If to read is to re-constitute *the* identical sense that the author first constituted, then interpretations can be of only two kinds: right or wrong.

On the other hand, as Stein argued (without swaying Husserl), constitution owes its possibility to a prior function: the empathic receptive following of another life. Construing Husserlian formulations in a way slightly different from the way *der Meister* wanted to arrange them, she threw the spotlight on the motivated coherence of thought itself as the target of understanding. This approach was congruent with Husserl's own position in the 1911 *Logos* article, but not with the *Ideen.*

Unlike Husserl, Stein formally addressed the problem of textuality (albeit very briefly). For her, written words and spoken words are but two out of a whole array of expressive modes that human feelings may discharge themselves into. We *understand* by letting our own i ride along with the i that lived-through the discharge of feeling into expression. Text, then, can facilitate empathic access to another's flowing life-stream. But among all the modes of expression considered by Stein, text is the one that can separate itself the farthest from live experience, and it is therefore also the one most susceptible to dissimulation and to loss of meaning in general. Besides written text, the other expressive modes invite interpretation as well. Among these are gestures, facial expressions, deeds, and even sublimation, according to Stein. By enumerating those modes, Stein anticipated a principle that is widely acknowledged in contemporary hermeneutics: that the human body itself is text-like, and it participates in reciprocal interpretive relationships of intertextuality. An individual life, a culture, a nation, even a church may be regarded as "texts" in this expanded sense. All require reading, that is, interpretation. For example, Stein's autobiography was a work of interpreting the "text" of her life and her social identity.

Thus, in accordance with Stein's own inclusive list of "expressions," our survey of interpretations of Stein will be broadly drawn. It includes social practices as well as interpretive texts. In fact, the very act of making a survey is itself an interpretive practice. As we did in the last chapter, here we will once again juxtapose a conventional way of sorting the texts with an alternative sorting. Part A runs through receptions of Stein chronologically and topically. Part B then re-sorts the texts phenomenologically, that is, by the various ways in which the interpreting i follows what Stein's authoring i has laid out.

A. Interpretations of Stein by Period and by Topic

Who was Edith Stein? What did she have to say? What is the significance of her work within the various academic disciplines, or for social and religious causes beyond the academy? There seem to be nearly as many "Steins" as there are commentators. Even during her life, she was something of an inkblot upon whom people projected their imaginations. Receptions of Edith Stein were determined partly by her own personality, deeds, and philosophy, and partly by what others needed her to be.

1. Interpretations of Stein before Auschwitz

We can infer that the men who profited from Stein's anonymous text production, like Husserl and Hans Lipps, regarded her as a resource but not as a colleague. Husserl described her thus in his 1919 letter[5] supposedly supporting her attempt to secure an entry-level teaching post at Göttingen:

Fräulein Dr. phil. Edith Stein, meine langjährige Schülerin an den Universitäten zu Göttingen und Freiburg, hat im Sommersemester d. J. 1916 in Freiburg summa cum laude das Doktorat der Philosophie gemacht, und zwar mit einer ausgezeichneten wissenschaftlichen Abhandlung über „Einfühlung", die sogleich nach ihrem Erscheinen das Interesse der Fachmänner erweckt hat. Sie wirkte nachher über 1 1/2 Jahre als meine Assistentin und leistete mir nicht nur wertvolle Dienste bei der Ordnung und Verarbeitung meiner Manuskripte für umfassende wissenschaftliche Publikationen, sondern nicht minder bei meiner akademischen Lehrtätigkeit. Sie hielt zu diesem Zweck regelmäßig philosophische Übungen für meine, nach tieferer wissenschaftlicher Ausbildung strebenden Hörer, an welchen nicht nur philosophische Anfänger, sondern auch Fortgeschrittene theilnahmen. Von dem ausgezeichneten Erfolge dieser Mitwirkung konnte ich mich im Fortgange meiner eigenen seminaristischen Übungen und durch persönliche Fühlungnahme mit meinen Hörern überzeugen. Fräulein Dr. Stein hat in der Philosophie eine weite und tiefe Bildung gewonnen, und ihre Fähigkeiten für selbständige wissenschaft-

Dr. Edith Stein, a student of mine for many years at the Universities of Göttingen and Freiburg, earned the doctorate in philosophy in the summer semester of the year 1916 in Freiburg, *summa cum laude*, with an excellent scientific treatise on *Einfühlung*, which aroused the interest of the experts as soon as it appeared. After that she was active for over one and one-half years as my assistant, and peformed valuable services for me not only through the arrangement and processing of my manuscripts for extensive scientific publications, but no less [by helping] with my academic teaching activities. For this purpose she held regular philsophical exercises for my graduate students who wanted deeper scientific instruction. Not only philosophical beginners took part in those exercises, but advanced students as well. Thanks to the excellent results of this collaboration, I was able to devote myself to the progress of my own seminar exercises and close personal contact with my graduate students. Dr. Stein has achieved a broad and deep education in philosophy, and her capacities for independent scientific research and

[5]First printed in Leuven (1983: 22-23) and now included in the *Briefwechsel* (3: 548-549). The letter was dated 6 February 1919.

liche Forschung und Lehre sind außer Frage. Sollte die akademische Laufbahn für Damen eröffnet werden, so könnte ich sie an allererster Stelle u. aufs Wärmste für die Zulassung zur Habilitation empfehlen.	teaching are beyond question. If the career of university teaching were supposed to be open for ladies, then I would be the very first to be able to recommend her enthusiastically for admission to habilitation.

Husserl damned Stein with loud praise subjunctively muted. True to his conviction that *the who* doesn't matter in the egoic function of object constitution, he did not recognize how much of what would be taken to be *his* thought was first *hers*. Two instances of this reception of Stein's work into Husserl's were discussed in the preceding chapter: the "hinge" chapters on the live human body in *Ideen II* that link its treatments of nature and culture, and the homogenization of the time-consciousness manuscripts. Both of those texts were to become hugely influential in the phenomenological tradition and in the existentialism stemming from it. A third important and unrecognized reception of Stein occurs in Husserl's lectures on psychology from the mid-1920's, which bear comparison with her *Jahrbuch* essays of 1922 on the philosophical foundations of psychology.[6]

My purpose here is not to regain for Stein the recognition that she deserved, much less to insist that individual ownership of philosophical work be strictly defended. My purpose is to examine the phenomenon of interpretation--not in pure consciousness, but as embodied through contagious *expression* in texts of various kinds. There is warrant for flouting Husserl's transcendental reduction by pursuing this phenomenon, "reception," beyond the cloister of consciousness and into real-world expressions: warrant is provided in Stein's own phenomenology of the "blending and tending" that carries an i through into whatever discharge concludes the expression of its feeling. On Stein's account, "reception" is *there* in the texts for immediate intuition. But it requires iteration. That is, only in my own receiving of the prior receptions do I gain the possibility of following how they went. The first disclosure occurring in a phenomenology of reception is that *no egoic following*--neither my own nor those expressed in texts--*is a perfect duplication.* Following is interpretation, and interpretation is variance. Of particular interest here, however, are the extent to which interpretation can also be effacement, and the ways in which such effacement is accomplished. In other words, *how Stein was disappeared* in the texts of others is the issue.

One can study Stein's textual disappearances only in instances where they did not completely succeed. For example, Max Scheler cited Stein by name in the 1923 revised edition of his *Sympathiebuch.* She had sent him a copy of her dissertation[7]; and, as mentioned earlier, the use he made of it was far greater than

---

[6]Compare Husserl's *Phänomenologische Psychologie: Vorlesungen Sommersemester 1925*, volume IX of Husserliana, with Stein's treatises on "Psychische Kausalität" and "Individuum und Gemeinschaft," published together as "Beiträge zur philosophische Begrundung der Psychologie und der Geisteswissenschaften" in Husserl's *Jarhbuch* of 1922. See also the 1917 manuscripts on psychology and epistemology, which Stein edited (Husserliana XXV: 82-206).

[7]According to her letter to Roman Ingarden of 27 April 1917. She wrote that Scheler

he acknowledged. Stein's contrast between in-feeling and one-feeling (*Einfühlung* versus *Einsfühlung*) seems to have inspired Scheler vastly to broaden the scope of the work. I have been unable, in a search of the literature, to find any other contemporary mention of Stein's dissertation.[8] Thus Husserl's statement that "the experts" took an immediate interest in it was something of an exaggeration. The earliest critical review of Stein's work that I have turned up is a mention in a yearly roundup review of German philosophy, where the 1929 *Jahrbuch* was discussed. That volume was the seventieth-birthday *Festschrift* for Husserl. Stein had contributed an essay comparing Husserl and Aquinas.[9] Martin Heidegger, editor of the *Festschrift* volume, had rejected Stein's first plan for the article--a script for a personal conversation between Husserl and Aquinas--and demanded a more "neutral" treatment.[10]

Stein's teaching activities at Speyer in the 1920's left lasting impressions with her students and colleagues there. After her death, Stein's Carmelite community collected statements from those who had known her at Speyer and later in the cloister. Those statements are on file at the Archive in Cologne, and portions have been published in various biographies. The following remarks by a former student are perhaps not untypical:

> (N)one of us has ever been able to forget the spell that her personality exerted. Her manner alone made her a model for us at that critical age. There's not a single remark of hers that I can repeat--and it isn't that her comments weren't memorable, but that she was a quiet, untalkative person who could influence us simply by who she was. Whenever she had to offer criticism, she always did it with the perfect balance of fairness and kindness. Nobody ever saw her as anything but tranquil, dignified, and calm.[11]

---

had sent her a postcard saying that he had marked up this copy quite extensively and would like to talk to her about it.

[8]An exhaustive search of German philosophical publications of the 1910's and 1920's would best be undertaken in Germany. We can look forward to having more information about early receptions of Stein's work as philosophical interest in her grows.

[9]See Faber (1930),

[10]See editor's note 5 for Stein's letter to Roman Ingarden of 1 November 1928. See also Leuven (1983: 56). Graef (1955: 50) cites the original opening of the dialogue.

[11]Cited in Herbstrith ([1971] 1992: 76-77). Posselt (1952) presents what appear to be longer verbatim recollections of Stein. Hilda Graef, who was given access to the convent's files and who independently interviewed many of Stein's acquaintances in the early 1950's, adds some contrasting accounts, particularly of the first years of her teaching at Speyer. Graef writes: "Her very perfection was too much for some of the girls--if they could have detected a flaw in her they would have liked her much better. The total absence of flaws extended also to dress. However unfashionable it might be, she never tolerated the slightest spot or speck of dust on it. Girls who were themselves exuberant and full of mischief were naturally somewhat exasperated by this almost superhuman perfection and would have preferred her to have lost her temper--or at least a hairpin--occasionally. Nor was this irritation, which her somewhat self-contained impeccability could inspire, confined to her pupils. One day a director of education came to inspect the school and assisted at one of her lessons. He gave as his opinion: 'She knows much, but

Stein stood stock still when she delivered public lectures and also while teaching. Her comments on student papers could be caustic. One woman recalls the following notation on a composition that had attempted to cite numerous sources: "The use of quotations proves that *other people* are clever."[12] An art instructor who worked at Speyer with Stein recalled an incident in which the philosopher correctly interpreted the woman's dream about her mother's impending death.[13]

Such stories, however, may convey less information about contemporary personal readings of Stein, than about the mechanisms of legend formation. More significant is the information that one can glean indirectly from documents involving several individuals who had mentoring relationships with Stein after her baptism in 1922 and immediately before it. How those trusted mentors interpreted Stein would have a formative influence upon the person and the philosopher she was to become in her mature years. The common thread in their "readings" of Stein, it seems to me, was this: Edith Stein was a channel, an adapter, a perpetual learner and mediator, but never an originator. Her words and deeds were to be representations of "something else" beyond herself. The value of her thought was to lie elsewhere than in her own live experience. Eventually Stein herself was brought to internalize this assessment. But how? This was the phenomenologist who had gone eyeball to eyeball with Edmund Husserl himself without blinking.[14] Now the insufficiency of *Evidenz* was made to appear to her with *Evidenz*.

How was this accomplished? Stein received the notion that Catholicism required a special kind of philosophy, other than phenomenology. Her acceptance of this view is puzzling, given her familiarity with Scheler's work in his "Catholic" period, which was thoroughly phenomenological. Herbstrith reports that the university chaplain in Breslau, Professor Günther Schulemann, suggested to Stein

---

she cannot teach'." See Graef (1955: 42-43).

[12]See Graef (1955: 42). This comment of Stein's reflects her philosophical training. Phenomenology aims to exhibit the evidence of phenomena, not enlist external authorities. This critical remark on a student's paper may be juxtaposed with what Stein reports about her first take on the writing of Thomas Aquinas himself (in *Endlich und ewiges Sein*, page 489, cited and translated in Graef [1955: 49]): "When I began to study the works of Aquinas I was constantly troubled by the question: according to which method does he proceed? I had been accustomed to the phenomenological method, which uses no traditional teaching but examines everything that is needed for the solution of a question *ab ovo*. I was baffled by a procedure that adduces sometimes scriptural passages, sometimes citations from the Fathers, or again sayings of the old philosophers in order to deduce results from them." Even Aquinas, it seems, quoted clever others too often for Stein's taste.

[13]See Leuven (1983: 64). When Stein was told by the art instructor that she had dreamed of her mother calling to her, Stein insisted that the woman immediately phone home and covered for her so that she could do so. As it happened, the mother had fallen gravely ill. The timely phone call enabled the teacher to reach her mother's bedside before she died.

[14]One recalls that she quit Husserl's employ over his attempt to impose a set of manuscript-processing instructions for her to follow.

that she read the works of Saint Thomas Aquinas in preparation for her baptism.[15] Thomas had been designated the alternative and antidote to Kant in nineteenth-century Germany, by Catholics who opposed certain political innovations that they associated with Kantian Idealism.[16] The papal encyclical *Aeterni patris* of Pope Leo XIII on 4 August 1879 had mandated the study of Thomism in seminaries. It was taken for granted, therefore, that the religious conversion of a philosopher should include professional re-tooling in Thomistic thought.

If the South German Catholicism to which Stein converted was scholastic in its philosophy, it was monastic in its ethos. Life was regarded as a spiritual path for which one needed a guide. Here Stein would have found a congenial notion, akin to her own earlier phenomenology of inner awareness as the following of flowing conscious life. She even found quasi-phenomenological descriptions of prayer experiences in the writings of the Spanish mystics: first Teresa of Avila, and much later John of the Cross. Moreover, the pursuit of prayer was supposed to be *led*, for it was only through obedience to another's advice that one could avoid deceptions and illusions. The leader would be one's "spiritual director."

Stein's first director was a cleric by the name of Schwind who was on the episcopal staff at Speyer.[17] No record remains of their conversations, because at Stein's request his family burned her letters to him after he died in 1927.[18] Soon afterward, Stein turned for direction to Raphael Walzer, abbot of the Benedictine monastery at Beuron where she customarily spent religious holidays and participated in the liturgy. Like Schwind, Walzer at first discouraged Stein from seeking entrance to the cloister, and he still had misgivings after granting permission in the wake of the political calamities of 1933. Walzer submitted a lengthy statement to the Cologne Carmel in 1946.[19] Among other things, he recalls Stein's tranquility and her ability to remain motionless for many hours during prayer in the monastery church. Before Stein entered the Carmelite cloister, Walzer had feared that she would suffer from Carmelite liturgical practices--tasteless by Benedictine standards--and from lack of intellectual stimulation in a small closed community of uneducated women. However, he states that she assured him after entering that she felt completely at home in Carmel. The third person who was an important spiritual guide for Stein was her mistress of novices, Sister Teresia Renata de Spiritu Sancto Posselt, who introduced her to the practices of the Carmelite way of life, and later would write the first biography of Stein.

All in all, the *content* of the direction that Stein received from Schwind,

---

[15]See Herbstrith ([1971] 1992: 68).

[16]For a discussion of the background of the neo-Scholastic movement in Germany, see McCool (1977). See also Inglis (1993).

[17]He is mentioned in her biographies as Dean Schwind, Canon Schwind, or Prelate Schwind, but I have been unable to learn his first name.

[18]See Leuven (1983: 58). But see Stein's comments on Schwind's manner of offering direction, excerpted from her obituary for him; Herbstrith (1985: 89).

[19]The text appears in Schlafke (1980: 15-19).

Walzer, and Posselt is perhaps less significant than the bare fact that it occurred. These relationships indicate that Stein had accepted the necessity of *receiving direction* for the flow of one's conscious life. What does that mean? For Stein's philosophy, it means that thought cannot autonomously reach being. For Stein's teaching and writing, it means that she now understands herself to be a *directed director* when communicating. In the classrooms at Speyer, in lecture halls, in letters and in personal counseling, Stein will be "directing" in the sense that she will be brokering a pattern of thought-movement whose validity or value she does not vouch for herself. She will be imparting coherences that never were originarily hers (if one may borrow the terms of her earlier phenomenology). Stein expects to *receive* those coherences herself through a direction-facilitated process.

Prayer now plays the role previously held by reflection and description. Prayer is the empathic reception (*Einfühlung*) of religious mysteries: a led following of the life-and-death story of Jesus of Nazareth, whose pattern is canonized in the gospel narratives, reprised in the lives of Mary and the saints, and dramatically rendered in the discursive structure of daily, weekly, and yearly liturgical observances. Empathizing those expressions, Stein's prayer aspires to ride along on the flowing i that first discharged its originary feeling in them. In general the universe, in particular the humanity and career of Jesus, are to be read in prayer as the personal expression of the divine creator. One's own life is meant to be a following of the life of Jesus, up to and through the death of Jesus. This is the intention signaled in the phrase "to carry the cross" and included in the religious name that Edith Stein chose: Benedicta a Cruce, "Woman Blessed by Means of the Cross." Prayerful following brings mystical union with the i of the other--Jesus--yet union is not an identification, a displacement, or an overlay. Relationship between numerically distinct i's remains, while what becomes "the same" is *the path*, called sometimes the cross. Where consciousness first goes in prayer, the events of real life are expected to go as one's life story unfurls. Edith follows Jesus.

The contrasts and the coincidences between the contemplative ambition of "cross carrying" and phenomenological descriptions of empathy are quite striking. Prayer preserves the relationship of two i's while assimilating one's path to the other's and while suppressing the variance that erodes any possibility of perfect coincidence between expression and understanding. More precisely, prayer must continually affirm that *divine* suppression of such variance is reliably occurring. In other words, Stein took it for granted that "the way of the cross" as Jesus of Nazareth first traveled it was *expressed without variation* in the gospel narratives, the liturgy of the Mass, and the lives of the saints--at least in its essential points. She saw no gap between these expressions and the ability of the human understanding to make a faithful copy of them when assisted by divine grace.[20]

Husserl's pre-Armistice account of empathy had not only insisted that i's remain separate, but had affirmed that the pathways along with they flow must

[20]Hermeneutical and historical scholarship has made this position untenable today.

remain separate as well. "No canals between lifestreams" for him. Only the forms of thought-paths could be shared, because the validity of the forms *already was* shared. Paradoxically, then, Husserl's transcendental phenomenology went on to displace alien i's with the non-individuated "pure i" who has no relationships. Husserl could dispense with other i's because any i already monadically held within itself the means to certify the coherence of a thought path. On the contrary, young Edith Stein vested the validity and value of followable paths in the person of the one who first blazed them. The worth of the path was owing not merely to its logical coherence (as for Husserl), but to its historical enactment *by someone*. The indispensable other person, not the path, was the prime target of understanding; but one's own person was the means and guarantor of understanding. Finally, Sister Benedicta abdicated the duty to certify that what determines the understanding stems from the other i *rather than from one's own desires*.

As a mature philosopher, then, Stein came to believe that understanding needs a power-assist in order to overcome its wayward creativity. Stein found that assist in Thomas Aquinas. Or rather, she was directed to find it there. Her own search for "Catholic philosophy" led first to the works of the English convert John Henry Newman. She had the 1921 edition of Newman's book *The Idea of a University*, and she began to translate it in the summer of 1923.[21] That coincided with the appearance of Erich Przywara's programmatic *Religionsbegründung: Max Scheler--J.H. Newman*. Przywara, a Jesuit priest, was a leader in the Catholic effort to counteract Kantian idealism.[22] He also was philosophy editor for the journal *Stimmen der Zeit*. Stein corresponded with Przywara about her Newman translation, and eventually met him. Przywara saw that Stein could be useful to his own effort to make Newman's thought more widely known in Germany. He asked her to translate Newman's letters and diaries from 1801 to 1845, and became the editor of the volume.[23] Przywara next urged Stein to undertake the translation of the *De veritate* of Thomas Aquinas. His favorable review of the work appeared in the very year of its publication, 1931.[24]

Przywara's own interpretation of Stein at that period exhibits a curious mix of racial stereotype and self-projection. He wrote of their 1925 meeting:

---

[21]According to her letter to Roman Ingarden of 19 June 1924. The manuscript of the translation is reported to be in the Carmelite Archive in Brussels, but seems never to have been published. Stein could of course easily read the English text for herself, so her activity of translating it indicates that she was intending to make it available to other German readers.

[22]Przywara (1923: v-xiii) examines the state of that program and reviews the relevant literature of his collaborators.

[23]Bejas (1987: 163) gives the publication data as *John Kardinal Newman, Briefe und Tagebücher 1801-1845*, translated by Edith Stein, edited by Erich Przywara (Munich: Theatinerverlag, 1928). It seems curious that Przywara favored the publication of the letters and diaries, rather than Newman's influential work on the university.

[24]See excerpts from this and several other contemporary notices of the Thomas translation in Gelber's "Nachwort" to volume 4 of Edith Steins Werke, 464-466.

[W]e had our first conversation in Speyer, under the auspices of the unforgettable, wise and kindly Vicar-General Dr. Schwind. Dr. Schwind told me at once that I was going to have a surprise: for he had never met anyone whose looks betrayed her race as little as those of Edith Stein. He was right; for the woman who met us might rather be compared to the statue of Uta in Naumberg Cathedral. . . . This was indeed a very special trait of Edith Stein--she came from absolutely pure Jewish blood and was yet a true German woman.[25]

If Przywara put Stein on a pedestal, he also set her as figurehead on the bow of German neo-Scholasticism--with himself as helmsman. His 1933 assessment of the past and future of Catholic philosophy in Germany, published in the American journal *The Modern Schoolman*, cites Stein's Thomas translation as the shining methodological exemplar, and brags that it coincides with his own approach.

The position and point of view which, as we ourselves hold, promises more successful results than any other, has been very clearly enunciated and expounded by Edith Stein, the former assistant of Husserl, in *Untersuchungen über die Wahrheit* (Breslau, 1931-2). This gigantic work is a compilation of the various translations and commentaries of the *Quaestio de Veritate* of St. Thomas. . . . [T]he work of Edith Stein confronts present-day views and objections with Aquinas' theory of knowledge; in this same spirit our own *Analogia Entis* (Vol. 1, Munich, 1932) is designed to visualize the methodic basic principles of Christian philosophy.[26]

Reading as adroitly as Pygmalion chiseled, Przywara extricates Stein from her ethnic identity and assimilates her into his own philosophical project and personal sense of mission.

Thanks in part to such readings, the Thomas translation established Stein as a Catholic philosopher, at least among Catholics. She was seen as someone who could help to realize Przywara's dream of a "reborn high scholasticism" (*Neugeburt der Hochscholastik*).[27] But her spurt of popularity as a lecturer and popular author

---

[25]Przywara's remarks appeared in an article enntitled "Edith Stein zu ihrem zehnten Todestag" in *a 1952 issue of Die Besinnung*, according to Hilda Graef, who translates them. See Graef (1955: 48).

[26]See Przywara (1933: 92). (The original text could not be located; one doubts the accuracy of the translation because Przywara knew better than to designate Stein's work a "compilation of the various translations.") Pzywara concludes that neo-Scholasticism must study and cultivate the classical tradition of Plato, Aristotle, Augustine, and Thomas "reduced and unified into a philosophical system and method which will embrace every point of view universally and positively and by all means critically." He asserts that Scholasticism has the potential "for striking off the shackles now holding truth in slavery --actively faithful to its whole function as liberator and defender of human reason."

[27]See Przywara (1923: ix). Implicitly, Przywara wanted to father a born-again Thomism with Stein as mother. Stein herself wrote to Ingarden on 8 August 1925 that Przywara had indicated in a letter that ". . . both of us look upon the same desideratum as presently an urgent task: namely, a comparison between the traditional Catholic philosophy and modern philosophy (in connection with which phenomenology is the most important for him, too) [. . . wir beide dasselbe Desiderat als gegenwärtig dringende Aufgabe betrachten: nämlich eine Auseinandersetzung zwischen der traditionallen

popular author was owing more to her appeal to Catholic women's organizations, as a professional who could articulate a religious rationale for women's participation in white-collar employment. Przywara arranged lecture tours for her through the Association of Catholic Women Teachers and the Association of Catholic University Graduates.[28] Stein's lectures were widely reported in Catholic publications. From 1927 through 1930 her popularity grew, and she was groomed for another attempt at university *Habilitation*. In the spring of 1931 she began composing another *Habilitationsschrift*, "Potenz und Akt" ("Potency and Act"). In order to work on it and to begin the round of visits to academic sponsors, Stein gave up teaching at Speyer. Meanwhile in France the Société Thomiste was planning a high-level conference to explore the possibility of establishing common philosophical ground with the Phenomenological Movement, known to the French in the works of Scheler, Hering, Koyré, and Husserl himself--who recently had delivered his *Méditations cartésiennes* at the Sorbonne. Stein was one of about 30 invited participants for this conference, held at Juvissy in September 1932.[29]

Nevertheless her second *Habilitation* attempt foundered. Stein was turned down by the university at Breslau, her mother's home city, where she had a brother-in-law on the faculty.[30] At Freiburg she received the encouragement of Martin Heidegger, who advised her that as a Catholic philosopher she should seek the sponsorship of Martin Honecker.[31] Honecker agreed to consider Stein's candidacy, but he seems to have found her credentials mediocre.[32] However the

katholischen und der Modernen Philosophie (wobei ihm auch die Phänomenologie die Wichtigste ist).]

[28]See Herbstrith ([1971]: 1992: 95).

[29]See Société Thomiste (1932). See also the recollections of Daniel Feuling, who participated in the conference, in Posselt (1952: 110-112).

[30]See her letter of 28 June 1931 to Adelgundis Jaegerschmid. The details of this attempt are not known, and it may not have proceeded very far.

[31]See her letter of 26 January 1931 to Adelgundis Jaegerschmid.

[32]As Honecker's son has pointed out, on Thomistic philosophy Stein had published only the 1929 *Festschrift* essay comparing Husserl and Thomas. Her two-volume translation of the *De veritate*, appearing in 1930 and 1931, did not count as original philosophical work, and her popular lectures and articles were irrelevant. See Raimund Honecker (1991). Hugo Ott has published Martin Honecker's notes concerning Stein's manuscript of "Potenz und Akt." See Ott (1993b). In a personal conversation in July 1993, professor Ott told me that the copy of the manuscript eventually returned to Stein by Honecker was probably marked up extensively with his comments, and is in the possession of the Archive in Brussels but unavailable for examination. Stein also received constructive comments on the manuscript in a two-hour conversation with Heidegger, as she reports in her letter to Roman Ingarden of 25 December 1931. Apparently Stein rewrote the manuscript soon thereafter, expanding it in light of those comments. A longer version is preserved with the papers of Hedwig Conrad-Martius in Munich, according to editor's note 4 on page 227 of Stein's correspondence with Ingarden, volume 14 of the Werke. Because this essay would form the basis for the principal philosophical work of Stein's mature years, *Endlich und ewiges Sein*, one would like to

greatest obstacles were economic ones, for Germany was suffering in the worldwide Great Depression. On 29 November 1931 Stein wrote to Ingarden that the question of her *Habilitation* at Freiburg had been decided negatively because of the general economic crisis. She said that both experts--presumably Honecker and Heidegger--informed her that she had proven her competence, nevertheless they had persuaded her not to submit an official application because it would not be put through by the ministry. Meanwhile behind the scenes, attempts were being made to obtain private funding for Stein's appointment to Freiburg. Support was sought from Catholic women's organizations and from the Görres-Gesellschaft, but it appears that her sponsors could come up with only 400 marks for the 1931-1932 academic year.[33] At the end of 1931 Stein had no more prospects for a university appointment.[34] She was recruited to join the faculty of the ten-year-old national Catholic teachers' college in Münster by its founder, the educator and women's leader Maria Schmitz.[35] Stein took up her duties there on 1 March 1932.[36] She had spent the last weeks of 1931 and the first of 1932 in Freiburg revising her manuscript "Potenz und Akt."[37]

---

see how Heidegger's and Honecker's comments may have guided Stein's reworking. It is to be hoped that Honecker's comments will be included in the promised publication of the manuscript. Honecker's notes as published by Ott (1993b) are keyed to the manuscript page numbers. (For example, Honecker jotted on page 105, that "the concept 'God' is not at all clarified.") Interestingly, a few years later Honecker would reject another manuscript destined to become enormously important in twentieth-century German Catholic philosophical theology: the 1936 doctoral dissertation of Karl Rahner, published in 1939 as *Geist in Welt.* See Vorgrimler (1986: 59-62). Rahner's philosophy has been characterized as transcendental Thomism, and Vorgrimler suggests that Honecker found too much of Kant and Heidegger in it. On the contrary Hugo Ott, with access to Honecker's notes in the archives of the University of Freiburg, says that Rahner simply fell short of the standards set by Honecker. See Gaboriau (1989: 161). For a brief account of Honecker's own philosophical work, see Knight (1927: 82-83).

[33]See Ott (1987: 272).

[34]But hope returned, briefly, shortly thereafter. Once Stein was teaching again at the Institute in Münster, she had informal contact with university students. In a letter to Martin Honecker of 8 July 1932 she mentions that she has established contact with three philosophy professors at the University of Münster but has not yet raised the question of *Habilitation* with them. On 24 February 1933 she writes to Hedwig Conrad-Martius requesting that this friend take a look at her philosophical writings to date. This seems to be a hint from Stein that Conrad-Martius might produce a review of Stein's work that would enhance its credibility in philosophical circles.

[35]See editor's note 3 on page 224 of Stein's correspondence with Ingarden, volume 14 of the Werke. The German Institute for Scientific Pedagogy was supported by the Society of Catholic German Women Teachers, to which Stein belonged, and by the Association of Catholic Teachers of Germany. Within walking distance in Münster were also the university and the Marianum, a school and residence for sisters and some lay students. Stein roomed and took her meals there.

[36]See her letter of 9 March 1932 to Roman Ingarden.

[37]It now seems clear that Stein was revising the manuscript after having learned that

Stein taught at Münster for only a year before being informed, on her return from the 1933 Easter holiday, that she was to cease her lectures in the wake of National Socialist policies barring Jews from teaching.[38] She entered the cloister in Cologne that autumn, over the objections and tears of her mother. Mrs. Stein interpreted her daughter's decision as an abandonment of monotheistic religious faith and a refusal to join in the life of the Jewish people through marriage and motherhood. But that interpretation did not prevail against the views of her Catholic mentors and colleagues, which Edith herself adopted: that a Carmelite nun was the bride of Christ and that her prayer bore fruit for God's new people. The ancient Christian myth that Jews were guilty of the murder of Jesus of Nazareth was fully internalized by Stein. It became the overriding theme of her prayer. On its terms she interpreted the Nazi genocidal policies and her own fate. Stein writes that, during a service she attended in the Carmel chapel in 1933 just before losing her job,

> I spoke to our Savior and told Him that I knew it was His Cross which was now being laid on the Jewish people. Most of them did not understand it; but those who did understand must accept it willingly in the name of all. I wanted to do that, let Him only show me how. When the service was over I had an interior conviction that I had been heard. But in what the bearing of the Cross was to consist I did not yet know.[39]

Stein's Carmelite religious superior attributes the following words to her in 1939, on an occasion when she had heard reports of violence against Jews:

> It is the shadow of the Cross which is falling upon my people. If only they would see this! It is the fulfillment of the curse which my people called upon its own head.[40]

---

it would not be accepted as a *Habilitationsschrift*. Publication of the Ingarden correspondence removes previous uncertainty about when the manuscript was first submitted to Honecker. Honecker's son mistakenly had recalled that the manuscript was not submitted until the summer of 1932. See Honecker (1991) and Ott (1987: 272 and 1993b: 140-141). Ott (1987) presents several letters that do not appear in the three volumes of Stein's correspondence in the Werke. At Freiburg in the winter of 1931-32 Stein stayed in the guesthouse of the Benedictine sisters at St. Lioba's monastery in Güntersthal, a pleasant streetcar ride away from the university. One of the nuns of that community, Adelgundis Jaegerschmid, had been a student of Stein's when Stein was teaching for Husserl, and she remained in close contact with Husserl throughout the increasingly difficult years of his retirement. Baseheart (1960: 12) reports that Husserl died at St. Lioba's, where he and his wife had taken refuge as Nazi persecutions became more severe; however I have been unable to substantiate this claim.

[38]Recollections of Stein's time in Münster, including her own account, are found in chapters 10 and 11 of Posselt (1952).

[39]See Posselt (1952: 118). By "the Cross" is meant the way things went for Jesus: the distinctive coherence discerned in his life and death, as set into narrative form in the canonical Christian gospels.

[40]See Posselt (1952: 184). Stein may be alluding to the Gospel of Matthew 27: 20-26, an episode in which a crowd in Jerusalem says of Jesus, "His blood be upon us and upon our children." Another possible allusion is Luke 19: 41-44, a scene where Jesus

In the cloister Stein received no special honor, of course, and the sisters were rather disappointed at her clumsiness and poor sewing. Her Jewish identity remained a problem because the Nazis took increasingly severe measures against anyone who sheltered Jews. On the last day of 1938 Stein was sent for safety to a Carmelite cloister in Echt, across the border to the west.[41]

The Jesuit John H. Nota, who visited Stein at Echt just before she was killed, penned the following recollection of her. Italics are added here in order to highlight the curious rewritings of Stein that already are under way in his text:

> When I arrived at the convent, Edith Stein asked me to deliver a homily at the Holy Hour. I felt a little nervous, having never preached in public since my ordination, but Edith Stein directed me to some beautiful scripture texts found in the Carmelite Office and *helped me to put the sermon together. In fact, she almost wrote it herself.* Yet she did it all in a friendly, unassuming way, happy to have me take her suggestions. It occurred to me that *Edith Stein's intellectual talents had in no way impaired the feminine side of her personality. She was anxious that I take back enough food for the return journey.* She loved to show me photos of her family, and of Husserl and Scheler too. . . . When I said goodbye to Edith Stein that July day, I had no idea that *it was the last time I would ever be able to do anything for her.*[42]

A busy text indeed! This Nota, this young sprout of an incompetent preacher, appoints himself arbiter of what it means to be a woman. His nostalgic words do a job of constructing Stein as *client*, even while they disclose her clearly to have been Nota's *patron*. We see that the future professor of philosophy and phenomenology at McMaster University has been given three gifts by Stein: information about Husserl and Scheler, food for his journey, and the substance of his sermon. Yet he portrays himself as having been *her* benefactor.

The German occupation forces in Holland regarded Stein as a pawn in a power game, because she was both a Jew and a Catholic nun. In July of 1942, the leaders of many Christian denominations in the Netherlands together protested German deportation of Dutch Jews. The Nazis then conceded that Jews who were Christian converts would be spared as long as the churches ceased their protests over the deportation of the rest of the Dutch Jews. The Catholic bishops did not accept that concession, and their denunciation of the Nazi policies was read from pulpits across the country at Sunday masses on 2 August 1942. That evening throughout Holland the SS rounded up all members of Catholic religious orders

---

says that Jerusalem will be razed because the city has not recognized the time of its divine visitation.

[41]This transfer was accomplished relatively rapidly after *Kristallnacht*, 8-9 November 1938, a violent and emphatic demonstration of the Nazi will to destroy real estate. The presence of Stein in the Cologne cloister in 1938 placed *its physical structures* at risk. In contrast, by 1942 the Nazis' policy had shifted. Now they were bent on destroying not the property but the Jews themselves. One cannot help remarking that in this different situation, the ecclesiastical bureaucracy moved much more slowly to secure a new shelter for Stein.

[42] Cited in Herbstrith (1971] 1992: 12-13).

who were of Jewish descent. This act of reprisal against the Dutch bishops was the occasion for Edith Stein's arrest, deportation, and murder.[43]

2. Posthumous interpretations of Stein.

Whether Stein died for being a Catholic or for being a Jew remains the key disputed point in readings of her life today, and we will return to this question below. Immediately after the Second World War, however, Stein was read as an exemplar of fidelity to "the way of the Cross" as construed in Catholic monastic spirituality. This hagiographic interpretive strand has been relatively constant in the reading of Stein's texts over the last four decades.

The three definitive early biographies--by Posselt, Miribel, and Graef--exhibit this interest very clearly, although all are careful to examine Stein's professional contributions as well. Posselt had lived with Stein in the cloister and, as her religious superior, had had numerous confidential conversations with her; Posselt also collected statements from Stein's former colleagues and friends. Posselt's biography and the archive of documents that she began in the Cologne cloister are remarkably free of sentimentality and ideology.[44] Miribel, writing anonymously as "une moniale française," also incorporated statements from people who had known Stein; yet one can hear the statue-carving chisel of the theological establishment echoing in the background of this text.[45] Graef interviewed Stein's sister, the gynecologist Erna Biberstein, as well as many of Stein's former students, colleagues, and acquaintances for her 1955 biography. She also examined materials in the Stein archive at Köln and the Husserl archive in Leuven, and complains only that she was denied access to the manuscript of Stein's autobiography "for I was not able to secure the co-operation of the person in charge of this material in spite of urgent representations to the authorities of the

[43]See Posselt (1952: 202-210) and (Schlafke 1980: 20-31). In the preceding months, Stein had very urgently sought permission to flee to another Carmelite cloister in Switzerland. The slow process of securing the invitation from the Swiss sisters had been completed, and Stein was awaiting only her Dutch travel documents. When the German officers appeared at the cloister to arrest her, the superior sent Stein out to them because she mistakenly assumed that the officers were merely delivering Stein's exit permit.

[44]I have been citing the 1952 English translation of Posselt's biography--which, according to religious customs of the time, identified its author only by her religious name, "Sister Teresia de Spiritu Sancto, O.D.C." This work grew through many editions. Apparently it began as a booklet prepared for distribution to people who inquired about Edith Stein after the Second World War. It expanded to book length as the Carmel collected more statements from Stein's friends and colleagues. The introduction to the translated edition is dated 1948.

[45]For example, a statement by the Thomist Aloïs Dempf on "La grande oeuvre d'Édith Stein" is appended, pages 214-216; and Erich Przywara's recollections are cited prominently. The name of Elisabeth de Miribel does not appear on the copy that I was able to examine; however, that volume carries a preface by the historian of education H.-I. Marrou. Baseheart (1960: 1) cites a 1954 biography with a Marrou preface, from the same publisher, as having been authored by Miribel.

Dutch province of the Carmelite Order."[46] (Graef also translated selections of Stein's works in a 1956 volume that introduced her to the English-speaking world, and in 1960 she published a full-length translation of the first volume of Edith Stein's Werke, *Science of the Cross.)*

These three early biographies--all authored by women, one each in German, French, and English--are loving but credible portraits even though they assimilate Stein's philosophical work to her devotional writings. Altogether different is the ideological tone of Oesterreicher's 1953 sketch of Stein as a converted Jewish philosopher, to which we will return in the final section of this chapter. Virtually all recent publications on Edith Stein, no matter what their angle, present some biographical treatment of her. The most worthwhile recent biographies are Koepke (1991) and Herbstrith (1971; English translation of the fifth edition, 1985). An exhaustive bibliography of works by and about Stein was published by Hauke and Dick in 1984, and it may be supplemented by the sixth (1987) edition of Herbstrith's biography, by Reifenrath (1985), and by Otto (1990). Lists of archival holdings in Brussels, Köln, Echt, and Speyer are given in Bejas (1987). Linssen (1991) offers an update on the publishing activities of the Edith Stein Archive in Brussels.

We turn now to consideration of various projects of reading Stein. Three of them are disciplinary: those of education science, theology, and philosophy; while two others are community based: feminist and Jewish appraisals of Stein, respectively.

(a) Education-science readings. We have seen that Stein argued in her 1917 work on empathy that a person literally cannot "be educated" because a person *as such* is impervious to influences impacting her from beyond herself. The young doctoral candidate had proposed that a person could only be invited to unfold her own possibilities: by first empathizing them in a more mature person "of her own type." Subsequently, as I argued above, in becoming a Catholic Stein accepted the necessity of "direction" for the flow of consciousness. The pattern for the flow should be imported into the soul, through the guidance of someone else, from the classic Christian patterns of Christ, "the Cross," Mary, the saints, the liturgy, or the monastic rule. When Stein took up her teaching duties at the pedagogical institute in Münster in 1932, she was challenged to formulate a scientific theory of pedagogy, consonant with the Catholic beliefs and psychological theories of that day. She did not have time to complete that task.[47]

---

[46]See Graef (1955: vi). Presumably the "person in charge" was Lucy Gelber, who formerly had been associated with the Husserl archive in Louvain and had become archivist of the Archivum Carmelitanum Edith Stein in Brussels. Graef's biography treats Stein sympathetically, but conveys critical details (such as the negative evaluation of her teaching performance, cited above) that do not appear in the other sources.

[47]Stein's lectures on "the structure of the human person," published in 1994 as volume 16 of Edith Steins Werke, attempt to come to terms with this task. It is important to read them in this context, without forcibly assimilating them either to her earlier phenomenology of empathy, or her later metaphysical and theological writings.

However, Stein's lectures and popular articles on women's education and role in society, from the late 1920's and early 1930's, had already presented a kind of patchwork coverage of pedagogical questions. (This work became widely available in 1959 as the fifth volume of Edith Steins Werke.) From Thomism it derives the principle that "the soul is the form of the body" and is prior to the body.[48] Hence feminine souls are the source of feminine bodies, and feminine souls require appropriate education that will be different from the education offered to men. All professions should be open to women, although women will practice them in characteristically womanly ways.[49] Education ought to be tailored to human characteristics at three levels: the individual person, the gender or "species," and common human nature. Besides soul-based gender differences, persons differ by "type" and therefore require different kinds of pedagogical guidance. Throughout these discussions runs the figure of Mary, construed (with absolutely no regard for historical fact) as the exemplar of Christian womanhood or, sometimes, of Christian personhood irrespective of gender.

Had Stein been permitted to work at Münster for ten or twenty years instead of just one, and had she been able to bring these insights into mature philosophical coherence in dialogue with colleagues of all faiths, then perhaps a viable pedagogical theory could have been produced from them. Stein herself turned enthusiastically to this new line of research, and her supporters in the German Catholic educational societies were eager to see her undertake it. Tragically, however, the work was cut off short by Nazi repression in the spring of 1933.

This broken thread of Stein's thought was the first to be picked up anew by her academic disciples after the war. In my view, however, the results have been disappointing because the tentative and preliminary character of what Stein left behind has not been acknowledged. The first studies to focus specifically on Stein's educational theories came from the United States. Verbillion, using principally the selections of Stein's works translated by Graef in 1956, wrote *A Critical Analysis of the Educational Theories of Edith Stein* to receive the Ed.D. degree from Loyola University in Chicago in 1960.[50] Madden achieved more philosophical precision and historical objectivity in her 1962 Ph.D. dissertation *Edith Stein and the Education of Women: Augustinian Themes.* Baseheart, whose philosophical work on Stein will be discussed below, published articles on Stein's

---

[48]This often quoted formula receives philosophical development in Stein's Münster lectures on the structure of the human person.

[49]Stein's views on the "essence" of womanhood bear comparison with those expressed in a comtemporary philosophy of sexuality, Ernst Bergmann's 1932 *Erkenntnisgeist und Muttergeist: Eine. Soziosophie der Geschlechter*; see the summary in Knight (1933: 96-97).

[50]This work, now 37 years old, is startlingly contemporary in some of its formulations. It surveys the history of the question of women's education, but refers to feminism always pejoratively *and in the past tense*. Many passages in the early chapters are lifted verbatim from the standard biographies.

educational theory in 1966 and in 1989. Oben produced an annotated translation of the fifth volume of Steins Werke, *Die Frau*, for her 1979 Ph.D. dissertation at The Catholic University of America; this became the basis of volume two of the American series of Stein's works, *Essays on Woman*.[51] Unaccountably, this earlier scholarship of American women is not taken into consideration in Reifenrath's 1985 publication *Erziehung im Licht des Ewigen: Die Pädagogik Edith Steins.* Like Madden, Reifenrath situates Stein's theory in its historical and political context, but he also compares her philosophical orientation to contemporary scientific pedagogies.

(b) Theological readings. As we have seen, from the mid-1920's onward, Stein's work was guided by and assimilated into the production of an officially sanctioned Catholic philosophy. The enthusiastic reception of her work by philosophers of the Scholastic persuasion brought it to the attention of theologians but did not lead immediately into a fruitful theological appropriation of it. Reviews of *Endliches und ewiges Sein* in the 1950's are remarkably similar in tone to reviews of Stein's Thomas translation appearing in the 1930's.[52] The period of the 1950's and early 1960's was a transitional one for Catholic theology. Scholasticism had failed to offer any constructive engagement with the religious and social questions confronting post-war Europe, just as it had been impotent to resist Hitler's rise to power. The growing edges of Catholic theological thought lay elsewhere: in the voices of the young churches of Africa, in the liturgical movement, in historical-critical textual investigations into Christian origins and early Judaism, and in the "Christian existentialism" and "transcendental Thomism" of a generation who took their philosophical inspiration from Heidegger and his followers.[53] These theological currents were moving toward the postwar ecumenical council that would be known as Vatican II. They bypassed the work of Edith Stein.

Apart from hagiographical and inspirational works, no serious theological study of Stein appeared until Ruf's 1973 dissertation on the treatment of trinitarian

---

[51]Oben states that she converted from Judaism to Catholicism in 1960, and learned German for the sole purpose of reading Stein's works. See Stein (1987: vii). See also Oben (1990).

[52]Compare Przywara (1931), Naber (1932), Koyré (1932-33), and Dempf (1934) with Dempf (1953), Collins (1952), Geiger (1954), and even Dubois (1973). But Stein's old friend Fritz Kaufmann tried to situate her *Endliches und ewiges Sein* in its phenomenological context in his 1952 review of the work.

[53]The co-called transcendental Thomists--Rahner, Lonergan, and Marechal--are claimed also by proponents of Scholasticism, however. The term "Thomism" can be used ideologically to connote a kind of generic Catholic worldview comprising both an *intellectual* component with a faith-assisted epistemology and a *practical* component embracing social activities devoid of any apparent link to the thought of Thomas. In this usage, Aquinas functions as a "symbol" of this integrated but insulated way of Catholic life. The term Thomism is used in this diffuse sense by Ralph McInerny; see McInerny (1966: 162-166). American manifestations of Catholic intellectual life at mid twentieth century may not coincide in all respects with the German, however.

images in creation in *Endliches und ewiges Sein.* The next year, Guilead's survey of Stein's philosophy (for the official French edition, Les Oeuvres d'Édith Stein) canonized the theological interpretation of her work: that Stein had progressed *from* phenomenology *toward* the science of the cross, and that she had broken completely with phenomenology when she was baptized in 1922.[54] Theological interest in Stein began to revive in German-speaking lands with two studies that appeared in the *Freiburger Zeitschrift für Philosophie und Theologie*: Secretan (1979) and Kuhn (1988). Barukinamwo, an African scholar studying in Rome and writing in French, examined themes of creation and transcendence in Stein's later work, relegating her phenomenology to a preliminary treatment as if it were a mere biographical factor. His 1982 dissertation was followed in 1990 by a work adapting Stein's thought for an African cultural context. Ziegenaus (1987) offered a reconsideration of Stein's comments on Judaism, in the context of criticism evoked by her beatification. By contrast, Müller (1993) eschews theological interests in his study of Stein's philosophy of religion. His exploration begins instead with phenomenological issues identified by Lembeck (1988, 1991, and 1993): specifically, Stein's account of "the pure i" and the basis she finds for the concept of the deity.

Meanwhile, theological interest in Stein had continued outside academic circles. Throughout the 1960's and early 1970's a member of Stein's Carmel in Köln, Waltraud Herbstrith, had been publishing essays on spiritual and inspirational themes in Stein's life and works. By the early 1980's, it seemed, the trends of popular theology had "caught up with" Stein once more. Post-conciliar reforms were fostering interest in practical and spiritual theology and in liturgical formation. Manhausen (1983) treated Stein's life and writings almost as a case study to illustrate the theories of "mystagogy" propounded by the academic theologian Karl Rahner.[55] Bejas (1987) examined Stein's work on mysticism in its own right, although he reinscribed the maxim that she had left phenomenology behind in a kind of life-pilgrimage toward higher truth. Herbstrith edited volumes on Jewish-Christian dialogue and on spirituality, and she emphasized Stein's significance for women's theological agenda.[56] When a new Carmelite cloister was founded in Tübingen by the Cologne community, Herbstrith moved there. This presence has helped to engender interest in Stein's work among theological students and faculty in that city and elsewhere. Otto (1990) develops a theological

[54]See Guilead (1974). Compare Dubois (1973).

[55]The term "mystagogy" means a teaching of the mysteries. In patristic theology, especially in the writings of classic fourth-century Greek theologians, this term denoted the period of teaching that followed a new Christian's initiation into the church through the liturgy of baptism, eucharist, and chrismation. Since Stein herself had presented profound theoretical treatments of prayer, the liturgy, and the Christian spiritual journey, and since she also had quite a bit to say about teaching, it is no small insult to use her as *a mere illustration* for some other writer's theories.

[56]Herbstrith also has edited a volume of philosophical studies on Stein, and has contributed to scholarly journals. Her wide-ranging work must be mentioned both here and below in the discussions of philosophical and feminist readings of Stein.

foundation for faith and spirituality out of Stein's work on the human person.[57]

Currently there are at least two theologically driven interpretive programs that bear watching. Each seeks to enlist Stein into its cause, and with some warrant. Kalinowski (1987) would align Stein with Karol Wojtyla as a philosopher of the person, pursuing the alliance between Thomism and phenomenology that was attempted in the early 1930's. Herbstrith and many others would position Stein as a bridge for dialogue between Jews and Christians--but on terms that are objectionable to many Jews, as we will presently see.

(c) Philosophical readings. Confessional interests spurred and guided the philosophical reception of Stein, at least until the 1970's. As we have seen, in the 1930's Stein's turn to Thomas was publicly applauded by the very proponents of Scholasticism who had engineered it in the first place. The great hope expressed by Przywara and others before World War II was that a synthesis of Plato, Aristotle, Augustine, and (especially) Aquinas would equip the twentieth century with the conceptual tools needed to address contemporary epistemological and social challenges. Scholastic philosophers of the thirties saw the Phenomenological Movement as an ally in their battle against the great nemesis, Kantian relativistic idealism. Stein's writings therefore were read as the tracks of a pilgrim making her way into a holy land yet carrying with her some valuable provisions harvested from Husserl's garden. As suggested above, this was the gist of philosophical reviews of Stein's Thomas translation in the early 1930's, and twenty years later the posthumous publication of her masterpiece *Endliches und ewiges Sein* evoked the same sort of reading. This fact is quite disturbing, given what happened to Europe during those twenty years. Scholasticism was unruffled by World War II and the Shoah. This seems to be *prima facie* evidence that Scholasticism was no longer a living philosophy.[58]

Nevertheless, one can identify the Scholastic appropriation of Stein as one important trajectory within the philosophical reception of her work.[59] Already in

---

[57]That work, written at Tübingen, could now be expanded in light of the publication of Stein's Münster lectures on the structure of the human person. Hauck and Dick, who produced the 1984 bibliography of Stein, did so at the Tübingen Carmel, which is named in honor of Edith Stein.

[58]Here is not the place to attempt to fix the date of Scholasticism's demise. (One would first have to find evidence that it had been a living philosophy at some point in the nineteenth century.) Ralph McInerny speaks of "the Thomistic revival" as having drawn to a close by the early 1960's, before the Second Vatican Council. Nevertheless he regards the critique of the Enlightenment in contemporary philosophy as grounds for rapproachment between post-modernism and Scholasticism. (Personal interview, 7 March 1995, Notre Dame, IN.)

[59]The following account is intended merely to identify trends of interpretation, and it cannot pretend to be an exhaustive survey of the literature. I regret that I could read only the works in German, French, and English; for intriguing titles have appeared in Spanish, Italian, and Polish as well. Also excluded from consideration are dissertations which apparently had no impact upon later Stein studies, and whose authors presented no further publications. However, the very existence of those isolated dissertations and academic theses on Stein may have some significance. A number of them are to be found

1942, the American philosopher James Collins published a detailed discussion of Stein's phenomenology--that is, her dissertation and her three *Jahrbuch* articles. Collins followed Przywara in praising the realism of her thought and what he regarded as its inevitable approach to Thomism. Baseheart's 1960 dissertation examined the five points of comparison between Husserl and Thomas that Stein had set forth in her 1929 *Jahrbuch* article, offering corroboration from Stein's mature work *Endliches und ewiges Sein*. By checking Stein's work against her sources--Thomas and Husserl--Baseheart identified the originality of some aspects of Stein's position, such as her account of individuation.[60] Three more dissertations on Stein's philosophy appeared in the 1960's from Roman universities.[61] Höfliger (1968) examined the problem of universals in Stein's *Endliches und ewiges Sein*, comparing it to treatments in Aristotle, Thomas, Duns Scotus, and other Scholastics; his work was published in Switzerland under Dominican auspices.[62]

Post-war French philosophy had embraced phenomenology in the form of existentialism, which had become the idiom for reflection on the social and personal disruptions of mid-twentieth-century European life. But Dubois's 1973 essay in the journal *Revue Thomiste* attempted to recover the earlier momentum toward a synthesis of phenomenology and Scholasticism, begun at Juvissy in 1932. Stein Studies received an unexpected boost in 1979 with the election to the papacy of Karol Wojtyla, a philosopher who had written his dissertation on Scheler and who was trained in the early or "realist" brand of phenomenology practiced by both Stein and her Polish friend Roman Ingarden. In method and in topics, there was much common ground between Stein and the new pope. For example, Kalinowski (1984) compared the treatments of personhood in their respective works. Wojtyla himself was an admirer of Stein, whom he regarded as a martyr. He raised her to the official status of "blessed" on 1 May 1987, with fulsome and ideologically driven praise. (Stein, it seemed, had attracted another Polish Pygmalion.) In 1988 Gaboriau published a four-part discussion of Stein's life and work in *Revue Thomiste*, which was reprinted as a book in the following year. This treatment employs the familiar device of portraying Stein as a pilgrim who left phenomenology, found Thomism, and passed on to us the building blocks for a fruitful contemporary Scholastic philosophy. But Gaboriau excels his predecessors in his frank assessment of Stein's masterpiece as an *unfinished* synthesis, and in his

---

in the Archive of the Carmelite cloister in Cologne, and I have appended a chronological list to this work.

[60]To my knowledge, Baseheart was the first interpreter to entertain an expectation that Edith Stein might have had some thoughts *of her own*; that is, some philosophical positions that deviated from those of Husserl, Thomas, or other canonical authors. This insight, and its fruits, had to be expressed very guardedly in her dissertation. See, for example, Baseheart (1960: 188, note 8).

[61]See Bajsič (1961), Bernbeck (1966), and Salmen (1968).

[62]The Dominican order includes many proponents of the thought of Thomas Aquinas, who belonged to it.

nuanced awareness of the plurality within both the Scholasticism and the phenomenology of her day. What would have happened, he wonders, had there been a colloquium in 1933 between the Scholastics and the phenomenologists, including Heidegger? Gaboriau proposes an agenda for such a discussion, and the second half of his book sketches how the issues would have to have been engaged, based upon extrapolation from Stein's writings. This attempt to "prolong the echo" of Stein's thought, he says, is the next step for someone who has studied her philosophy.[63]

In the United States, meanwhile, Graef's biography and translations of the 1960's had attracted the attention of readers who were interested in saints. Baseheart was struggling almost single-handedly to win recognition for Stein among philosophers. In 1981 she presented a paper on Stein before the American Catholic Philosophical Association. That event led to a conference on Stein's work in 1984 sponsored jointly by the Institute of Carmelite Studies in Washington, D.C. and the School of Philosophy and the Friends of the Library of The Catholic University of America.[64]

At that conference, McInerny formulated the question why Stein's philosophy had been largely ignored. His historical analysis differs markedly from the one that I have proposed. McInerny says that Thomism, including Stein's philosophy, is ignored because Catholic philosophers simply are no longer "docile" to the official ecclesiastical advice that they study Thomas. Stein herself, he says, first turned to the study of Aquinas because she followed church policy. Nostalgically, McInerny writes that

> Edith Stein reminds us of a better time when the believing intellectual, far from being apologetic about the faith and accepting the suggestion that it is intellectually suspect, glories in Revelation as a positive boon for the human mind. (1984: 85)

McInerny states that people will not be interested in Stein unless and until they are responsive to the Catholic magisterium. But he predicts that Stein will come out of the shadows and, with Maritain and Thomas, will "play a significant role in the future intellectual and spiritual lives of many" (1984: 79). In the decade since McInerny made that prediction, Stein has indeed moved back into the philosophical spotlight, but not in the way he expected. Scholastic philosophy has not come back to life; Stein's rediscovery is owing to other factors. The one exception that might count as confirming McInerny's prediction is MacIntyre (1993), who in the wake of his own acknowledged conversion to Catholic traditionalism now proposes Stein as an exemplar of Catholic philosophy.[65]

---

[63]See Gaboriau (1988: 440).

[64]See *Carmelite Studies* 4, where the papers are published, for a description of the conference.

[65]See MacIntyre (1993). McInerny and MacIntyre were colleagues at the University of Notre Dame. McInerny (1966: 48-56) argues for the importance of tradition in philosophy.

What I have termed the Scholastic trajectory has been an important and constant interpretive strand in the fabric of Stein Studies; but there are others. They run with the trends in post-war European philosophy, soundly defeating any claim that interest in Edith Stein waxes and wanes in exact correlation with "docility" to Rome. To begin with, Stein's phenomenology fell into shadow when Husserl's own mature transcendental phenomenology led him away from their common ground, and even more when Heidegger's project put Husserl himself into the background. Existentialism responded to questions arising from war experiences; Scholasticism did not. Moreover, at least in the United States, Scholastic philosophy in the post-war era neglected to engage with the issues that were enlivening the Analytic tradition: questions of metalogic, philosophy of science, and language analysis. This was owing in part to its academic confinement within Catholic colleges.

The timing of the posthumous publication of Stein's works also affected the degree and kind of interest that they evoked. In Germany the later, mature works were published first; while the American series emphasized Stein's personality.[66] These publication strategies are compared in Figure 5.1.

FIGURE 5.1: COMPARISON OF PUBLICATION STRATEGIES

EDITH STEINS WERKE		COLLECTED WORKS OF E.S.	
1950	*Kreuzeswissenschaft*	1986	the autobiography
1950	*Endliches und ewiges Sein*	1987	*Essays on Woman*
1952	new edition of the Thomas translation (2 volumes)	1989	new edition of the dissertation
		1992	*The Hidden Life* (inspirational essays)
1959	*Die Frau* (lectures and articles)		
1962	*Welt und Person* (philosophical essays)	1993	*Self-Portrait in Letters* (combining the two German volumes of correspondence)
1977-8	*Selbstbildnis in Briefen* (2 volumes)		

With the American series, appearing under the sponsorship of the Carmelite order, priority seems to have been given to fostering devotional use of Stein. Therefore

[66]Both series are continuing. The German series now numbers 17 volumes; the American, 5.

anglophone philosophers cannot yet read her most important philosophical works if they cannot read German. Both philosophical and religious interest in Stein was heightened by her beatification in 1987. In that year, Imhof's publication of the first half of his dissertation also established the basic outlines of an intellectual biography of Stein. His analysis places into perspective the relative value to philosophy of the various works by and about Stein. For the American reader, it underscores the urgency of making Stein's philosophical works available in creditable English translations.

However the single most powerful factor to affect the nature and degree of interest in Edith Stein has been the women's movement that revived in the early 1970's and profoundly transformed the academy across the disciplines. The quest to recover *both* philosophies of woman *and* the philosophies of women has led many scholars to rediscover Edith Stein. (A distinct feminist interpretation of Stein is emerging, and will be treated separately below.) Encyclopedia articles by Herbstrith (1988) and by Baseheart *et al.* (1995) bear witness to this trend in scholarship, and insure that Stein has entered the academic canon of philosophy. The increase in publications on Stein since the late 1980's is owing in part to the general heightening of interest in women's philosophy, and in part to the discrediting of Heidegger because of his support of National Socialism and the consequent broadening of interest in other figures from his era. However several direct initiatives also fostered philosophical study of Stein. In the United States, Baseheart established the Edith Stein Center for Study and Research at Spalding University, with a growing archive and a program of annual lectures. Several philosophical conferences on Stein's thought also have been organized on both sides of the Atlantic. Besides the 1984 symposium at The Catholic University of America, conferences were held in 1991 at Rolduc and in 1993 at Eichstatt, with publications following.[67] Herbstrith (1991) collected papers on Stein's philosophy that also belong to this new generation of scholarship. Increasingly, Stein is placed in her own philosophical context and her arguments receive critical comparative attention.

Stein Studies in the 1990's, then, moved beyond the genre constraints that seemed to be operating in earlier philosophical readings of her work. Stein was no longer interpreted formulaically as the "type" of a Christian philosopher, or as the shining example of the Catholic professional woman. Nor was she continually "introduced" as an oddity. One did not ubiquitously encounter the facts of her life (or texts from her autobiography) wielded as though they were keys to unlock a body of thought that otherwise would remain unintelligible. Nor was one asked to view her death as a martyrdom and therefore as a sacred seal rendering her philosophy unassailable. The genres of Stein Studies at last became those of *philosophy in general*: historical recovery, internal and comparative analysis, and application to new problems. Three book-length general introductions to Stein's philosophy were available: Secretan (1992), Gaboriau (1988), and Matzker (1991).

---

[67] See the volumes edited by Elders and by Fetz *et al.* These recent papers represent a diversity of approaches that cannot be examined individually here.

Although the last is rather journalistic in tone, all three present straightforward accounts of Stein's thought in its intellectual, academic, and historical contexts. A comparable English-language work is to be wished for.[68] In Germany, an Edith Stein Society was established in 1994 and reportedly has begun to publish a journal entitled *Stein-Studien.*

It is tempting to conclude that ideological readings of Stein continue now only on two fronts: in feminist interpretation, and among those who desire to position Stein as facilitator of religious understanding between Jews and Christians. (These options are considered further below.) But such a conclusion would be premature. The canonization of Stein *by practicing academic philosophers* should be regarded with no less suspicion than her expected canonization by the pope. What I have just termed "the genres of philosophy in general" are conventional practices of reading that must be called into question. "Historical recovery, analysis, and application" were the terms that I used just now to denote the activities constituting mainstream philosophy. But we have already encountered a rather wide variance from those practices, even within the relatively homogenous stream of interpretation that I designated the "Thomistic trajectory" of Stein readings. On one hand, McInery praises Stein's "docile" acceptance of Thomas and regards her as a signpost on the way back to authentic Scholastic principles. On the other hand, Gaboriau favored "prolonging the echo" of Stein's own work by carrying forward with the lines of thought that she herself laid out.

Both of these programs are ways of working *with* Stein, not *on* her; yet they differ in their reading strategies. The "docile" reader demands perfect coincidence between Thomas and Stein, and then between Stein and her contemporary disciples, all enforced by loyalty to the teaching authority of the Catholic Church. The "echoing" reader wants to think the thoughts that Stein herself wanted to pursue, but with adaptations for new situations.[69] Those egoic partnerings of the reader with Stein, though they engage Stein's subjectivity in different ways, contrast sharply with other philosophical practices that would treat Stein as an object. What kind of object can a philosophical text become? Stein's enrollment in the philosophical canon means that she now is in danger of turning into a portrait in the rogues' gallery of philosophers, a vignette for a women's history course, a stockpile of arguments to be mined, or an arsenal of weapons for defending one's own opinions. How to position the philosopher's reading i vis a vis Stein's authoring i is a question to which we must return in section B of this chapter.

(d) Feminist readings. As a woman philosopher and as a philosopher

---

[68]Publication of an English-language introduction to Stein's work, by Mary Catharine Baseheart, was announced for late 1996 by Kluwer.

[69]Gaboriau presented this option in philosophy, and Verbillion presented it in education science. Ziegenaus, a theological interpreter, suggests a looser version of this option. He says that readers of Stein today must not apply her actual words to present situations, but must discern what Stein *would have said* had she lived through the historical events of the 1940's through the 1980's.

who wrote about womanhood, Stein arouses interest within the contemporary feminist movement. Stein called herself a feminist in her student years at Breslau and Göttingen, before women achieved the right to vote in 1919. As a university student she was an advocate for working women's education, and had once intervened when she saw that a male teacher was taking advantage of women students. Later, however, she wrote that women were essentially different from men and required special education. She declined to marry, and as a nun she allied herself exclusively with a closed community of women and directed most of her correspondence to the women whom she loved. While this complex profile does not fit easily into the agenda of any segment of the women's movement of the 1990's, contemporary feminist readings sift through Stein's life and works to reconstruct a useable history.

As mentioned above, Verbillion (1960) actually had positioned Stein as an opponent of feminism, a movement that Verbillion was able to describe in 1960 in the past tense. Oben (1979 and 1990) continued to emphasize the *contrasts* between Stein and later feminists. Many commentators, however, seek and find important *continuities* between the thoughts and deeds of Stein, and the objectives of the women's movement in the 1908's and 1990's. McAlister (1989: 9) writes:

> This then is the evidence of Edith Stein's feminism: her youthful support of the women's rights movement; her friendship and support of other women in philosophy; her willingness to fight for the rights of herself and others in the face of discrimination; her efforts to effect a synthesis of feminism and Roman Catholicism after her conversion and particularly on the subject of girls' education and women's careers; and the congruence between her religious thought and that of contemporary Christians who seek to emphasize an explicitly inclusive and feminist theology.

McAlister is editor of *Hypatia*, which is the leading English-language feminist philosophical journal. Baseheart's 1989 article in *Hypatia* brought Stein to the attention of feminist philosophers in the United States.[70] In the same year in Germany, Gerl published a selection of Stein's writings on "the women's question" which reintroduced Stein's voice into the feminist philosophical conversation in that country. The beatification of Stein in 1987 directly promoted her canonization within the literature of feminist theory and women's studies. Biographies by Koepcke (1985 and 1991), Gerl (1991), and others sought to recover Stein as a piece of women's history. Not all women's history is written from a feminist perspective, however. Herbstrith's earlier biography, translated into English in 1985 but soon out of print, was reissued in 1992 by a conservative Catholic publisher.[71] Koeppel's hagiographical biography appeared in 1990 in a series on Christian mystics. Cross (1989) suggested that Stein offered a path of compromise

[70]Including the present writer. However I had earlier met Stein's work through Graef (1955) and Graef's 1956 translation of excerpts from Stein, which I read in the 1960's as a high school and college student.

[71]The first English edition was brought out by Harper and Row. After it went out of print, it was reissued by Ignatius Press.

to repair "the shattering impact of feminism on the Church."

Herbstrith (1993), however, offered positive proposals toward understanding the being of women on the basis of Stein's thought and life. Miles (1991) attempted to use Stein as a kind of interpretive key to unlock obscure portions of Simone de Beauvoir's thought, by positioning Stein and Beauvoir as proxies for an encounter between the philosophies of their respective mentors, Husserl and Sartre.[72] McAlister (1993) proposed a more suitable way of positioning Stein philosophically. She placed her within a constellation of feminist options on the question of whether there is an "essence" of womanhood, and examined those options in the context of the larger issue of how sameness and difference may be theorized. McAlister also urged philosophers not to be put off by the religious tone of Stein's mature writings:

> I believe that Stein's philosophy of women can be studied by secular feminist philosophers and that it has a potential contribution to make to their exploration of feminist philosophy. There is no essential difference--just differences in style, vocabulary, and point of departure--between Stein's efforts to understand the extent to which women are socially constructed and contemporary feminists' efforts to explore the same issues.

McAlister also helped to place Stein within the feminist philosophical canon by co-authoring the entry on Stein for the series *A History of Women Philosophers* (Baseheart *et al.* 1995).

Thus we see that while feminism is a "topic" found amid interpretations of Stein, it is a topic in a very different way from that in which education-science, theology, and philosophy were found as topical areas. This peculiarity is owing to what McAlister identifies as the socially constructed character of womanhood, and to the co-participation in that construction by Stein and her readers alike. Stein not only *writes about* women, she *is* one; and this very being-as-woman becomes an explicit component of Stein's theorizing. The same goes for many of her readers. All of her readers, whether men or women, feminist or not, have a stake in determining what womanhood is. "Objective" dispassionate interpretation of Stein on this score becomes impossible.

(e) Jewish readings. Like gender, race too is a social construction. Stein not only *writes about* Jews, she *is* one. What it means to be a Jew is treated explicitly in her writings, especially her autobiography and her writings on spirituality.[73] Moreover, all of Stein's readers, gentiles and Jews alike, have a stake in determining what Judaism is. Jewish and Christian believers have

---

[72]This doctoral dissertation from Claremont suffers not only from its curious rationale, but also from the fact that Miles relied upon flawed translations and neglected to consult the secondary literature. The work is 89 pages long in typescript, and lacks a bibliography. Nevertheless it must be mentioned in this section because it purports to read Stein within "a philosophical analysis of feminism."

[73]However, she explicitly refuses to discuss race in her Münster lectures on the human person. There she argues that the more significant category is "the people" (Volk) as a community (Volksgemeinschaft) in which persons participate. See Edith Steins Werke 16: 177-178.

opposing views on religious grounds, of course, because of their disagreement over the status of Jesus of Nazareth. Yet even non-believing readers of Stein must care about the practices of racial identification; for no human being who hears of the Shoah can avoid considering how it can have happened and how we are obliged to live in its aftermath.

The Jewish identity of an author took on increasing hermeneutic significance in the 1930's under National Socialism. Any work of fiction or non-fiction that promoted liberal social views might be designated as decadent "asphalt literature," supposedly churned out by Jews to pave over the authentic expressions of the pure Aryan soul. Even works of gentile authors were included in the condemnation of "Jewish literature"; you didn't even have to be a Jew to write it. But all writing by Jews past or present was suspect, and its publication was forbidden. Practically speaking, it was difficult for Germans to tell which of the works that they had previously enjoyed now were to be avoided as "asphalt." Handbooks had to be issued.[74] Like their texts, the bodies and faces of Jews were not distinguishable from those of other Germans, Poles, Czechs, or Dutch.

The single most decisive hermeneutical act to affect the texts and the bodies of Jews occurred in antiquity when Christians designated the Hebrew scripture as an "old testament." The obsolescence of the Tanak had to be asserted if it were to be replaced with a new testament, or agreement, between humanity and the deity. The continuing existence of the people of the original testament continually reminded Christians that their founding hermeneutical act had not succeeded and therefore could not be attributed to the initiative of the *almighty* being whom both Christians and Jews worshiped.[75]

Stein, like many in her day, distinguished between the Jewish race and the Jewish religion. She still considered herself to be a Jew after she had given up Jewish faith in her teens and accepted baptism at the age of 30. On this view, Jewishness as a racial identity is no impediment to accepting--and being accepted into--the Christian church. The pejorative and condescending attitude toward Judaism in this view is striking. Jewishness is regarded as a racial flaw that can be generously overlooked in someone who gives up the religious expression of that racial heritage.

As we saw above, this was the attitude that Christians held toward Stein during her lifetime. Her Scholastic philosophical mentor Przywara was glad that his protégée didn't look Jewish; while Nazis in Holland offered to leave baptized Jews alone if the Catholic bishops would acquiesce to the deportation of the unbaptized Jewish population. This motif--Judaism as a water-soluble racial stain--is repeated in one of the earliest posthumous textual interpretations of Stein's life and work. Oesterreicher included her among the seven thinkers whom he discussed in a widely read and reprinted work, *Walls Are Crumbling: Seven Jewish*

---

[74]As mentioned in chapter one, the 1943 edition of a philosophical handbook meticulously identifies which thinkers were Jewish.

[75]This was acknowledged by the Second Vatican Council.

Philosophers Discover Christ, first published in 1953.[76] He says in the preface:

> For almost two thousand years, Jews have denied that Jesus is the Messias. For almost two hundred years, modern thought has been a philosophy of denial and contradiction. Finally, rancor has invented the slogan, "Jews lack the power of spiritual thought, they are to blame for the destructive philosophies of materialism." . . . For centuries Jewish teachers have rejected Jesus' claim to be the Christ, the one and only Savior of the world; for centuries there has been a conspiracy of silence, so that His Name was not uttered, neither in home nor in house of learning. Jewish unbelief and, no less, Gentile persecution built a wall which kept the Jews from Christ, but now this wall is giving way. (1953: xiii)

Oesterreicher enlisted Stein posthumously into the ranks of a final wave of conversions that were supposed to bring the end of the Jewish religion, once and for all.[77] Jewish religion was "unbelief," and it had to end.

The same rationale underlies the more benign expressions of some who have asserted that Stein's life and works can provide the basis for dialog between Christians and Jews. Schandl (1990) collected Stein's statements on Judaism and highlighted what he saw as "Jewish elements" and themes in her works, as if to assert that (and illustrate how) one could remain Jewish while becoming Christian. Böckel attempted something similar in a more popular presentation in 1989, the second edition of which (1991) included some Jewish responses. Yet despite Christian fantasies, Stein's way is not one that Jews wish to follow. Her beatification in 1987 provoked negative responses in the Jewish community. A dialog has begun, but the conversion of Jews is not up for negotiation. Herbstrith (1990) edited a volume of statements representing both Jewish and Christian perspectives on Stein and on the significance of her beatification. She wrote:

> Durch Edith Stein lernen wir, wie behutsam wir als Christen mit Vereinnahmungen und triumphalistischen Tendenzen in unserer Sprache umgehen müssen. Wirkliche Ökumene, und Edith Steins Gestalt ist ökumenisch, heißt, daß wir das Denken, den Glauben der Anderen ernst nehmen und in ihrer Eigengesetzlichkeit stehen lassen. (1990: 17)

> Through Edith Stein we learn how carefully we as Christians have to deal with inherited phrases and triumphalistic tendencies in our language. Real ecumenism--and Edith Stein's profile is ecumenical--means that we take the thought and the faith of others seriously and let them stand just the way they are.

---

[76]This work was trimmed and republished in 1967 as *Five in Search of Wisdom*. Its new preface repents of the earlier subtitle insofar as it seemed to imply that those seven Jewish philosophers had made no other discoveries. But Oesterreicher dismisses most other criticisms of the earlier work as attributable to misunderstandings by the critics.

[77]Oesterreicher also interpreted Stein as a convert *from feminism*. The opening lines of his essay illustrate that he worked just as hard at the construction of gender as at the construction of race: "An unmotherly woman--which is not at all the same as a woman who has no children--is a spectre, almost all the world admits. Only a feminist, in the fever of the extreme, will deny that woman's calling differs from man's that for all their common human nature, the souls of men and women have each their own design, declared Edith Stein, who herself had known this fever." See Oesterreicher (1953: 288).

This volume signals a new willingness to allow "entirely different answers" to be expressed side by side, and indeed, a new recognition that Stein's life and work remain a compelling challenge for all sides. Among the more than sixty statements about Stein collected by Herbstrith for that kaleidoscopic volume, none attempts to impose a definitive interpretation.

There are many points where Jewish interpretations of Stein diverge from those of her Catholic adherents, but two questions stand out with particular significance: whether Stein should still be regarded as a Jew, and the meaning of her death in the camp at Auschwitz. In the estimation of her surviving Jewish relatives, Edith Stein remains a Jew. A niece, Suzanne Batzdorff, was a girl of twelve when she last spoke with her aunt before Stein entered the Carmelite cloister.[78] Batzdorff has contributed several essays to discussions of Stein in German and in English, and she has translated some of her aunt's work. She writes:

> I feel, as did my mother, who died in 1978, that Edith Stein was a human being who accomplished much, contributed to philosophical and religious literature, won the love and admiration of many and died a horrible death. Though she was a Catholic who embraced her chosen faith with joy and devotion, she was not, in the end, separated from those who had remained Jews and were killed because they were Jews. (1987a: 70)

Batzdorff contends that Stein died *because she was a Jew*. There is a report, perhaps legendary but found in some biographies, that when the officers were taking them away from the cloister Stein said to her sister, "Come on, let's go die for our people." Christians understand this remark as expressing Stein's intention to expiate for the supposed guilt of the Jews. But Batzdorff takes serious exception to this interpretation. She writes:

Das Judentum erkennt das Martyrium als stellvertretende Tat für den Mitmenschen nicht an. Jeder Mensch muß vor Gott für seine eigenen Taten Rechenschaft ablegen. Das gibt es keinen Vermittler. Daher ist die Idee, daß Edith "für ihr Volk" in den Tod ging, nach unserem Glauben unhaltbar. Dazu kommt noch, daß sie ja in den letzten bitteren Tagen ihres Lebens klar sehen konnte, daß von einer stellvertretenden Rolle ihrerseits keine Rede sein konnte; es wurde ihr Schicksal, **mit ihrem Volk, nicht für ihr Volk** zu sterben. Ihr Opfertod war schließlich ein Tod wie der von Millionen von Juden, die mit ihr starben. Er kam nicht als Konsequenz einer freiwilligen Entscheidung, sondern der Willkür. . . . Eine Wahl oder einen Ausweg gab es für sie nicht. (1991: 135-6)	Judaism doesn't recognize martyrdom as an act of standing in for a fellow human being. Every human being has to render an account before God for his or her own deeds. There is no middleman. Therefore the idea that Edith went to death "for her people" is untenable according to our faith. As it happens, in the last bitter days of her life she could see quite clearly that there could be no talk of her standing in for anybody else. It became her own fate to die **with her people, not for her people**. Her sacrificial death was finally a death like that of millions of Jews who died with her. It came as a consequence not of a free decision, but of a brutal ruler's caprice. . . . For her there wasn't any choice or any way out.

[78]See Batzdorff (1987a) for an account of that last conversation. Batzdorff is the child of Edith Stein's closest sister, the physician Erna Biberstein.

If Stein died for being Jewish, and not for her Christian faith, then the question arises how anyone can interpret her death as a martyrdom for Christ. This point is made by Baaden (1987) as well as by Batzdorff. Fuchs-Kreimer (1991) finds qualities to admire in Stein's life and work, but expresses anger at Stein's decision to abandon the Jewish religion without attempting to explore the rich depths of her own tradition.

The most cogently argued critical interpretation of Stein on Jewish premises comes in an "open letter to Cardinal Lustiger" published by Raphaël Draï, a political scientist. Some revisionist historians, says Draï, have sought to minimize the horrors of the Shoah or even deny outright that it ever took place. He sees this program of denial as a kind of second, rhetorical holocaust. The deliberate and calculated non-recognition of the Shoah is a double negation (*dénégation*) that annihilates the annihilation of the Jewish people. Draï reads several of the official actions of Pope John Paul II as complicit with this program. Among them are: the appointment of Lustiger, a convert from Judaism, as cardinal archbishop of Paris; the beatifications of Stein and Maximilian Kolbe, who died in place of a fellow prisoner at Auschwitz; and the universalization of Auschwitz by homiletically designating it as a place where people of many nationalities died.

These moves are cited by Draï to show how papal policy has sought to efface Jews from twentieth-century history through a rhetorical strategy of equivocation. In each instance, Draï's charges have a bearing upon the interpretation of Stein. First, Lustiger's conversion raises the question of how Jewish identity is to be legally determined. Draï says that Lustiger remains a Jew according to traditional religious law (*halachah*), and he cites historical cases in which people who apostasized during troubled times were still recognized as Jews. But by the Israeli constitutional Law of Return, however, Draï cites precedents to indicate that Lustiger would not be judged eligible today for citizenship in the State of Israel. By implication, then, although Stein's relatives might continue to regard her as a Jew, it is unlikely that an Israeli court would have granted her citizenship had she survived the war and sought to emigrate to Israel.

Second, Draï also detects a rhetorical re-annihilation of Jews in the decision to single out Kolbe's death as an example of self-sacrifice. Kolbe, a Catholic priest, traded places with the father of a family who was about to be isolated and starved to death in the camp. But the man spared through Kolbe's heroic generosity was a fellow Christian, although the camp was full of Jews in peril every day. Draï does not doubt that there may have been many acts of heroism by Christians on behalf of Jews; he simply asks why those acts were overlooked in order to throw the spotlight of acknowledged sanctity upon something that happened between Christians.

Third, in effect the pope is trying to take Auschwitz away from the Jews, according to Draï. He points out that the word "Jew" was hardly mentioned by the pope when he spoke at Auschwitz. In the Catholic imagination, the camp is becoming assimilated to Golgotha, the hill where Jesus died, and Poland is turning into the new Holy Land. Poles, who one might think would bear some responsibility for what occurred on their territory, instead regard themselves as having a christological vocation of which Auschwitz is the confirmation. Thus celebrations of the two Christian martyrdoms alleged to have occurred at Auschwitz--Kolbe's and Stein's--romanticize the place and obscure the fact that it nearly succeeded at what it was designed to do: wipe out the Jewish people.

Draï has one more complaint against Stein herself. She was, and deliberately remained, ignorant of the philosophical traditions of her own people. Her vaunted quest for truth took a detour around the Mishnah and the Talmuds.[79] She could compare Husserl with Thomas, but not Maimonides. She would translate classical works from English and Latin, but not Hebrew. I will chime in with Draï and close this section by adding an observation of my own. After her baptism, Stein would spend many hours each day in prayer using the Roman breviary, which is largely made up of translated Psalms. The original Hebrew version of those prayers is quite lovely--all the more so for a Christian who realizes that those very syllables once came alive in the mouth and heart of Jesus, too. The poetry of the Psalms does not survive in the translation that Stein used. By choice, Stein fed her spirit on texts whose primal zest had been boiled down to bland correctness. The hardest thing for me to understand about Edith Stein is not why she left philosophy, but why she prayed in Latin.

B. Interpretations of Stein, According to Deployment of I's

The discussion in Part A of this chapter established that Stein's work has evoked academic discussion in several disciplinary areas, that the academic discussion is amplified by discussions in popular media and by social movements responding to Stein's life and work, and that these discussions are rapidly expanding. Our review of the profusion of Stein literature could barely dip below the surface of numerous issues that invite further research. Our concern was to understand the surface itself: the terrain of a new philosophical subfield--Stein Studies--for which we have provided a first and very tentative mapping.

However, we alluded also to the danger inherent in launching into what might be termed "philosophical business as usual" in the academic exploitation of this newly discovered "field." Ironically, the canonization of Stein within the list of recognized important philosophers, which signals a kind of *success* for her philosophy after some 70 years of obscurity, may also portend the *failure* of her own phenomenological project, which was intended to overhaul how we understand understanding itself.

Recognizing that danger, in this section we will take another look at projects of interpreting Stein. Instead of classifying them conventionally by topic

---

[79]Fuchs-Kreimer, like Draï, mentions the comparable case of Franz Rosenzweig. He was a young philosopher on the verge of abandoning Judaism, but discovered its spiritual depth during a visit to an Orthodox synagogue on the eve of Yom Kippur. Why could Stein not have had such an experience? Fuchs-Kreimer, from a Jewish feminist perspective, remarks: "Would Franz Rosenzweig have remained a Jew if he had been hurried upstairs to sit behind a *mehitza* that fateful night? Frankly, I doubt it." (1991: 7). Rabbi Fuchs-Kreimer's reference is to the tradition of separating women from men at worship either by means of a screen or in a balcony.

and by their chronological development and historical context, as we did above, now we will suggest a different sorting. One can identify several different manners in which the i of the interpreter allows itself to be led by the i of Stein the writer. Conscious life is a flowing stream, according to Stein's hermeneutic, and the living i rides its feelings through into expressions of various kinds. Texts are one variety of expression. Any expression makes a live and immediate presentation *both* of a meaning *and* of the particular individual i whose meaning it first was. Through empathy (*Einfühlung*), Stein held, the i of the one who understands an expression follows along with the i who expresses. The one who understands thus receives *both* meaning *and* intimate inner subjective contact with another i--not just as any old generic i differing only numerically from oneself, but rather as the unique active originator without whom this received meaning would be different than it actually is.

Herein lay the roots of Stein's divergence from Husserl on the issue of constitution. Husserl assumed that the conscious object whose unity was enacted by one subject would be identical and interchangeable with the conscious object constituted by any other subject. *Who did* the constituting didn't matter. In fact, Husserl defined constitution by abstracting whatever might be owing to the individual character of any given existent subject. He postulated a "pure i" as being that object-unifying capability shared by any subject whatsoever. Husserl did not claim that the pure i *exists*; quite the contrary. But the functions of the Husserlian pure i are exactly those expected of the ideal reader of a text. The premise of reading is this: that meaning will be reconstituted from the text irrespective of the individual characteristics of particular readers. The text is built to enforce identical replications of meaning in all possible readers. A well-built text excludes ambiguities. A well-trained reader tries to coincide with the thought of the writer. The act of reading makes the words of the writer flow through the conscious life of the reader.

Stein saw that real-life interpretation of expressions never matches this ideal scheme. Nor should it. The success of such a program of reading would efface the artifactual character of the text as well as the factual character of its maker. But in fact, the actual occurrence of understanding discloses something *prior to* the constitution of more-or-less identical objects for consciousness. Constitution of same or similar meanings is founded upon constitution of *the other subject*, with his or her personal peculiarities, and furthermore upon constitution of *one's relationship with* that other subject. Even in the limit case--when the relationship approaches coincidence, when the led i merges into identity with the leading i, when the reader forgets the alien authorship of the text and complelely appropriates its meanings--the act of empathy (*Einfühlen*) still has preceded and founded the act of constitution. Object-constitution concludes and occludes the act of ablative empathy, yet still includes it anonymously within the intention of the object *as real*--that is, as available in principle to vague "others" whose constituting acts would do nothing other than corroborate one's own.

Husserl was concerned, throughout his career, with how science is epistemologically secured. His conception of "science" was determined by his own culture and historical era. Thus, his account of constitution and of the "pure i" that accomplishes it was focused narrowly upon establishing the replicability of "the same" meaning regardless of who might entertain that meaning. He concentrated on the constitution of *objects*, neglecting the role played *within object-constitution itself* by empathy of other i's as constitutors and by creation of relationship among them.

Stein's hermeneutic prescribes a fuller picture. Science is just *one* kind of understanding. But *all* kinds of understanding comprise egoic activities coordinated on three fronts: empathizing an alien i, contracting a relationship of responsible following with that i, and constituting objects affirmable by that i. Moreover, each of those activities can be done more or less adequately, while their coordination can be more or less well balanced. This complexity suggests a way to diagnose certain trends in how Stein's life and works are read.

1. The authority syndrome, and "docility."

As we have seen, McInerny regarded Steins' adoption of Thomistic philosophy as evidence of her docility to ecclesiastical advice. Having discovered Thomas (or having had Thomas discovered to her), Stein relinquished (or abdicated) the capacity of her own flowing i to register the coherence of an argument. She "docilely" allowed herself to be led along the paths of Scholastic philosophy and christic vicarious suffering. There can be no phenomenological intelligibility to such a choice. It is free and mysteriously personal, even though many would consider it a sin.

This choice of Stein's elevates her, in the eyes of some interpreters, to a place of authority indistinguishable from that of Thomas of Aquin and Theresa of Avila. Stein is invested with the mantle of unquestionable authority. This is equivalent to an alteration in how Stein *can be read.* Thus among some Stein scholars one encounters a pathology of interpretation in which the reader abdicates the responsibility to evaluate what the author has expressed, but simply accepts it on the author's word. The purpose of interpretation then becomes merely the establishment of Stein's doctrine. One asks what she thought, but does not go on to ask whether she was correct. After baptism, Stein herself read Thomas and other classic Christian texts in this way--quite unlike the way in which she earlier had read Husserl.

Among Stein's philosophical and theological interpreters alike, there is widespread reluctance to criticize her. This docility may be owing in part to the exaggerated awe with which many regard Husserlian phenomenology. It seems to be viewed as a great dark forest where thinkers get lost without the help of an adept to lead them by the hand. Stein is positioned as an expert in the ways of this inscrutable and imposing system. Another factor that dampens criticism of Stein is the tragedy of her death and the aura of beatification which now surrounds her legend.

Thus even the most astute philosophical commentaries on Stein rest content with giving a historical account of what she wrote. They do not discuss its social genesis, nor do they redirect any of Stein's investigations or amend her results. This has the unfortunate effect of sealing Stein's thought into the past, for without being bent and pruned it cannot grow into a living resource for today's and tomorrow's challenges. Stein is, in effect, idolized by scholars who read her in the mode of docility. She becomes the target of study, rather than a partner in the study of other issues. The trouble is that the relationship between Stein's philosophical i and the i of her interpreter has been wrongly construed. The one read and the one reading are not partnered as leader and led. They are juxtaposed as object and subject. The commentator studies Stein as a finished entity, a completed set of works.

Overtly docile, such commentators are covertly aggressive. Interpreters who cite Stein's writings authoritatively may seem at first glance to be subordinating their own judgment to the superior wisdom of a great thinker. They may seem to be transmitting truths established by someone more capable than themselves. This is a ruse. The activity of scholarly citation wields its own kind of power. It follows a line of thinking that is analytic and historical; it does *not* follow the line of thinking to which Stein's own philosophical thought beckons. To summarize Stein for Stein's sake may serve a worthwhile pedagogical purpose, but it also postpones the start of philosophical thought, that is, the kind of thought that Stein herself did and taught--at least, in her early phenomenological period.

To regard Stein as an inimitable expert and a source of unassailable truths makes it impossible to pose a key question in Stein Studies: how did Stein bring herself to accept the spiritual and professional direction that turned her toward an authoritative reading of Scholastic philosophy? One does not ask, because the answer has been pregiven: since Stein was a much better philosopher than any of us, she must have had a reason, but we cannot know it. Or more simplistically, as one often reads in more popular commentaries: because Thomas and Theresa had the truth. This is not an answer acceptable to phenomenologists, philosophers in general, or responsible thinkers at large.

2. Type, the drift toward biography, and "echoing."

As we saw above, the sharing of social stereotypes between writer and readers can serve as a shortcut to understanding. But there are costs. The reader's i may be too quickly led through the moves that Stein makes, lulled by assurances that she is "the right kind" of woman or teacher or German or philosopher or saint or whatever. The "click" of recognition for the exhibited type substitutes for the "click" of recognition within one's own conscious process that the line of thinking offered for following is indeed coherent. The echo is not a new sound but the repetition of an old sound. Nothing originary comes to presentation in it.

Type is an object whose constitution *should be* identical for all i's, but isn't. Type is supposed to appear as an item in the natural world, yet for a few of us its artificially constructed character will be glaringly apparent. Two of Stein's type-constructions--the woman, and the Jew--illustrate this. She was quite successful on the lecture circuit when she appealed to notions of bourgeois femininity that she happened to hold in common with her audiences. Yet later generations object to them, or turn away from her work altogether because they are so repelled by these constructions; the echo rings false to them.

Stein's notions of Judaism were more nuanced and complex. Her autobiography was designed to defend the Jewish people by portraying them as coinciding with the type of the modern German urban professional. Along the way, however, in projecting that type she repeated ethnic slurs against "Eastern" Jews and Yiddish speakers. Those remarks, together with scattered comments on Jewish scriptures in her spiritual and theological writings, put up a serious barrier for Jews today who might wish to study her philosophy. In my own case, Stein's offhand insults to Poles and cabinetmakers impeded my ability to follow her thought and sapped my will to do so.

Such instances when Stein's appeal to type backfires are very valuable from the standpoint of hermeneutical analysis, however. They serve to illuminate the use of type in counterfeiting the connection between the writer and the reader. They also disclose an even more serious impediment, which is amply represented in the literature on Stein. Not just Stein's type, but her whole constructed life story, often has been made to be the disclosive principle for her philosophy. The intelligibility of her thought is vested in the events of her life. There is a creeping tendency in Stein Studies to subordinate the philosophy to the biography. Interpreters seek not only to trace the chronological unfolding of Stein's work in parallel with her career and human relationships, but also to *explain* her philosophy as a kind of epiphenomenon issuing out of personal or emotional upheavals transpiring in her life.

This cannot be avoided altogether, if Stein is to be made available to the various constituencies who have a right to claim her; for example, young women and men in the Christian churches. But when found in philosophical commentaries, this tendency should be recognized as a pathology of interpretation. Following Stein should not be reduced to imitating her life decisions while mining her writings for citations that may be "applied" as stock answers to questions occurring in a historical situation quite different from her own.

3. The Pygmalion syndrome, and "receptive adaptation."

We have had occasion to complain that Stein was susceptible, both in life and after death, to being redesigned in somebody else's image. Like Przywara before him, Pope John Paul II became enamoured of an effigy of Stein designed by himself, and then set about bringing it to life. Przywara cultivated a Scholastic philosopher; the pontiff is creating a saint. Yet they are doing little more than any interpreter does in constituting the alien i with whom one's thought is partnered. Stein, too, breathed life into Thomas and Husserl when she scripted their lines for a philosophical dialogue (which Heidegger made her rewrite before publishing it in the 1929 *Jahrbuch*). She groomed Hans Lipps for *Habilitation* at Göttingen and all but wrote Nota's homily for him at Echt in 1942.

Thus besides the options of "docility" and "imitative echoing," a third way of egoic following opens up for Stein Studies. This way attempts continuity with Stein's philosophical practices, rather than replication of her words and deeds. This is the way of *Einfühlen*, not *Einsfühlen.* In simplest terms, it aims to think and do as Stein *would have* thought and done, were she alive among us. A spirit of receptive adaptation characterizes the readings of Herbstrith, Ziegenaus, and several other recent interpreters of Stein. One must credit her with sense enough to have changed her views had she learned what is now known about the historical circumstances of the lives of Jesus and Mary of Nazareth, and had she perceived the motivational links between Christian theology and Nazi policy. Stein certainly would not want to be an authority docilely cited or a voice nostalgically echoed--especially on issues where she turned out to have been wrong.

There must be a little of Pygmalion in every interpreter. Love, desire, and self-projection essentially characterize the movement of the following i as it chases after the i that leads it. Pathology sets in when empathy crowds out the independent otherness of the loved leader. Thinking he has caught and subdued the other, the interpreter mistakes himself for her and winds up like a puppy chasing his own tail. This is an unavoidable danger in any interpretive project. The antidote is this: make sure there are many interpreters rather than one authoritative reading. The greater the number of statues set to dancing, the less the chance that any one of those idols will be mistaken for the real Edith Stein.

In conclusion, we see that Stein's insistance that *the who matters* in the constitution of objective knowledge has opened the door not to chaos, but to a closet full of denied facts. The healthy and balanced interplay between i and i, leader and led, can introduce delightful variance, creative growth, and productive adaptation in Edith Stein's philosophy, as in any other human enterprise. Without it, her work is finished and sterile. In the final chapter of this study, we will consider Stein's own account of creativity in the essay "Psychische Kausalität," written immediately after her dissertation. We will then be in a position to review the implications of her work for contemporary discussions in feminist epistemology and philosophy of science.

Chapter Six

Science as Literacy

The task of this concluding chapter is to make a beginning. What begins here is a constructive philosophy of empathy that will address several contemporary questions in philosophy of science while drawing on the legacy of Edith Stein. The two preceding chapters have identified key components of that legacy: the priority of bodily life for any hermeneutic endeavor, and the capabilities of egoic following within any program of reading. Wielding those keys, Stein herself unlocked the gate to the goal of establishing the unified basis for natural and cultural sciences alike--the very goal which had eluded Husserl's grasp in the *Ideen*. Stein's solution appears in her 1919 essay "Psychische Kausalität." That essay funds my critical reappraisal of contemporary psychoanalytic and materialist interpretive programs, as well as my constructive alternative. Before examining psychic causality (*psychische Kausalität*) according to Stein, let us briefly recall Stein's accounts of bodily life and of empathic egoic following (*Einfühlen*).

A. Reading Life

In chapters two and four above, we saw that the second book of Husserl's *Ideen* embraces two contrasting accounts of the phenomenological foundation of the sciences. The strategy that we termed "the second solution" was the one ultimately favored by Husserl himself. It accords priority to *transcendental object-constitution*, and it was to undergo development in Husserl's subsequent writings. What we termed "the first solution" accords priority instead to *felt bodily life*. Husserl tried out this strategy in various writings before 1917. In Stein's hands, those manuscripts were crafted into the central portions of the text of *Ideen II*, to be published only posthumously in 1952. Stein thus is the principal architect and proponent of the Husserlian phenomenology of the living body.

Science is possible at all, according to her account, simply because the bedrock that grounds all cognition is bodily life, immediately available to consciousness. Life is felt *as* the reciprocal influence and responsiveness of material, sentient, and soulful levels in their openness to intellect. "Blending" is what Stein called this transmission of influence across the intrinsically permeable borders of the physical, the psychic, and soul within human being. Empathy (*Einfühlung*) presents such life as i-drenched. As we saw, empathy is logically more primordial than is individuation into the juxtaposed objects "my body" and "body not mine." In other words, the plurality of individuals is owing to empathy. After individuation, empathy remains the basis for communication among persons. Space, as possibility for plural locations in relation to plural bodies, arises subsequently to egoic individuation. These structures all are presupposed by object-constitution--which Husserl's "second solution" would (mistakenly) take to be the primordial starting point for science.

Thus the differentiation of i into i's is logically posterior to empathy. The content of empathy is *i* before it can be I or you. Determination of i into I, you, she, he, they, we occurs as empathy ceases. Yet I find that individuation can come undone. Sometimes I spontaneously succumb to empathy, surrender my individuality, and am caught up ecstatically into a stream of live experience that (I will later discover!) originates with someone else. Moreover, I can even deliberately turn over my i to another, partially and temporarily. This is precisely what I do when I read. With narratives, my i rides along the constructed experience-strands of the various characters and it lives-through their choices; while with nonfiction my i rides the reasoning and the evidentiary trails as they unfold. We saw in chapters three and five that Stein offered a rudimentary theory of literacy in her description of the leading and following of i through a flowing live experience. To write is to offer one's i as a guide; to read is to follow an i-drenched live-experiential stream that one does not originate. This possibility is owing to the chief characteristic of i: *i flows*. What i is, is live experience streaming along in a great variety of modes, each with its own lawfulness. Stein calls this "tending."

I "tends" when it spontaneously moves toward some goal that it already, in some manner, has. For example, in perception the i flows forward through a series of partial appearances to arrive at a unified object. In memory, the i flows from a non-live content toward a revival of that content in a live presentation. In expectation, the i flows from an unreal content toward its realization. There is no guarantee of success; the i may fail to arrive at live coincidence with its goal. For example, if a geometric theorem is misremembered, the i will never be able to recover just that faulty formulation *as* the content of a live assent. Nevertheless, the i's capacity to retrace and re-live egoic flow-paths is the foundation of literacy. Both writing and reading require the ability to suspend individuality and register an experiential flowpath that i--anyone's i--could follow and enliven.[1]

With geometry and logic, as with technical manuals and recipes, *whose i it is* that originates the pattern of the flow *hardly matters* to the content. Such texts are written with the expectation that any i can perfectly replicate the validating insight in all relevant respects. But Stein saw that with most instances of communication, the in-felt content retains the fingerprints of the unique individual person with whom it originated. In fact, she took this to be the general rule. Communications always offer access to i-hood: if not to that of a recognizable individual, then to that of *someone* ("the author"), or at least to non-individuated i. Even the Pythagorean Theorem has no claim to validity apart from the possibility of its being worked through by an i.[2] Therefore Stein held that alien

[1]Insofar as science relies on writing and reading research reports, we see that it presupposes the possibility that one i can follow another's act of validating confirmation. Teaching and learning depend upon the same capacity, but cannot be investigated here.

[2]The limit case for this argument would not be logic formulated in text, but the logical circuits imprinted on a microchip in a computer processor. Their formal validity, so to speak, inheres in their configuration, which must be recognized and certified by

i was immediately accessible within *Einfühlung*, piggybacked onto whatever content was being re-enlivened in the activity of understanding. If I live something as pre-lived, then I receive alien i and ride with it. I am led. I am led through the "tending" and "blending" of another life.

There are many kinds of dances in which readers and writers can engage. In the previous chapter we examined three modes of reading: (1) docility, (2) echoing, and (3) chiseling like Pygmalion. These modes were instantiated by the strategies of Stein's interpreters; but we can now see that they were practiced by Stein herself as well.

1. Docility to an autographed text.

Docile readers deny that their own subjectivity--their i's--actively contribute to the content that is re-enlivened as they follow the lead of the author. The docile i pretends that its following is slaved to a leading i, perfectly replicating every move and every pre-lived content. The coherence of the meaning-content is not independently validated (that is, enacted anew) by the docile reader, but rather is accepted on the authority of the author. McInerny reads Stein docilely. Stein read Thomas docilely.

Docility needs authority, and authority needs docility. Stein herself was capable of writing and teaching authoritatively, as we saw in chapter four. In her "autograph" mode, Stein wrote as if her text would become a clear window into the subject matter. Yet paradoxically, an autographed text itself frames the definitive portrait of the matter under consideration and displaces it. The authored text makes this displacement in the name of the author. The author's name--for example, "Stein" or "Thomas" or "Husserl"--then comes to stand in for a complex configuration of affairs. So, for example, phenomenology becomes what Husserl signed his name to rather than what Husserl actively lived through. The docile reader surrenders the right and abdicates the duty to peak around the frame and check out the validity of the configuration engineered in the text. In authored text, the content becomes unavailable for re-enlivening precisely because it is so handily packaged and pre-certified. Perhaps the most striking example of this paradox is Stein's own attempt docilely to "follow Jesus" by correctly mouthing correctly Latinized Psalms at the correct hours during her cloistered religious life, all the while accepting the "cross" of anti-Semitism on behalf of "guilty" Jews.

There is idolatry in docility. The autograph puts dead letters in place of living subjectivity. The docile reader idolizes an author, in a futile quest to merge with that which should merely lead. Both docile reading and authoritative autographical writing are pathologies of empathy. That is, they are faulty deployments of i.

---

something other than a chip. Their validity does not inhere in their functional ability to produce "correct answers" in calculations, for the correctness of correct answers is not self-certifying.

2. Echoing an autobiographical text.

The echoing reader is more active than the docile one. Rather than pretending to coincide with the live experience of the writer, the echoing reader aspires to prolong it. Though active, the echoer is lazy and uses "type" as a shortcut. Echoing depends on prior sharing of pre-established stereotypes between writer and reader. Those types are kept insulated from re-evaluation; the reader does not take the trouble to re-enliven them to discover whether they are valid. Stein herself, in her 1916 dissertation, had criticized racial stereotypes on precisely this score.[3] Anti-Semites, she wrote, have been infected with prejudice by someone else. Were they to try to re-enliven the originary experiences in which they assigned a negative valuation to Jews, they would find that they could not do so. Nevertheless, in her autobiographical writings of the 1930's, Stein herself relied heavily upon stereotypes in an attempt to make Jews seem less alien and more appealing to a German readership. She designed herself as a *German* Jew. Judaism of "the right kind" was just another way of being "the right kind" of German.

Among Stein's interpreters, Gaboriau and Ziegenaus exemplify the echoing mode of reading. They aspire to prolong Stein's philosophy by extending its applications into fields that she herself did not examine. If the docile reader regards Stein's philosophy as complete, the echoing reader regards it as needing elaboration. Stein's thought becomes a content that can be replicated and exported. But the echoing export of "Stein's philosophy" is lazy in the same way that Stein's own use of "type" was lazy. Both cut corners. Type was employed to evoke rapport or revulsion without the reader's having to engage anew in the originary constitution of the loved or hated object. Similarly, echo-philosophy incorporates prefabricated chunks of thought to be assimilated as contents, not processes. They are swallowed whole, so to speak. The echo is nostalgic, because it purports to return home to the ground of originary experience without actually doing so. The reading i does not invest itself in following the leading i of the text. It may mime the moves, but it declines to enact them within itself. Therefore it cannot vouch for the validity of the insights that it brokers. It vouches only for the accuracy of their replication.

---

[3]Stein cited the examples of race hatred and pornography in order to illustrate how the activity of reading brings into play the phenomena of "tending" and "blending" (to be discussed further below). When a racial stereotype is encountered on the page, there is a "tending" in which the i assembles the pieces of the image and constitutes a whole object that it already has: *the kyke, the nigger, the pollack.* Thus the "tending" succeeds and terminates in a click of recognition. Moreover through "blending" the intellectual registration of the stereotype is accompanied by a psychological registration of hatred and perhaps even a visceral shiver of revulsion. The intellectual, soulful, and psychological levels of the human person are open to one another, and influences cross among them. Pornography works similarly. The i flows through the series of elements and thus constitutes the object. At the same time, the constituting act reverberates across all levels. The body registers physical arousal, and the soul floods with lust. Stein's acount of "tending and blending" would support a theory of cinema; but that would carry us beyond the confines of our present interest.

3. Chiseling anonymously.

The chiseling reader, by contrast, is perhaps overly active. Chiselers breathe their own life into the beautiful form of someone else. Like Pygmalion, Przywara tried to give Stein a Thomist soul identical to his own. Stein, too, chiseled when she read Husserl. She was obliged to chisel because as Husserl's assistant, she was supposed to check out Husserl's transcendental phenomenology and vouch for its validity. In order to do that, she had to let her i ride through the moves made in his manuscripts. At many points, Stein found that she was able to re-enliven the "tending" of Husserl's thought within her own i. But as we have seen, Stein's i-ride through *Ideen II* brought her up against several places where no egoic validation was possible because Husserl's argument ran in circles. Stein detected the problematic question-begging in Husserl's doctrine of constitution through her own inability to assent to the coherence of his thought. At those points in the manuscript, as we have seen, Stein literally cut and pasted the paragraphs to gloss over the logical glitches. By all indications, she also drafted a positive statement showing that felt bodily life is logically prior to constitution. That statement, partly original and partly inspired by Husserl's earlier work, became what was termed above "the first solution" to the problem of founding the sciences.

Stein also breathed her own phenomenology into nascent texts of Hans Lipps, Roman Ingarden, and others. We have labeled Stein's editorial activity as ventriloquism or ghostwriting when it produced words or texts to go under the name of someone else. But the capacity to ghost-write presupposes the general competence to ghost-read: that is, to pump one's own live experience stream (*Erlebnisstrom*) through the lines of a text laid down by somebody else. A book acquires a new author every time it is read in this way. While being led--that is, while living through a stream of experiences that originate with someone other than myself--I am at the same time living in a stream of my own experiences. My reading is *receptive*, which it to say that it is passive and active at once.

In principle, it is possible that a text would offer no glitches, so that my thought as led by the text would perfectly coincide with my thought as actively tracking and checking where I go. In practice, my own originary experiences diverged from my non-originary experiences at many points during reading Stein's texts (as Stein's diverged from Husserl's over the priority of object-constitution). My own reading "jumped the tracks" when Stein's text invited me to replicate her disparagement of Poles, cabinetmakers, and "Eastern" Jews. But let us not yet call this divergence "chiseling."

A reader chisels only when he or she endeavors to displace the subjectivity of the author rather than co-exist in dissonance with it. Thus, if you will, Stein ultimately abstained from chiseling Husserl out of *Ideen II* because she retained "the second solution" in the text. (Nor did Husserl chisel away Stein's "first solution" during his subsequent revisions of those manuscripts.) Variance is inevitable when the act of reading is an act of responsible receptive adaptation. Let us then reserve the pejorative term "chiseling" for those instances where the reader denies that she or he has departed from the i-stream marked out by the author. Such cases reprise Schleiermacher's ambition: to understand by taking the place of the other. The chiseling reader effaces the subjectivity of the writer and substitutes one's own.

Because Edith Stein's scholarly work during her mature years was devoted to *following Christ*, one may ask about the depth of engagement between Stein's i and the patterns laid down by the gospels or by Jesus of Nazareth himself. Is the Christian religious life best characterized as docile replication, echoing prolongation, receptive adaptation, or chiseling displacement of Jesus? What degree of variance is tolerable?[4] Exactly how does an i "take up the cross"? These questions cannot be pursued here.

B. Writing Science

Husserl had no reader more astute than Edith Stein in the decade of the 1910's. She intimately understood the purpose and the method of his *Ideen zu einer reinen Phänomenologie und phänomenologischen Philosophie*. The first book of the *Ideen* had appeared in 1913 just as Stein began her studies under Husserl. Materials intended for the second book had passed through her hands and been shaped by her own subjectivity during the months of 1916 and 1917 while she worked as Husserl's assistant. Stein had reason to expect that the rest of the *Ideen* manuscripts would be published soon after the Armistice. She also had intimate knowledge of Husserl's writings on psychology and epistemology from the spring of 1917.

Thus Stein's postwar essay "Psychische Kausalität" is intended as an organic part of Husserl's larger project: providing a unified account of how all sciences achieve their validity. By addressing the human cultural sciences (*Geisteswissenschaften*), Stein's essay completes a portion of the work that Husserl had left undone. Written in 1919 as a *Habilitationsschrift*, Stein's essay on psychic causality was published in the *Jahrbuch* in honor of Husserl's 60th birthday (belatedly in 1922, for financial reasons). It was the first of two treatises labeled "contributions toward the philosophical foundation of psychology and the cultural sciences."[5] However Stein's essay is no mere echoing extrapolation of Husserl's phenomenology, much less a docile repetition of it. She pursues her own original contributions to the elaboration of what I have called "the first solution."

The 1919 essay enlarges upon Stein's earlier insights into the priority of bodily life and its availability through *Einfühlung*. But there is a startling new assertion. All objects are *values*, and thus every act of object-constitution must register within the constituting i as an inclination toward or against the object. This insight, derived from Pfänder's account of motivation, reinforces Stein's

---

[4]The canonical gospels vary among themselves. Thus the questions posed here in relation to Edith Stein are also relevant to the egoic following that held between Jesus and members of the early churches, some of whom became authors of the gospels.

[5]The other treatise was Stein's "Individuum und Gemeinschaft" ("individual and community"). The designation *Ideen IV* is my own suggestion. In fact, *Ideen II* and *Ideen III* had not been published by 1922, and would not appear until 1952.

contention that individuation logically precedes objectivation. The valuing of an object depends on the individual i who constitutes it. Strictly speaking, Husserl's nonindividuated "transcendental i" would be impotent with regard to constitution, because it could not confer value. Thus *there are no intersubjectively identical objects*. They are not merely nonexistent by virtue of being ideal essences; rather, they are eidetically inconceivable, and therefore science cannot be founded by appealing to such. Instead, says Stein, science is possible because *Einfühlung* is the capacity to follow motivations of various kinds. Not objects, but the inclinations of subjects with regard to them, are what become intersubjectively available through *Einfühlung*. The possibility of predicting human actions rests on the possibility of understanding the complex interplay of causality and motivation in human decision. Let's look at Stein's arguments in greater detail.

1. "Blending" and psychic causality.

Stein starts her treatise "Psychische Kausalität" with the observation that what we call causality in the natural world ordinarily involves multiple elements. Causing happens in chains. A causing event is easily distinguished from the corresponding caused event. Thus it would seem that there can be no proper causality within consciousness, inasmuch as the stream of live experiences is not chain-like, not articulated into segments. The contents of consciousness are continually subsiding into the past, but they can also be recalled and revived because they remain accessible to the same streaming life. Yet there are unities within the stream; for example, discrete experiences and fields of experience. These are not mechanically external to one another, like links in a chain, but are overlaid and are capable of co-variance. For example, a feeling of lassitude will imbue whatever transpires in the stream of live experiences. Stein terms this "causality of life" (*Erlebniskausalität*). The lassitude is the source (*Ursache*) that conditions or determines (*bedingt*) the live experiences transpiring under its influence. Stein asserts that the causality operating within the sphere of live experiences (*Erlebnisse*) is analogous to the mechanical causality that operates in the natural world. According to this analogy, the role of the causing is played by "life-feelings" (*Lebensgefühle*) such as fatigue or freshness, while the role of the being-caused is played by the rest of the live experience as it runs its course (*Ablauf des sonstigen Erlebens*).[6]

But this is not yet what Stein means by "psychic causality." Life-feelings belong to consciousness, not to the psychic. *The psychic* in this essay, as in Stein's dissertation, is the sentient level of human being, parallel to the physical level and to the soul and sandwiched between them, so to speak.[7] The psychic does not flow like consciousness. Rather, it passes through successive states, with the later states determined by the former states as in any biological process. The states of the psyche at various times--its *Lebenszuständen* or "live states"--may or may not

---

[6]This much is established in the first chapter of the treatise. See Stein ([1922] 1970: 7-18).

[7]See Figure 1.2.

register within the "life-feelings" (*Lebensgefühlen*) of consciousness. But as the psyche processes through its states, it has a relatively stable capability that Stein terms its "lifeforce" (*Lebenskraft*).[8] That psychic lifeforce announces itself to consciousness in such "life-feelings" as lassitude or enthusiasm. Stein conceives of the lifeforce as an energy source that fuels the state-changes that the psyche undergoes and is depleted by them.

Psychic causality, therefore, is like natural causality in that it is governed by the law of conservation of energy. The psyche changes its state only if energy is expended. Psychic processes "cost" the lifeforce something. (By contrast, the life-feelings within consciousness do not draw energy from the lifeforce at all.) Stein thus portrays the psyche as a self-regulating mechanism. It spends lifeforce to intensify the voltage of its living (*eine Steigerung der Spannung des Erlebens*), or it ratchets down its experiences to conserve lifeforce.

When the lifeforce is spent down, what replenishes it?[9] This is a *quantitative* question, the only kind of question that can be asked of the psychic level. *Qualitative* questions must be deferred to the sphere where the lifeforce "announces itself" in life-feelings to consciousness. This *conscious* level is provisionally designated as the "life sphere. . . a lower stratum of the experience stream" (*Die Lebenssphäre bildet eine Unterschicht des Erlebnisstroms)*. It is what Stein formerly designated the soul. The *quantity* of lifeforce available in the psyche registers as the *quality* of life-feeling permeating the soul. Stein implies that the soul, as the field of pervasive life-feelings, is therefore a field of conscious experience. There could be consciousness and a flow of conscious experiences without such a soul as underlayment; but in a soul-less consciousness there would be no causality operating. However a consciousness undergirded with soul can receive causal influences--not as an energy drain or gain, to be sure, but rather as colorations continually renewed moment to moment or shifting through a spectrum as the psychic lifeforce wanes and waxes.

While psychic states do not change without an expenditure of lifeforce, the character of the psychic alteration depends on additional factors besides the quantity of lifeforce expended. Stein asserts that the content of psychic processes is determined not only by the amount of lifeforce allocated to the reception of data, but also by whichever data happen to present themselves for reception. This input obviously cannot be predicted in advance, so it is not possible to forecast the future states of a psyche from its present state (much less the future achievements of any individual whom one might encounter today). Nevertheless, Stein holds, the causality operating on the psychic level is sufficient to support less ambitious inferences.[10] This possibility of (limited) prediction through inference lets

---

[8]The life-power or *Lebenskraft* seems to be akin to what Sigmund Freud would later term the libido.

[9]Later in the treatise, Stein suggests that human love or divine grace may refill the depleted lifeforce, but she does not pursue that intriguing line of argument. See Stein ([1922] 1970: 76-77).

[10]See chapter 2 of "Psychische Kausalität," Stein ([1922] 1970: 18-34).

psychology be truly a science.

This account in the 1919 essay puts a fine point on the analysis of "blending" (*Verschmelzung*) offered in the 1916 dissertation. Blending, as we saw, was Stein's label for the permeability of each level of human being to the other levels. Physical events that befall the body reverberate in the psyche and the soul, while psychic and soulful events express themselves physically. The psyche can initiate processes that affect the physical body and the soul, just as events originating in the soul can affect the psyche and the physical body. With the 1919 essay, however, Stein connects this "blending" to consciouness by positioning the soul as the sublevel of the stream of live experiencing (*Erlebnisstrom*).

Stein was attempting to work out a new vocabulary in which to address an old philosophical problem: the problem of how the body connects the mind to the world. The originality of her work can be appreciated when her essay is contrasted with the "Project for a Scientific Psychology" drafted by Sigmund Freud in 1895. Freud did not publish that sketch, and by the 1920's he had repudiated its approach because he failed to find a way to integrate consciousness into it.[11]

Freud's aim was "to furnish a psychology that shall be a natural science: that is, to represent psychical processes as quantitatively determinate states of specifiable material particles."[12] He postulated three systems of nerves--"neurones"--operating in the human body. In his manuscript these systems are designated by three Greek letters: the φ system, the ψ system, and the ω system. From the context, one gathers that φ stands for φύσις (*physis*, material nature) and ψ stands for ψυχή (*psychē*, mind or animating principle).[13] The translator suggests that ω was chosen to designate the third or mediating system that supposedly comes into play with perception (German *Wahrnehmung*) because omega, ω, resembles the letter W.

Freud says that the systems differ in how their neurones handle the passage of stimuli across them. The φ neurones offer no resistance, and they spring back to their original state after the stimulus passes. They are "permeable." By contrast, the ψ neurones are "impermeable" because they are changed by the passage of stimuli. Thus they "remember" by existing in a different state if they have carried the stimulus than if they have not. Curiously, in this proposal Freud is attempting to give an entirely physical accounting of the difference between something physical and something psychic. In a similar vein, he goes on to offer

---

[11]The "Project" was first published in 1950. A critical edition has not appeared. The translator, James Strachey, worked from a photocopy of the manuscript and warns that the German edition is not to be trusted. Therefore my citations are taken from the Strachey translation. Although Husserl and Freud were about the same age and had been students in Vienna at the same time, I have found no evidence of contact between them or familiarity with one another's writings. Stein of course had no access to Freud's unpublished manuscripts, so her account was developed independently of his.

[12]See Freud (1966: 295).

[13]However the translator does not offer any interpretation of these abbreviations, and I have not found any corroboration of my view elsewhere.

a purely physical account of the difference between influences entering the organism from the outer world and influences originating within the organism itself. He says that the φ system is reached by external stimuli, for it comprises the grey matter of the spinal cord. The ψ system receives only endogenous excitations; it is the grey matter of the brain, and it has pathways only into the interior of the body. "To the best of our knowledge," Freud writes, "the system ψ is out of contact with the external world."[14]

This scheme created more problems than it solved. Freud needed to invent a third system, the ω neurones, to account for perception--the process in which sensory impressions are assembled into conscious objects. He suggested that consciousness ought to reside in the ω neurones. (Unfortunately, the ω system also was conceived as operating in a physical, mechanical way like the other two.) As Stein would later do, Freud postulated force-gradients within the systems, to be governed by a law of energy conservation. But Freud described optimal function as the maintenance of *low*-energy states through strategic discharges, where Stein regarded *high*-energy states as optimal. Stein postulated the integration of the physical, psychic, and soulful levels through "blending" that facilitated transmission of influences across them, but Freud kept his φ and ψ systems insulated from one another. Most importantly, Stein managed to integrate consciousness with the levels of somatic life through the underlayment of "soul." Freud's 1895 proposal failed to do so. Eventually he began again to build his science of the human mind from the side of consciousness; but that work would reduce bodily manifestations *to consciousness* just as the 1895 proposal had reduced the psychic to *the physical.* Freud's psychoanalytic theory never achieved the coordination of consciousness with organismic functions, as Stein's phenomenology of *Einfühlung* did.

The ego, according to Freud's early proposal, was a configuration among ψ neurones that interfered with the passage of certain stimuli across them. It acquired stability owing to the repeated passage of those stimuli along the same pathways. But the ego's existence amounted to little more than the inhibition of selected psychical processes. The ego was a gatekeeper; in no way was it an agent or the originator of impulses. By contrast, Stein brought to her psychology the complex phenomenology of subjectivity developed by Husserl, enriched by her own studies of egoic following (*Einfühlung*). Stein's account of "tending" (*Hervorgehen*) within the i has no parallel in the so-called depth psychology of Freud.[15]

2. "Tending" and motivation.

In her dissertation, Stein had established that an i *can* follow the motivated coherences once followed by another i. In her 1919 essay, she asks whether an i

---

[14]See Freud (1950: 303-4).

[15]I have not found mention of Freud in any of Stein's work. There are a few general comments on "depth psychology" in Stein's theological anthropology, written in the early 1930's, but the bulk of that work is based on Thomistic philosophy. See Stein (1994: 26-28).

must follow. She explores various sorts of coherences that an i may refuse to follow altogether, or may elect to follow partially and creatively.[16]

Motivated coherences invite the i to follow them, without constraining it to do so. Engaging with such a coherence, the i registers within itself a tendency to flow forward. It recognizes that several possible directions beckon, any of which would "fit in" with the unfolding coherence. Any of them would be warranted. If the i happens to be following a coherence that another i has pioneered, then at each moment the led i recognizes that the leading i *chose* the direction in which the coherence actually unfolds. Perhaps the simplest example is a melody. As it plays, the tune *could* move to any note in the scale. We understand while listening that any of a certain range of options would complete the melody coherently; but given that range, we are eager and delighted to discover which way the tune actually goes in its flow toward wholeness. We understand the melody as a pattern precipitating out of someone's free choice at every moment. Its contours are comprehensible (and enjoyable) as they are, because they could have been otherwise. Life is like that.

If the freedom of choices were entirely arbitrary there could be no understanding, that is, no leading or being led for the i. In fact, choices are bounded within a range of possibility stemming from i-hood itself. The range is quite narrow in cases of what Stein terms *rational necessity*; for example, when the i recognizes the form of a valid logical inference. In such cases, choice ends at the moment when the i first latches onto the logical problem. From then on it is constrained--by its own rational being--to flow into either an assent (if the form is proper) or a rejection (if the form is flawed). But in cases of *valuation*, the necessity of the flowpath is relative to the intentional target. A great variety of actions could make sense as motivated responses to any given valued object. In either kind of case, however, motivation is perspicacious (*einsichtig*). The i finds the necessity within itself; it "sees the way." This is how motivation differs from causality. The chains of causes and effects are "blind," that is, not available for the i to inhabit. Extensive regions of human being itself are "blind" in this way. These are precisely the regions where psychic and physical causality operate.

Stein says that motivation is the general structure for the whole range of intentional live experiences (*Erlebnisse*). That means that the varieties of motivation previously examined by herself or by Husserl now are to be subsumed into the intricate webwork of conscious life. Let us recall some of those earlier examples of motivation:

- combining adumbrated appearances into one object;
- completing the series of appearings of an object;
- memory: recalling to presentation a past experience;
- expectation: presenting a future state of affairs;
- expression: outward embodiment of feelings;

[16]The discussion in this section is based on §1 of the third chapter of "Psychische Kausalität"; see Stein ([1922] 1970: 34-41).

- logical inference;
- making explicit what was implicit;
- object constitution;
- empathy (*Einfühlung*) of another's live experience;
- inclination (*Strebung*) toward or away from an object.

There is no causality in any of these. To be sure, causal chains connect into these motivations at various points, thanks to the "blending" of the physical, psychic, and soulful levels in human being. Blending is causal; tending is not.[17]

3. Objects as motives.

Motivation and causality transpire together within human being. The vitally important difference between the two is that causes produce effects "blindly" and in a one-way sequence from which i is excluded, while motives are posited by an i "before" it registers within itself an inclination to flow--optionally--toward their completion. Causal series are infinite in principle; motivation springs from a beginning and arcs out toward a whole. The *motive*, for Stein, is merely the objective correlate of an act of meaning. (As Husserl would say, the motive is noematic.) However that objective meaning exercises no reciprocal effect or influence on the i. Instead, the i is motivated *by its own prior meaning act* itself (that is, by what Husserl would term the noetic intentionality of that act.) Here lies the ground of motivation for Stein. The act that intentionally grasps the motive--grasps it as a determinate unity--is the *motivant* in her terminology.[18] This act of apprehending the motive does two things: it posits an object, and it also generates a decision with regard to that object. The *motivated*, then, is whatever decisional act may flow out of the act of apprehending the motive.

---

[17]"Tending" is motivated. But "motivated" does not necessarily mean "conscious." There are some conscious events that are not motivated. For example, association is a blending rather than a tending. The flowing away of live experience into the past also is designated as a mere blending. Stein writes: "Motivation in unserm allgemeinen Sinn ist *die* Verbindung, die Akte überhaupt miteinander eingehen: kein bloßes Verschmelzen wie das der gleichzeitig oder nacheinander abfließenden Phasen des Erlebnisflusses oder die assoziative Verknüpfung von Erlebnissen, sondern ein *Hervorgehen* des einen *aus* dem andern, ein Sichvollziehen oder Vollzogenwerden des einen *auf Grund* des andern, *um* des andern *willen*. (Motivation in our universal sense is *the* connection that acts generally contract with one another: not a mere blending like that of simultaneous or sequentially subsiding phases of the live experience flow, or [like] the associative linkage of live experiences, but rather an *emerging* of the one out of the other, a self-consummating or becoming-consummated of the one *on the ground* of the other, *for the sake* of the other.)" The i "tends" when it actively consummates in this way, whether originarily or nonoriginarily. "Blending" conscious processes are passive registrations for the i, in which the i cannot do otherwise than it does. See Stein ([1922] 1970: 35).

[18]The rendering of "motivants" for *Motivanten* and "motivated" for *Motivate* comes from the partial translation of Stein's essay by Sister Mary Catharine Baseheart. I was able to consult Baseheart's manuscript through the kindness of Father Steven Payne, editorial director of ICS Publications, the publisher of the English translation of Edith Stein's Collected Works.

FIGURE 6.1: MOTIVATION OF ACT BY ACT

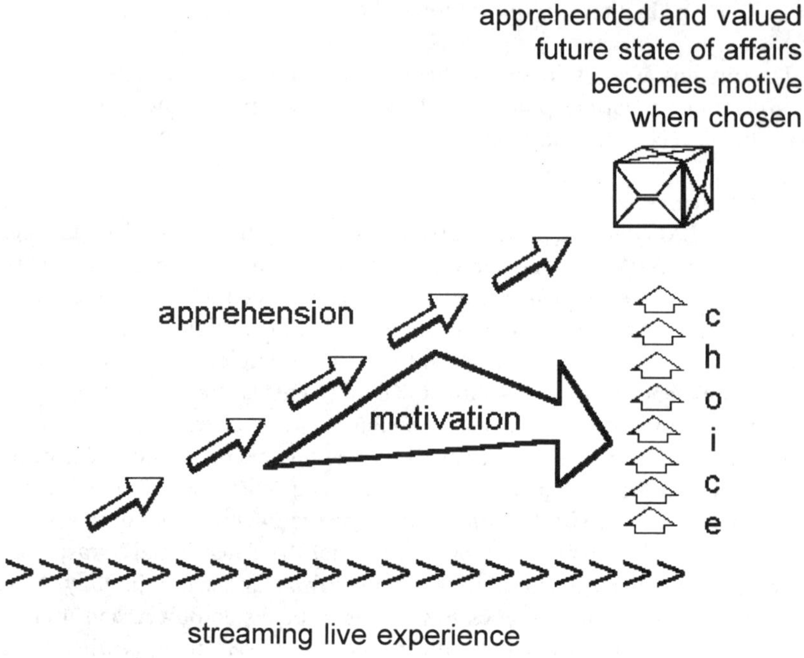

There is no causality in this picture. The motive does not "cause" *either* its apprehension *or* the choice made in regard to it by subjectivity. *Nor* does the act of apprehension "cause" the decision. The motivant act just rolls over into the motivated act, but it does so *as* an insightful live move of i. This i-drenched and *einsichtig* transition of act into act is what Stein terms "motivation" in the general sense.

In the diagram the apprehension is an act of object-constitution, the "motivant" act.[19] It is a tending that presupposes blending--or more precisely, presupposes contributions from those fields of human being where causal blending operates. The data to be apprehended must be received into consciousness somehow. It is the live body which delivers them, through causal chains of influence involving the physical, the sentient, and the soulful levels. Influences stemming from the soul--for example the brightening of enthusiasm or the dampening of depression--affect the way in which an object is constituted, although

[19]The identity of Stein's "motive" and Husserl's "object" is my own inference. It is not plainly asserted in her text.

they neither cause nor prevent its constitution. "Life-feelings" such as optimism and depression impart value to the object, because they affect the sensitivity of sentience at the psychic level within which sense data must register. Such causal influences are unique to the individual who undergoes them. They cannot be "understood" from the inside by others. Because of the individuality of the body, objects are constituted only idiosyncratically, *never intersubjectively*. Moreover, each object is constituted with a kind of valence for decision. The object is always desirable or repulsive for the i whose object it is, since it has been constituted and valued in a single stroke.

Thus I cannot share your objects, because I cannot displace your body in the causal or "blending" sequences upon which your objects depend. However, through *Einfühlung* I can in principle follow you in a re-enactment of any of your egoic acts. I can live-through your act of object constitution as well as your act of decision that arises out of it. Those "motivant" and "motivated" acts of yours are available for me, as is the "motivation" that obtains between them. As I understand, I live-through the motivation after you, but I can only infer the causal influences that shaped your experience. Those causal influences--your sensitivity to physical stimuli, your mood, the repertoire of stereotypes that you imported without originarily constituting them--all are elements that I cannot get inside of. But then, neither can you.[20]

4. The sequence of motive-constitution and choice.

To understand another's motivation is to understand that the motivated decision was warranted by, but not required by, the motivant apprehension. The very constitution of an object determines the latitude within which responses to it are warranted. Certain stances toward an object are *required* by its constitution. For example, I cannot help believing in the existence of certain objects, and I cannot help admiring a person who has certain virtuous qualities. The belief and the admiration are attitudes (*Stellungnahmen*) that are inescapable. They have been caused in me, not motivated. Yet they in turn will become motivants for subsequent optional acts. Denial of my admiration or ignoring the existence of the object are options, and may be chosen in appropriate circumstances.[21] While I

[20]Chapter three already provided an example of how objects fall short of being intersubjectively constituted. In reading Stein's autobiography, I found that I could not empathically follow her constitution of the objects "Pole" and "cabinetmaker." I lacked certain prefabricated stereotypes that Stein was taking for granted, and that she was using as short-cuts in her portrayals. Those stereotypes were opaque to me; I could not originarily enact their constitution. Their opacity to my i disclosed that they were causal influences on Stein's objectification of certain persons--her blind spots, as it were.

[21]For example, Husserl's "epoche" or suspension of belief in the existence of the world would be seen now as having been *motivated* by his conviction that the constituted object "world" is infallibly and Anselmianly endowed with existence. (Had he no such conviction, he could not elect to turn it off.) Similarly I could decide to suspend my admiration for someone to insure that I did not act upon that feeling; but such a decision could occur only as an implicit acknowledgement that the admiration was an inseparable component of my apprehension of the person in question.

cannot refrain from admiring or believing, yet I can render those attitudes inoperative by acts of suspending them: acts that are motivated (in part) by the very acts in which I first constitute the objects as admirable or existent. The denials that suspend belief or affection *are not caused by anything*. They are chosen--ordinarily for the sake of some other motivant. For example, a phenomenologist suspends belief in the existence of things for the sake of science, or a husband suspends his affection for one woman because of his vows to another.

Stein holds that attitudes (*Stellungnahmen*) are not chosen, although they can be simulated when absent just as they can be rendered inoperative when present. Attitudes inhere in egoic acts but they register in the soul. As long as they are refused and rendered inoperative there, they never come to expression.[22] But attractions and repulsions, if not refused and rendered inoperative in the soul, can also register in the psyche; then I may have an inclination (*Strebung*) toward chocolate, for example, that I can choose to indulge selectively.[23] An attitude that I choose to suspend still registers in the soul, but my decision can eradicate the corresponding psychic appetite or revulsion.

On the other hand, there are inclinations that seem to originate at the level of sentience, such as hunger, sleepiness, or instinctive impulses. Here, psychic processes are supplying data to consciousness. As those psychic data are apprehended, objects are constituted: "my hunger," "my sleepiness," "my sexual interest." These are objects in their own right, not value-valences on other objects. Thus they are unfailingly accompanied by attitudes (*Stellungnahmen*) that will register in the soul and will suggest to the i a range of appropriate "motivated" acts--among which will be the act of refusing the attitude and thus inhibiting the corresponding inclinations. Granted, some psychic states are such that they directly cause other psychic states without bringing motivation into play. In such cases the i merely observes the events from the sidelines; it does not live them. But human acts are not events of that kind.

Stein has described the priority of the motive-constituting act to the decision-taking act. Neither is entirely determined by any psychic state, although psychic states and soul-qualities do partly condition them. Furthermore, a motive is not a cause, but a meaning-content of consciousness.

5. The dynamics of creative choice.

Stein's account makes possible a distinction between attitude and inclination (*Stellungnahmen* and *Streben*). When attitudes are refused, they persist in a neutralized version. When inclinations are refused, they can be extinguished.[24] There are three ways in which to rid oneself of an inclination: (1) by turning one's

---

[22]See Stein ([1922] 1970: 44). Obviously Stein disagrees with Freud and the psychoanalytic tradition, which holds that repressed content finds expression through dreams, fantasies, mistakes, and slips of the tongue.

[23]Stein does not consider cases in which a denied affection for another person could resurface, so to speak, as a craving for chocolate or some other socially permitted object.

[24]The following discussion is based on Stein ([1922] 1970: 82-87).

attention away from the object whose constitution arouses the inclination; (2) by deliberately interrupting the influence of the causal factors involved in the delivery of the data out of which the object is constituted; or (3) by deliberately altering those causes. The first way extinguishes the inclination temporarily, but the inclination returns whenever the object is recalled to consciousness. The second way breaks the chain of data transmission; for example, when a dieter takes the long way home in order to avoid seeing the pastry in a bakery window. The third way alters the state of the region from which causal influences are emanating; for example, when wholesome food and exercise are given to the live body in order to break its addictions.

Stein also considers unmotivated inclination.[25] Some impulses (*Triebe*) that arise in the psyche are not directed toward any particular goal; for example, the impulse to run and leap. These impulses register passively with the i, and the i feels itself impelled by them without investing itself in them. Maintaining such impulses is a drain on the lifeforce, and when the energy alloted to them has been used up, they burn themselves out. The soul may register a qualitative change in "life-feelings" as the quantity of lifeforce is spent down by impulses. Thus these unmotivated impulses work like a safety valve for the lifeforce in what Stein terms "the psychic mechanism." When the battery is charged up, it emits activity-impulses (*Betätigungstriebe*); when it is depleted, it emits need-impulses (*Bedürfnistriebe*) for an infusion of new force.[26]

Thus we have a way to distinguish between two phenomena that both go by the name of "desire." The first variety of desire is the unmotivated "impulse." A need-impulse originates in a psychic deficit and, although it registers with the i, involves no object-motive and therefore no i-drenched act. The same goes for activity-impulses, except that they originate in a psychic overload. By contrast, the second variety of "desire" is the motivated inclination, which arises from an act of motive-constitution enacted by the i. The i is the source of the i's inclinations but not of the i's impulses. Furthermore, the inclination is directed toward a realization while the impulse aims merely to equalize pressure. What "I want" is that "I accomplish" something. The accomplishing is wanted precisely as something whose realization is possible and lies within my capabilities, at least passively. I am "inclined" to realize only those states of affairs whose realizability-by-myself figures in to my constitution of them as motive-objects.[27]

Impulses are blind because, lacking objects, they lack also any inherent involvement of an insightful i. Motivated inclining is not blind, for the i actively constitutes its object. Willing presupposes inclining, because the decision of the

---

[25]See Stein ([1922] 1970: 87-92).

[26]See Stein ([1922] 1970: 61). The refill of lifeforce is supposed to come from some unspecified source, perhaps from divine grace or human love.

[27]See Stein ([1922] 1970: 64). One may hope to realize a state of affairs that one cannot procure by one's own initiative, but only through the initiative of someone else. For example, my wish that someone love me can be fulfilled only if she does so freely without my having coerced or earned the love through my own efforts.

will must be framed for or against an inclination. The i that wills invests itself in the inclination; this act of self-investment unfolds out of a prior act of motive-constitution. So the number of *acts* required for willing is two; for inclining, one; and for impulse, zero.[28]

Because of this complexity, it is no easy task to give a scientific account of psychic states and processes. In natural science, when the state of a system at T_2 can be predicted from its state at T_1, then the later state is said to be determined by the earlier state. Psychic processes are not determineable in that way. The psyche is continually receiving extra-psychic influences from the i-acts involved in object-constitution, inclining, and willing (not to mention influences stemming from the two other levels of human being, the soul and the material body). Stein therefore rejects:

psychischen Determinismus . . . die Auffassung, die einen psychischen Zustand als bestimmt durch die Reihe der vorhergehenden und als aus ihnen berechenbar ansieht. ([1922] 1970: 105)	*psychic determinism* . . . the conception that regards a psychic state as determined by the series of what went before and as calculatable out of that.

Thorough description and exact prediction of psychic states would require more than a history of previous states; it would require knowledge of an individual's core personality (*Persönlichkeitskern*) and will.[29] Motivations and willful choices can be empathized; however that empathy will retrieve only the rationality of the choices, not their necessity. In other words, the led i of the scientist can determine the meaning of someone's choices by determining that the choices have actualized options within a motivational range. What eludes empathy altogether is that which is utterly personal, including the value-assignment involved in object-constitution and the executive prerogative of willing itself (*Willensvorsatz*). These intellectual (*geistig*) acts are realizations of one out of a range of possible motivated acts. After realization, they can be understood through empathy. Before realization, they cannot be exactly predicted.

Stein concludes by postulating a threefold split in "lifeforce," that is, a branching of the lines of force feeding into the psyche. (1) Sensory lifeforce fluctuates according to the sense data that are received, and gives rise to the "blind" impulses. The history of fluctuations in sensory lifeforce would be the first determinant of a psychic state at any given time. But sensory lifeforce sustains (2) intellectual (*geistig*) lifeforce, which in turn makes egoically constituted and valued

---

[28]See Stein ([1922] 1970: 63-64). Stein acknowledges that she has refined Pfänder's distinction. He regarded inclination as "blind" and will as insightfully goal-directed, without distinguishing between impulses and motivated inclinations, as Stein has done. However the Husserlian account of constitution has demonstrated the i's active involvement in the constitution of any object. That discloses the double activity of i within willing.

[29]See Stein ([1922] 1970: 84). This seems to be the first mention of a personality core, and the notion is left undefined.

objects available to the psyche. The psyche's intellectual lifeforce has additional sources of support in extra-psychic regions such as the intellects of other persons, the world of values, and the divine intellect. Taken together, this would comprise the second determinant of any psychic state, the determinant which Stein designated as "core personality." Finally, there is (3) a force of will that taps into the i itself (*Willenskraft . . . die das Ich anscheinend aus sich selbst schöpft*).[30] The will gives the go-ahead to initiate a course of action whose meaning and value may already have been affirmed for some time past. (This creative stroke is what Stein had termed the *fiat* in her dissertation.)

These three varieties of force, involving the three determinations of the psyche, thus yield descriptors of the "what," the "why," and the "when" of psychic life. *Voilà.* Psychology.

6. Phenomenology of *Einfühlung* as the science of science.

Of what use is Stein's phenomenology of the psyche for establishing a unified foundation for the natural and cultural sciences? Stein's phenomenology succeeded where Husserl's bogged down, at two crucial points. She demonstrated that object-constitution is motive-valuation, and she described empathy as the means of intersubjective access to motivated coherences of all kinds.

A unified theory of the natural and cultural sciences must avoid reducing either side to the other, as Husserl argued already in his 1911 *Logos* manifesto. The methods of natural science have to do with causal chains, and would be inappropriate for history, literature, or music. On the other hand, the methods of cultural science have to do with decisions, desires, and values, and would be inappropriate for chemistry. Husserl sought to bring these divergent methodological practices together through a third science, phenomenology. But Stein found, through her practice of phenomenology, that natural science is a subsidiary component of cultural science. More precisely, natural science is that phase of *Geisteswissenschaft* in which processes that cannot be empathized are detected and observed from outside. Causal chains opaque to empathy lace throughout every instance of motivated experience. In Stein's terminology, "blending" interlocks with "tending." Human being is shot through with complex causal influences. The causality operating in the natural world is continuous and isomorphic with that operating within live human bodies. The "outer" world is no more opaque to us--and no less--than are the impulses of hunger and habit registering within consciousness.

Thus, the "natural sciences" are paradoxically the most *unnatural* of sciences. To pursue physical sciences in isolation from the live human body requires the enactment of an arbitrary abstraction. The openness of life to life through *Einfühlung* must be placed under suspension, so that the fiction of a solitary transcendental i can be entertained. Only then could one conceive of a transcendental constitution of objects in which the noetic activity owed nothing to differentially embodied life. In order to highlight causal chains, the bracketing of

---

[30]See Stein ([1922] 1970: 105-6).

the plurality of empathizing bodies might be a helpful technique. It cannot, however, remain in place throughout the practice of science, without landing the scientist into the sort of question-begging that Stein detected in *Ideen II.* This bracketing certainly cannot endow the natural sciences with a status "separate but equal" to that of the *Geisteswissenschaften.*

Stein saw that the science of science--phenomenology--*ought to coordinate the two strands of science rather than oppose them.* It should do that by examining the interplay of causal and motivational factors in the very arena where their mutual engagement is most intense: the individual human psyche. There, all objects are constituted with value-valences: they are attractive or repulsive in measures unique to the individual who intends them. The valuating act of object-constitution *motivates* further decisional acts. Historical sciences understand past choices by empathy, that is, by living through their motivations. But *natural sciences also* must feel-into and live-through the valuing aspects of object-constitution, for that is the only way in which two individual scientists can focus on *relatively* identical objects in their studies.

Moreover, all sciences are textual. Their practitioners must write reports of their findings to be read by other scientists. Scientific reading should avoid docility, echoing, and chiseling, just as scientific writing should avoid authoritative assertion, autobiography, and anonymity. More will be said below about "receptive adaptation" and "realization" as the preferred modes of the reading and writing that comprise scientific literacy.

Before that constructive proposal is presented, let us briefly examine two of its competitors in the contemporary marketplace of theory: psychoanalytic feminist philosophy of science, and materialist feminist philosophy of science. It is well for us to introduce these two contemporary theoretical programs at this point in our exposition, because they illustrate what a "science of science" must accomplish even though they fall short of accomplishing it. Each of these programs draws on a larger body of theory. Materialist feminism derives its orientation from the work of Marx, Engels, and their disciples, while psychoanalytic feminism hearkens back to Freud.

As Paul Ricoeur established already in 1965, Freud's psychology and Marx's economics have implications that reach beyond their respective scientific disciplines.[31] They contribute hermeneutic principles capable of application in any discipline whatsoever, as long as *the disciplinary subject matter is taken to be a text.* Unlike more traditional programs of interpretation, which aimed to understand or explain, the materialist and psychoanalytic investigative enterprises are "hermeneutics of suspicion." They look for meanings that are hidden, and for the practices and processes that hide those meanings. On one hand, appearance is to be distrusted. On the other hand, when given a critical diagnostic reading, appearance is trusted to disclose displacements of desire. The unmasked displacement is then taken to be the real meaning of the cultural or scientific text. Furthermore the unmasking is taken to be therapeutic in itself, with no further ado.

---

[31]See Ricoeur (1965: 29-44).

Diagnosis and therapy become the objectives of all the sciences. Thus these diagnostic-therapeutic programs divert scientific attention away from the disciplinary object-fields, in the service of ill-defined "liberation" or free play.

One welcomes any articulation of a generalized scientific method and a general goal for the sciences. Yet one becomes suspicious of suspicion when every scientific question receives the same answer. As we shall see, the method of diagnosing desire's displacements seems to reduce motives to causes. The goal of liberation for play seems to disable creativity. Provisionally let us suggest that *a science of science ought to support future scientific practice and the realization of improvements in the conditions of human life.* Because the two initiatives in philosophy of science that we are about to survey are both avowedly *feminist*, they would endorse this criterion.

There are several further reasons for selecting strands of *feminist* philosophy of science for consideration here. This selection saves us from having to consider *all* materialist and psychoanalytic philosophy of science, while providing a reasonable sample of the capabilities of the two hermeneutic approaches. Furthermore, the efforts to be considered here share with Husserl and Stein the ambition of anchoring all sciences, whether natural or cultural, with the same general interpretive principles. They also complement our suggestion that sciences so secured will comprise a program of reading and writing--a literacy. Most importantly, the feminist psychoanalytic and materialist initiatives call into question the element of *gender* in science, along with *race*. Insofar as female gender and Jewish race are expressed as individual personal bodily characteristics and behaviors, they have been touchstones in our engagement with Stein's phenomenology up to this point. Nevertheless we have expressed dissatisfaction with Stein's own handling of them in terms of "types." We have yet to establish the relative priority of gender and race within the individuation of i's and the constitution of objects. We would like to learn whether they are boons or blocks to empathy. Both materialist and psychoanalytic hermeneutical programs have a great deal to say about "the subject."

C. Psychoanalytic Feminism as the Science of Science

We have already mentioned an early effort by Sigmund Freud to establish psychology as a science. His 1895 "Project" treated psychic phenomena as if they resulted from physical causal processes, and therefore it failed to integrate consciousness and meaning. Freud's later works focused on consciousness and symbolic meaning, often sacrificing real links to events within the live body and its outer world. Like Husserl, Freud regarded the object as an egoic accomplishment. However Freud's approach to objectivation was developmental rather than noetic.

In the psychoanalytic tradition stemming from Freud, the first years of life are of crucial importance. They pass in the well-known stages: the oral (birth to about 18 months), the anal (about 18 to 36 months), the genital (about 3 to 5 years), culminating in the so-called oedipal crisis as a transition into psychosexual latency and the relatively calm childhood period of readiness for formal education.

Object formation, in which the child becomes able to recognize the independent being of the mother, is considered the key achievement of the oral phase. Arguably it is also the beginning of gender identity. The personal individuation of the baby is the psychological correlate of noetic individuation, or the transition from i to i's and I.

Psychoanalytic object-relations theory, taking its name from that first achievement of the infant human life, explains reason and gender through the processes of their co-origination in the individual. Accordingly the capacity to reason, like personal gender identity, emerges out of the unfolding relationship between mother and child, especially during the oral stage. On its affective side, this relationship enables the baby to balance conflicts: the desire to return to a less differentiated state, against the fear of being swallowed up; the desire to use the mother ruthlessly, against an emerging rudimentary concern for her as a separate person. On its logical side, this relationship enables the child to balance contradictions: the object is created in the mind by baby's desire, but was there previously waiting to be created; the baby in fits of rage strikes out against the object, but the object survives destruction to become truly useful for the first time.[32]

The clinician and theorist D.W. Winnicott has proposed that between mother and infant there is, as it were, *a transitional space* that is neither subjective nor objective. This space opens up in the process of individuation as a kind of testing ground in which the baby safely experiments with relinquishing the fantasy of omnipotent control on the part of, alternately, baby or mother. As mother more often declines to respond instantaneously to baby's every demand, the availability within the transitional space of so-called transitional objects (for example, a blanket or a toy) helps the baby to cope. The transitional space is intermediary between the illusion of omnipotence and the unyielding resistance of reality.[33] This space, like other ego phenomena of infancy, persists into adulthood. It becomes the breeding ground for all cultural productions, according to object-relations theory.

The oral, anal, and genital phases are taken to be species-universal for human beings. They parallel the normal pattern of bodily growth and emergence of basic physical relations without which the baby body cannot live. Yet these stages also are culturally contingent, for they unfold in different versions among peoples with different childrearing practices. Nevertheless the child's first object, the mother, is the prototype for all subsequent objects of knowledge or love--be they people, molecules, plants, history, or literature. Different infancy experiences lead boys and girls to construct this object differently; but *how*? Nancy Chodorow theorizes that the baby's biologically determined sex is recognized by adult care-givers and evokes differential nurturant behavior from them. She assumes that, in every culture, people treat their sons differently than their daughters. Chodorow's hypothesis is that when the infant and the care-giver are the same sex, the nurturing produces slower and less complete ego differentiation in the child (for

---

[32]See Flax (1991: 107-132).

[33]See Winnicott (1965).

example in a girl whose more salient parent is the mother). When the infant and the care-giver are opposite sexes, then the nurturing produces faster and more complete ego differentiation (for example between mothers and sons). Since the results of this process of ego formation persist throughout life, there are exactly two distinct genders whose principal differences consist in emotional and epistemological competences. Western science and philosophy deem "objective" those ways of knowing and feeling that constitute masculine gender identity; these have been valued as rational, reliable, strong, and mentally healthy. By the same token, those ways of knowing and feeling that constitute feminine gender identity in Western societies are branded "relativistic," and they have been disvalued as irrational, deceptive, weak, and sick-making.[34]

Evelyn Fox Keller also associates the characteristics of Western "science" and "reason" with early-life events. According to Keller, the baby starts from an original ego unity with mother, and has to push back far enough to construct its self as not-mother; this self in turn constructs the mother as not-self; that is, as the first object. For boys, it is a double negation: I am not mother, and I am not even female like mother. The separation remains fragile, and the baby's principal ally in resisting a re-engulfment in the mother's ego is the father. As Keller describes the process,

> It is the father who comes to stand for individuation and differentiation--for objective reality itself; who indeed can represent the 'real' world by virtue of being *in* it. ... (F)or all of us--male and female alike--our earliest experiences incline us to associate the affective and cognitive posture of objectification with the masculine, while all processes that involve a blurring of the boundary between subject and object tend to be associated with the feminine.[35]

Yet for boys as well as girls, both emotional resiliency and cognitive creativity depend upon access to the in-between "transitional space" that belongs exclusively to neither self nor object.

Citing Winnicott, Keller proposes the ideal of a *dynamic objectivity* in which the autonomous knowing self can risk itself in forays into "transitional space" where there are phenomena "about which it cannot be determined whether they belong to the observer or the observed." Such phenomena include the indeterminacy that physicists face in quantum mechanics when they attempt to characterize the relation between a system and its description. This indeterminacy appears to be a feature of the objects of knowledge themselves (for example, the

---

[34]See Chodorow (1974: 51), emphasis added. Chodorow was calling for a cross-cultural anthropological research program, which has failed to materialize. Her theory was elaborated in a dissertation, published as Chodorow (1978). Despite the lack of empirical corroboration, it has become feminist dogma and is routinely cited as having been "demonstrated." See, for example, Bordo's (1987) uncritical adaptation of Chodorow's work.

[35]See Keller (1985: 85-87).

behaviors of subatomic entities) rather than an inadequacy in the knowing.[36]

The recognition of the intractable indeterminacy of reality throws a new light upon the conventional posture of science, whose quest for perfect objectivity now is disclosed as unrealistic and emotionally driven. Keller says that traditional science is founded upon two beliefs: that nature is objectifiable, and that it is knowable. The first tenet "embodies the radical dichotomy between subject and object," and it assumes "that reality outside us is composed of objects." The second tenet gives assurance that the separation is not total, and that in fact there is a special connection "that permits us to read the laws of reality without distortion, without error, and without omission." Keller's proposed "dynamic objectivity" would enable "the very real indeterminacy in the distinction between subject and object to function as a resource rather than as a source of confusion and threat."[37] Yet there are going to be enormous emotional obstacles to that kind of objectivity. Because the negation that founds both self and male gender continually threatens to dissolve, tremendous psychic energy is invested in maintaining the distinction between object and self that was established at such a cost in infancy. The scientist feels compelled to insist upon the autonomy of the object in order to defend his own psychic integrity, although he does not recognize the real source of the necessity that he feels. A dynamic objectivity therefore would require a *dynamic autonomy* on the part of the knower: one that takes pleasure in the interdependence of knower and known.[38]

Jane Flax's work represents object-relations theory in a methodologically reflexive moment. Her effort to find a voice is a kind of authorial self-analysis in which she coaxes herself (and readers) to face and understand the anxiety operative in the forgetting and camouflaging of information. In effect, she psychoanalyzes certain curious avoidances and incoherences in Freud's own account of how he conducted his therapeutic sessions and in his explanations of the mechanisms by which cures were effected. Flax finds that Freud misdescribed the main tool of his trade, namely, the analytic situation. While his narrative accounts of cases relate details that portray him as a warm and caring partner in an intimate relationship with his patients, his theoretical prescriptions demand that the analyst become a blank facade and an empty mirror, emotionally unavailable to the patient. The consulting room is supposed to be a controlled setting "like an empiricist's laboratory," Flax says. The so-called transference, the therapeutic alliance between patient and therapist that is the implement of healing, is misdescribed and left as a riddle.[39] Flax unravels the riddle by showing how the transference employs

---

[36]See Keller (1985: 85, 139-149).

[37]See Keller (1985: 141-149, 99).

[38]To illustrate the emotional stakes for the scientist, Keller cites researchers' vivid descriptions of their work in which they employ such metaphors as stalking, hunting, attacking, torturing, forcing, and conquering nature--with nature taking a female pronoun, like as not. Such brutal rhetoric has intergenerational consequences, in that it tends to attract into the sciences those individuals who have similar emotional needs.

[39]See Flax (1991: 66-73).

certain relational processes characteristic of the first two years of life, when the ego of the child gradually emerges from the mother's and learns to hold her apart cognitively and affectively as a separate object. Flax observes that Freud claimed to have difficulty understanding those early experiences. Although he articulated his object-relations theory in an attempt to account for basic human needs, he also proposed a rival, contradictory account known as instinct theory or economic-drive theory.[40]

These two glitches in Freud's work--his camouflage of the transference, and his contradiction of the significance of object relations--are read by Flax as nonaccidental, that is, as symptoms of something. What they betray to her is Freud's own attachment to the ideal of positivistic science, and the anxiety brought on by his recognition that his very own discoveries were eroding it. In Flax's view, the psychoanalytic account of how reason, objectivity, and gender emerge in the human infant closes the door irrevocably on the viability of empiricist and rationalist accounts of science and knowledge. She calls for an epistemology of the transitional space, the region of reality that is neither/both subject nor/and object. Flax frames her own project of "thinking fragments" as an example of transitional thinking.

Thus for Flax and the other feminist object-relations theorists, objectivity is not self-evidently equivalent to perfect knowledge. It actually owes its possibility to two prior states retained psychologically in adulthood: the undifferentiated ego of mother and child in early infancy, and the transitional space in which imaginative re-merging and re-emergence of subject and object were still permitted in later infancy. Utter objectivity is a fantasy impossible to realize, since there could be no knowing at all if the knower were completely autonomous with respect to the known. This fantasy of scientific objectivity, therefore, should be understood to function as an anxiety-reduction mechanism designed to secure the male ego, that is, as a gesture of negation toward and control over all that is other. But a male ego so secured is pathological and parasitic; by no means does it represent the "natural state" of male human beings.

As a science of science, object-relations theory diagnoses scientific practices and claims alike as neurotic symptoms in need of therapy. This theory argues for the *contingency* of: (1) what is culturally sanctioned as objective or reasonable; (2) the contours of maleness and femaleness; and (3) the emergence of gender and reason in any individual. These three regions of contingency in human life are co-determinative diachronically and synchronically. But the theory harbors latent *essentialist* assumptions as well: (1) that a human being results *causally* from processes constituting the oral, anal, and genital stages; and (2) that interpretive strategies deriving from those processes can in turn produce something like a true account of them. These assumptions are flaws that cannot be resolved by psychoanalytic theory, because it does not adequately distinguish between cause and motivation. Psychoanalytic dogma requires phenomenological critique.

---

[40]See Flax (1991: 66-73, 52-56).

D. Materialist Feminism as the Science of Science

The political and economic theories of Marx and Engels include epistemological claims that may be summed up in two axioms: that what human beings can know is determined by their work, and that correct knowledge of class society is available only to those who are its workers. Feminist philosophers of science adapt these tenets by associating women and workers. Some theorists regard this association as an analogy (women are oppressed *as if* they were workers). Others hold for the literal origin of all economic oppression out of the primordial and continuing oppression of women. A few voices question whether women are universally oppressed and whether "oppression" has any essential transcultural meaning.

For Nancy Hartsock, one's economic engagement in society constitutes a "standpoint" in the sense that it makes some things visible and others invisible. Hartsock asserts dogmatically that "women's work in every society differs systematically from men's." The specific work of women everywhere, she assumes, is the reproduction of children and the production of subsistence food and services. If Marx is correct in his claim that the character of one's participation in production shapes the character of one's knowledge, then women's special productive activities ought to have epistemological significance--even though Marx himself did not notice it. Therefore Hartsock calls for a theoretical account of the extraction and appropriation of women's productive and reproductive labor.[41] Hartsock holds that the sexual division of labor is the first social fact. In this respect, rigorous feminist standpoint theory departs from earlier feminist appropriations of marxist thought, which had accepted the myth of a primordial egalitarian society.[42] But like earlier theorists, Hartsock believes in the possibility of constructing a unitary generic standpoint for all Western women, and she declines to test out her theory with application to cases.

More fruitful feminist applications of marxist theory, leading in turn to theoretical reformulations, have occurred in anthropological and sociological fieldwork, particularly in Africa. Karen Sacks's 1979 book *Sisters and Wives*, a study of six traditional societies, contested the social-darwinist assumption of a biologically determined essential womanhood. She argued that "nature" is a metaphor or mask concealing the social relations characteristic of late industrial capitalism. Sacks denied the then-common anthropological assumption that babies and culture must be made in separate spheres. In no society, she wrote, are the carrying, bearing, and nursing of babies defined as the totality of women's work. Moreover, people ingeniously organize their childrearing activities to fit around their productive activities, not vice versa.

Sacks has demonstrated the arbitrary and obscuring character of distinctions ordinarily made between waged and domestic labor, between kinship and labor relations, and between home and workplace. Her data indicate that the

---

[41]See Hartsock (1983).

[42]See, for example, Millett (1969).

exchange of labor for wages between workers and capital does not constitute the closed system typically portrayed in marxist analysis. Rather, the formal (and often largely male) labor market is continually supported by the environing informal market and subsistence labor of non-wage-earners (many of whom are women).

The terms "sister" and "wife" for Sacks do not primarily connote either gender or affective personal relationship. Rather, they indicate two different nodes within the network of production and ownership. In the partilineal patrilocal societies that Sacks studied, a "sister" is a household member who is an owner, a decision-maker among others, and a controller of her own sexuality, all because of having been born into that household. By contrast, a "wife" is someone brought into the household who must work there but does not control the fruits of her own labor. (The same individual will be a wife in one household, and a sister in another when she goes home to visit.) All the sisters of the household share the right to the labor of all the wives of the household; for example, the wives' farming, weaving, meal preparation, and other commodity production. Thus it is clear that "wife" in these societies is first and foremost a kind of *laborer*; only secondarily does wifehood have anything to do with gender or with sexual relations.[43]

In a 1989 anthropological review article, Sacks presents a major feminist revision of marxist theory. Recording the consensus of feminist marxist ethnographers, she asserts that "women's unwaged domestic labor is a necessary condition for the existence of waged labor." Moreover, the standpoint of women within the economy has epistemological significance:

> When working-class women are the subjects and narrative voices of case studies, [then] class membership, gender and kinship organization, class-based mobilization, and class consciousness look very different from the way they have been portrayed in nonfeminist marxist analysis. . . . Feminist theory applied to the study of working-class women's lives has birthed questions like: What are the social relations by which the working class sustains and reproduces itself? How do women conceptualize their unwaged labor and community-building activities? . . . How do women's constructions of their sexuality relate to issues of class and kinship? Embedded in these questions, I would suggest, is a definition of the working class in which membership is not determinable on an individual basis, but rather as *membership in a community that is dependent upon waged labor, but that is unable to subsist or reproduce by such labor alone*.[44]

In Sacks's conception, then, it is as if the economic relationships of the workplace are enclosed within a permeable membrane afloat upon a nutrient sea of domestic reproductive labor. "Reproductive labor" includes generational reproduction (the raising of future workers) as well as the daily physical and psychic refreshment that enables a worker to return each day to his job. Women's

---

[43]This finding discloses the mistake upon which the social-darwinist or "essentialist" position is built. It construes the title "wife" as gender information equivalent to information about labor status: "wife" equals "female" equals no significant involvement in productive labor.

[44]See Sacks (1989: 543).

uncompensated labor thus benefits capital in two ways: by freeing it from paying the full cost of reproducing workers, and by maintaining the women themselves as a reserve army of wage laborers should they ever be needed. *Knowledge of* these facts becomes available when research design taps into the perspectives of investigators and respondents who occupy "women's places" in society.

However, rigorous pursuit of feminist-standpoint theory transcends the notion of a standpoint altogether when the marxist fiction of the solitary wage-earning individual is unmasked. The significant agencies for economic relations in the material world now seem to be networks and communities of support. Thus only an oxymoronic "diffuse standpoint" is possible. In Donna Harraway's call for "situated knowledges," the metaphor of location is profoundly altered. Feminists are to reject *both* the pretense of the seer who, seeing all angles at once, is never himself seen, *and* the relativism of social constructionism. Feminist objectivity will acknowledge the partiality of any knowledge while taking responsibility for what is known and how it has become known. The problem

> is how to have *simultaneously* an account of radical historical contingency for all knowledge claims and knowing subjects, a critical practice of recognizing our own "semiotic technologies" for making meanings, *and* a no-nonsense commitment to faithful accounts of a "real" world, one that can be partially shared and friendly to earth-wide projects of finite freedom, adequate material abundance, modest meaning in suffering, and limited happiness. . . . We need the power of modern critical theories of how meanings and bodies get made, not in order to deny meaning and bodies, but in order to live in meanings and bodies that have a chance for a future.[45]

Objectivity is partial connection. Haraway wants a doctrine of objectivity that is usable, although not innocent. The feminist knower is not to be self-identical, but split and contradictory. Because one does not coincide perfectly with one's social position (or even one's species), one is able to engage partially with others. Haraway's criticism of the visual rejects only a seeing *that cannot itself be seen.* She calls attention to the disappearing prosthetic "eyes" of contemporary popular techno-science; for example, those that digitally construct stunning color photographs out of telemetry from interplanetary probes or from probes of submicroscopic living structures. She calls the sense of discovery and conquest that such photos evoke the "Land, ho!" effect. The photos give us "things" that, it seems, we could reach out and grab or explore and colonize.[46]

In another critical study of visualization, Haraway investigated Carl Akeley's use of taxidermy and safaris to produce the dioramas of the American Museum of Natural History. The construction of these displays in the early part of this century was meant to offer a realistic vision of each species within its usual habitat. Obscured were the facts that animals were killed to be "saved" in the dioramas, that large male specimens were featured as typical, and that the exhibits were funded by wealthy industrialists to teach swarthy immigrants the virtue of

---

[45]See Haraway (1991: 187).

[46]See Haraway (1991: 221-222).

hard work.[47] The labor of hunting, shooting, skinning, and stuffing was hidden so that the contrived museum display could be seen as "natural."[48]

The *naturalization* of gender, race, and class is achieved by hiding the means of their production. Haraway turns to something quite explicitly *un*natural to dramatize her epistemological proposal: the cyborg. Fantasized in science fiction, a cyborg is a composite of human being, brute, and machine. Cyborg components come from both sides of three culturally significant boundaries: those between organism and machine, between human and animal, and between physical and non-physical (or hardware and software). Cyborgs were contrived in science fiction before they became social realities in medicine, in war, in publishing, and in many other fields. As fiction, the cyborg displays the possibilities and the pleasures of uniting across boundaries. Haraway says that this icon entices toward a kind of unity that is counter to three other, culturally privileged but counterfeit kinds of unity: (1) holism, or the fantasized original organic oneness with the mother; (2) unity imposed through domination; and (3) unity through incorporation. Cyborg unity is constructed, consensual, and partial.

The cyborg's insistence that unity in future will be constructed and partial is what undermines the possibility of a unified feminist standpoint as such. Since there is no generic woman, whose standpoint could it be?[49] "Woman" has not been an innocent category. The inclusion of women of color into the unified category "woman" has been a coerced and dominative inclusion. Even the key analogy that put the tool of marxist analysis into the hands of feminists--the analogy between reproductive and productive labor--is suspect for Haraway as an essentializing and naturalizing move. Before effective unity can be constructed, the dominating category of "woman" must be allowed to decompose back into "women" plural.[50] Thus the standpoint is dissolved, diffuse, porous, and hotlinked. It is not a home but a homepage.

Cyborgs are made, not born. As bio-artifacts they have no infancy, they do not undergo the gradual emergence of a fragile ego, and so they also escape the turbulence of gender identification along with the emotional commitment to objectivity that it entails. Flax and the object-relations theorists placed an epistemological premium upon competence to merge with the other in a pre-objective communion that would reprise the original mother-child unity, and

---

[47]Despite the rhetoric, it was clear to Haraway that the "endangered species" to be preserved through this instruction was the white male Yankee. See Haraway (1989: 36-42).

[48]Another sort of work is camouflaged behind the writings that Akeley supposedly authored. For the most part they were taken down by a stenographer as he told after-dinner stories, or ghostwritten outright by his wife. Photos in the museum files indicate that she also killed some of the animals that are claimed as his trophies in his publications. See Haraway (1989: 46).

[49]Haraway's critique is not considered by Seyla Benhabib, who employs space as a metaphor for discourse; see Benhabib's (1992) book *Situating the Self.*

[50]See Haraway (1991: 151-158).

she disparaged the objectivity of science for foreclosing that possibility. Haraway for her part looks askance at all myths of primordial unity, but prizes the possibilities for connection in the global integrated circuit now provided through techno-science.

As a science of science, feminist materialism denounces the practices and claims of science as oppressive ideology in need of unmasking. This theory argues for the *contingency* of (1) production upon reproduction, and especially upon the work of gender maintenance; (2) knowledge upon economic "location"--or, with Haraway, diffusely situated prosthetic vision; (3) "naturalness" upon erasure of productive activity. In each case, the contingency is denied and the genesis is hidden. "Goods," "truths," and "nature" all obscure the labor on which they depend. Moreover, materialist standpoint theory assumes that (1) the conditions of labor determine the knowledge that it produces; (2) survival is the essential imperative and defining criterion for human action; and (3) with the dissolving of "natural" boundaries, survival depends on coalitions among humans, other species, technology, and earth herself. But labor, survival, and coalition cannot be understood on materialist premises alone. Phenomenological critique is needed.

E. Realist Feminism as the Science of Science

By conventional standards, both psychoanalytic and materialist feminist philosophy of science fall short of their ambitions to be supersciences: *the science of* science. Neither presents an adequate empirical basis. Neither reliably predicts the individual cognitive style of any particular man or woman. Neither has been cross-culturally validated, although some impressive non-Western applications have been attempted. Neither demonstrates the sort of gender determinism craved by the scientific establishment, on the one hand, and by feminist ideologues, on the other. Object-relations theory deals with psychological phenomena; and although it handles them causally, it largely ignores the constraints that the material world imposes upon knowledge. Standpoint theory deals with material relations without due regard for the mind and the emotions.

But even by conventional standards, there are obvious strengths in these traditions as well. They regard as scientific knowledge that which equips one to protect and cherish the fragile contingency of life. The cyborg network of materialist theory and the transitional space of psychoanalytic object-relations theory alike have survival value in that they project a new kind of objectivity, one in which affectivity is key. These epistemologies highlight how desire has determined knowledge for Western science: either through the enforcement of rigid objectivity as a gender-identity project, or through the avarice of capital. At the same time they entice toward the intimacy of transitional space and the pleasure of cyborg transgression. Both epistemologies re-theorize "woman's position" as the competence for connection at the heart of cognitivity. For object-relations theory, knowledge shelters in the "between" where the scientist connects with the object to be studied (or rather, accepts the interdependence that always has been there). For diffused standpoint theory, knowledge shelters in the "among" of online variant technologies, ethnicities, species, and earth. Perhaps most importantly, both theories try to account for how knowledge *hides*: through repressed memory, or through a "naturalizing" camouflage of the labor that has constituted worldly realities. Thus the epistemological utility of these feminist proposals inheres in their access to the mechanisms of desire and deception.

In this concluding section of our study, then, we must indicate how that utility is retained in our constructive proposal, even while we overcome the troubled positivism of the psychoanalytic tradition and the blind constructionism of the materialist tradition. This is no easy task. Several influential feminists have given up entirely. Doyennes of academic feminism such as Susan R. Bordo and Elisabeth Schüssler Fiorenza regard science as beyond redemption--and good riddance to it. For them, any attempt at establishing the natural and cultural sciences must be a symptom of "anxious flight" (for Bordo) or "kyriarchy" (for Schüssler Fiorenza). Either symptom indicates fatal error. Since these critiques are axes laid a priori at the root of my proposal, I must remove them before proceeding.

1. Bordo's diagnosis: science as anxious flight.

The transition from the medieval to the modern worldview, culminating in the work of the seventeenth-century philosopher René Descartes, is the focus of Bordo's widely acclaimed 1987 book *The Flight to Objectivity*. She gives Descartes's *Meditations* a "psychocultural" reading through the categories of object-relations theory. This reading is intended to disclose "hidden psychological truths" that explain why Descartes would theorize that the material objects in the world are radically different from the human mind that knows them. Before Descartes, the medievals supposedly experienced no gulf between themselves and their world. Bordo likens that original unity to the unity between mother and fetus before birth. Thus in the stirrings of early modern science, a newly aggressive human mind opposed itself to the newly passive natural environment--much as babies are opposed to their mothers during the time of object-formation and identity-establishment, according to psychoanalytic object-relations theory. Analogously, Bordo holds, science emerges wrapped in "parturitional" anxieties like those that accompany the infant's separation from the mother. Science remains necessarily flawed because it must neurotically conceal the real character of its efforts. Science is not really about the world. It is about maintaining fragile masculinity. It insists on the necessity of objectivity in order to guarantee the integrity of personal identity, including gender identity. The practices of science continually conceal the historical, developmental, utterly contingent character of the twin achievements of objectivity and subjectivity.

Fortunately, the blade of Bordo's ax is dulled by history and phenomenology. *Historically*, one can point to numerous ancient and medieval writings in which the world is anything but a congenial all-enveloping womb. Qohelet, Epictetus, Augustine, Maimonides, Beruriah, and the Sages of the Talmud hardly viewed human beings as blissfully merged with an organic female universe. By the same token, one can name historical figures in philosophy such as Boethius and Schopenhauer who insisted that reason and objectivity were decidedly feminine. In our own century, object-relations theory itself had a historical origin and was in fact produced by the very sort of objectifying scientific practices that Bordo means to indict with it. Thus this theory cannot be accorded the status of transhistorical truth that Bordo gives it notwithstanding her frequent disclaimers that she wields its categories only as hermeneutical tools. In practice Bordo largely ignores the historical and cultural contingency of gender itself, and of the profiles of "masculinity and femininity" as various cultures understand them--although she gives lip service to this contingency. *Phenomenologically*, then, we can see that both the naive periodization of history and the psychoanalytic categories of object-relations have been received as dogmas into Bordo's argumentation. These dogmas were imported whole without undergoing critical examination. Bordo's work thus is what we have termed an "echoing" reading of the evidence. It prolongs interpretations offered by others, without bothering to check out their viability originarily. Great chunks of Bordo's work cannot be egoically followed--cannot be validated by having their coherence empathically re-enacted. Bordo has read her psychoanalytic sources *as authorities*. Many in turn read Bordo herself as an authority, and import her assertions without asking for any further certification than the author's name. But we cannot. Much as we might like to harbor in the security of authority, phenomenology renders us unable to do so.[51]

2. Schüssler Fiorenza's denunciation: science as kyriarchal oppression.

On Bordo's reading, natural science is neurosis. For Elizabeth Schüssler Fiorenza, cultural science is oppression. The marxist thesis that ubiquitous class conflict is the hidden core of history has been transposed by Schüssler Fiorenza into a principle for feminist history of the earliest Christian churches. She takes it as axiomatic that Jesus of Nazareth and his first adherents were engaged in the struggle against gender oppression. This premise is an axiom because it is held to need no historical evidence to support it; rather, it becomes the lens by which any claims about the past are to be validated or discredited.

Schüssler Fiorenza assumes that all past societies were structured to support the patriarchal oppression of women, and that women always somehow resisted that oppression. Her more recent work refines the problematic terms "women" and "patriarchy." The ideal category of "wo/men" now includes oppressed and marginalized men along with all women.[52] Not all males and fathers, but only those who are lords, are the perpetrators of oppression--now termed "kyriarchal" rather than patriarchal.[53] In any event, proto-Christian

[51]Because Bordo advocates "sympathy" or "empathy" as an alternative to Cartesian epistemology, readers may wonder why I do not look to her work for corroboration of Stein's philosophy of *Einfühlung* and of my own proposal. See Appendix Two for my argument that Bordo's supposedly empathetic reading of the "medieval" world view has in fact displaced and effaced its *living* proponents through acquiescence to the Western philosophical and literary canon.

[52]See Schüssler Fiorenza (1995: 191 and 89). On this view, Jesus of Nazareth is one of the wo/men who constantly struggle against their social oppression.

[53]From Greek κύριος (*kyrios*, lord). See Schüssler Fiorenza (1992: 7-9).

resistance to kyriarchal oppression is not an empirical finding, but an explanatory principle held to be beyond the reach of any empirical challenge.

Schüssler Fiorenza disallows any scientific use of this principle to frame hypotheses that could be confirmed or disconfirmed through empirical research. Indeed, she explicitly rejects all empirical research as positivist and "kyriocentric," that is, complicit with oppressive structures of social domination by elite males.[54] The charge of positivism seems to stem from the difficulty of distinguishing between data and model in historical research. Obviously, there can be no social science if heuristic models *entirely* determine the data that are gathered by means of them. Conflating data and models, Schüssler Fiorenza regards them all as *texts* with equal and arbitrarily assigned value. Having only texts, she argues, we have no extrinsic way to choose which of several contradictory texts give the most adequate representations of the past. That being the case, one may simply elect to favor the texts that promote some liberating project in the present.

This indictment of all historical sciences as inescapably positivist and kyriarchal is an ax that looms over our proposal at its inception. Fortunately this ax, too, is dulled by historical and phenomenological considerations. *Historically*, while it is true that all data are (in part) determined by the models through which they are produced, data are (in part) *not* determined by the means of their production. One finds that the data sometimes disconfirm prior assumptions and overturn the heuristic models that had been in force. This disconfirmation happens for two reasons. First, texts need not be read in docile, echoing, or chiseling mode. Readers can rely on their own i's either to certify coherence or to jump the tracks. The second potentiality for disconfirmation of textual models stems from the fact that there are more things in the universe *than texts*. Archaeology gathers extra-textual evidence. Therefore we have the possibility and the obligation to test and thereby to confirm or disprove the tentative conclusions drawn from an examination of texts facilitated by materialist social modeling. In historical sciences the alternatives facing us are not "pure facts" versus "theory-drenched facts." Rather, the alternatives are: a thread of facts spun from one kind of evidence and dyed with one theory, versus a fabric of facts woven of threads from many kinds of evidence and spun out through various social modelings. The "fabric" will be logically stronger than the "single thread." Its strength comes from the complex corroborations among data that have been produced through theories that are independent from one another, from the investigator's interests, and from the fields of evidence themselves.[55]

Schüssler Fiorenza's refusal to consider the wealth of material evidence

---

[54]For her rejection of empirical historical research, especially historical-Jesus research, as positivistic, see Schüssler Fiorenza (1994a: 80-86; 1995: 9, 73-74, 87-88, 156, and 108-9; 1992: 79-96). Many feminist theologians part ways with her here, while acknowledging the value of her pioneering work in feminist hermeneutics of scripture.

[55]What I call logical strength is termed "security" in the literature of science studies, which is too extensive to review thoroughly here. Wylie (1995a and 1995b) describes this strength as the convergence of many evidentiary lines upon one coherent model. See the discussion below.

emerging from the ancient world commits her to *constructionism* in practice. But here is a dull ax indeed. Jesus and the early churches are up for grabs; they may be portrayed in any manner that suits contemporary interests. Yet Schüssler Fiorenza wishes to avoid the appearance of historical constructionism. She asserts that certain events occurred in the past and produced effects that continue to affect the present.

> Recognizing that the past is only known to us through textual traces 'is not however the same as saying that the past is only textual as the semiotic idealism of some forms of poststructuralism seem to assert' [quoting Hutcheon]. Historical representation gives *meaning*, not *existence* to past events. Although in epistemological terms we can know the past today only in and through historical discourse, past events have occurred. By underlining the fact that . . . access to reality is always mediated through language, one problematizes and denaturalizes references to the real. Such a demystification does not, feminist theory insists, excuse us from giving a more adequate account of reality. . . .[56]

This disclaimer seems calculated to ward off criticism, but it rings hollow against the operative constructionism that runs throughout Schüssler Fiorenza's interpretive practices. Jesus of Nazareth is designated a "wo/man." Schüssler Fiorenza adopts this neologism without attempting to discover whether his own culture might offer indigenous categories of interpretation to highlight whatever resistive practices were undertaken by the early churches.[57] The twin theses of women's oppression and women's resistance to oppression are invoked equivocally: sometimes as a heuristic model to guide investigation of texts, but often as universal truths to fill in gaps in the texts. The slippage between these two intentions seems as deliberate as it is provocative. What is lacking is coherence. One does not find in Schüssler Fiorenza's published work any account of the relation between reality and text.

Phenomenologically, this creative slippage betrays the fact that Schüssler Fiorenza reads the New Testament with chisel in hand. She carves Jesus into a "wo/man," breathing her own political commitments into his form. What we have termed a "chiseling" mode of reading is one that ghostwrites and throws one's own voice into the text. In this case, indigenous first-century categories of interpretation are obliterated, and the early Greek churches are made to rail against "kyriarchy." Therefore, astoundingly, the woman's testimony that inaugurated Christian faith must be chiseled out. Mary Magdalen can no longer be permitted to say "ἑώρακα τὸν κύριον (*heōraka ton kyrion,* I have seen the Lord)," John 20:18. Whatever the title "Lord" might have meant to first-century Christians becomes irrelevant and unrecoverable. After Jesus is chiseled into a "wo/man," he cannot have been called what they in fact did call him: *Kyrios*, Lord.

Chiseling denies the creativity of the chiseler, even as it effaces the experiences that are read and renders those who lived through them anonymous. There may very well be reasons why today's interpreter might not wish to re-

---

[56]See Schüssler Fiorenza (1992: 91; cf. 1994).

[57]See my own recovery of indigenous resistive terms for securing the identity of Jesus, in Sawicki (1994a: 95-118).

enliven the first-century interpretive acts that designate Jesus of Nazareth as "Lord"; however on phenomenological grounds, she should make such a refusal in her own name.

3. Toward a realfeminist alternative.

We have succeeded, then, in deflecting a diagnosis and a denunciation that had threatened an a priori dismissal of our proposal. Bordo's diagnosis of natural science as *neurotic repression* was shown to arise from her "echoing" reading, and Schüssler Fiorenza's denunciation of historical science as *kyriarchal oppression* was shown to arise from her "chiseling" reading. Both critiques neglected to take historical facts adequately into account. Although these two eminent academic feminists recoil from the task of establishing the grounds for the natural and cultural sciences, one cannot help but remark that the coherence of their own analyses has suffered from this demurral.

It remains now to indicate the path toward a science of science that Edith Stein's phenomenology offers--when read in the mode of receptive adaptation. The philosophy of science that I wish to articulate should be both realist and feminist, and it should address gender and race as factors in all scientific projects. These realist aspirations signal an orientation toward *realization* of possible scientific and political objectives. Reality exceeds texts. No philosophy can claim to be feminist without commiting itself to the reality of a world in which real objectives are to be achieved. All varieties of feminism must answer to this criterion; and as we have seen, materialist and psychoanalytic feminisms do not answer well enough. At most, they have shown themselves capable of unmasking *un*realities, but that is not yet a commitment to reality and realization.

Although reality exceeds texts, texts cover reality. Scientific practice entails a complex engagement of the textual with the nontextual. On which side does the body fall? Is it textual: is it "inscribed" with social texts of race, gender, class, age, and ability? The metaphor of inscription has been widely employed in paraphilosophical discussion of how social identity propagates. Akin to it is the metaphor of the "discourse" as circumscribed space of socially disciplined human interaction.[58] According to these metaphors, the body is socially inscribed with the cultural practices of gender and racial identity, and its activities are delineated by them. Moreover, everything a body does contributes toward the reinscription of gender, race, and other social texts.

Curiously, this provocative metaphor of "body as text" has not yet produced a complementary theorization of body *as writer*: that is, a theory of creative transgression through variant reinscription of gendered and racial texts.[59]

---

[58]These metaphors are commonplaces of French psychoanalytic postmodern literary theory. The designation "paraphilosophy" arose in a personal conversation with Barry Smith in New York in December 1995.

[59]Elsewhere I have proposed that human bodies are capable of both autoinscription and heteroinscription. That is, we actively receive social texts into our own bodies and practices, and we also administer them to others. Race and gender, strictly speaking, are

Nor has it prompted an exploration of the dialectical intertextuality obtaining among bodies, landscapes, and artifacts. Thus the literacy of the social sciences so far has been understood to be a rather truncated literacy. Social texts seem to propagate themselves automatically; we are their dupes, their *tabulae rasae*. The mechanisms of this automatic copying are left unexplicated. This is owing to a failure to distinguish between the interwoven causal and motivational factors at play in the copying. To copy is to follow, to "read." But as we have seen in our study of Stein's phenomenology, what follows is i and *what* it follows are the motivated acts of i. An i tracks another i's moves intimately but never exactly. In tracking, the reading i registers not only causal factors and their ensuing effects, but also inclinations and the choices that they occasioned. Reading i's are at the same time writing i's, expressing their own choices, feelings, and registrations of causes, impulses, and indications.

The creative variations that i's introduce are the untheorized crucial factor in social and physical sciences. The so-called scripts of gender and race are not simply causes; they are templates for choice. Thus they cannot be "explained" by positivistic science. They cannot even be adequately "interpreted" by any hermeneutic program that lacks a poietic dimension. But a fully poietic hermeneutic, like Stein's, models a literacy that is at once both a reading competence and a writing competence.

The challenge of devising scientific practices for *knowing* the world *as* a written and rewritable reality already is being met. Contemporary archaeology offers us examples, and two illustrations have been selected for presentation here. The discipline of archaeology employs methods and data of both natural and cultural sciences. Thus these illustrations provide concrete instantiations of scientific practice that invokes a poietic hermeneutic.

(a) Leone's archaeology of class and race in Annapolis. The methodological problem that confronts practitioners of historical archaeology is to devise a way to coordinate their reading of conventional texts with their "reading" of other material remains, including artifacts and landscape configurations in the excavated earth. Archaeologists should cultivate a hermeneutic practice capable of recovering the creative choices of the people who inhabited the site that they are trying to understand. Mark Leone, Parker Potter and their colleagues employ a "recursive" approach to this task.

In recursive interpretation, neither the historical texts nor the material remains are taken to convey straightforward information "innocently." Both are read suspiciously and then those tandem readings are critically correlated. The archaeologists' own stance toward the tradition under investigation, and their own susceptibility to the persuasive power of artifacts and texts, are taken into account as well. Leone regards material things as carriers of meanings and relationships. In other words, as Stein would say, the material implements of everyday life, along

---

not texts but rather genres for the production of various instantiations of races and genders. As genres they are historical and they change over time. Most importantly, they are frameworks for selective defiance. See Sawicki (1994a).

with urban space itself, were being constituted as value-objects in colonial times. Material artifacts convey meanings because their physical making involved acts of valuation. They were made to serve purposes, and they did serve them--although the purposes may not have been present to anyone's consciousness.

If artifacts became subtly persuasive as soon as they were placed in use and were being "read," archaeologists wish to know how that "reading" happened. The intentional valuation of the artifacts figured into every activity and every project in which they were used. The constituting act repeatedly became a component of all subsequent encounters with the artifact; its constitution *motivated* all deliberate uses to which the thing was put. The valuation intrinsic in the object's constitution was more or less complied with in every realization to which its use contributed. Besides compliant object-constitution, archaeologists committed to a "recursive" interpretational program expect to encounter traces of "resistive" readings of various sorts in the past as well. Both texts and artifacts, particularly the landscape, are regarded as (potentially) strategic interventions in the negotiation of social relationships, whether for the sake of maintenance or for change.

Leone describes artifacts as active vectors for the cultivation of virtues and viewpoints in the population. Potter writes, "the recursive quality of material culture is the capacity of objects to teach their users ways of thinking and behaving."[60] Leone argues that

> using things substantiates and reproduces all the same social actions that went into the artifact in the first place. Just as language reflects and in use creates, so things that are made reflect but also substantiate and verify, and thus reproduce the processes that led to making them.[61]

Archaeological investigation and reporting, too, produce textual artifacts that are strategically powerful in the service of various academic and political enterprises today. At Annapolis, the foundation myths of the United States are corroborated in the spatial rhetoric of the reconstructed ("preserved") Statehouse where George Washington resigned his military commission and so fathered his country as a democracy. The site stabilizes republican values.

But Leone and his colleagues attempt to interrupt the present-day social cultivation of received values by teaching laypeople, as consumers of archaeological displays and reports, how to discern the covert mechanisms of persuasion at work within their own civic involvements--just as they worked in the past--and how to resist those messages when appropriate. In Leone's program there is a phase called "Archaeology in Public" (or "on the street") that is designed to teach tourists about these factors.[62] Leone's site--Annapolis and its historic

---

[60]See Potter (1992: 117).

[61]See Leone (1992: 131; and 1982). See also Shackel and Little (1992). Compare Scarry (1985), who describes phenomenologically how creativity redounds from the artifact to act upon its creator.

[62]See Potter (1992).

Statehouse--was progressively redesigned in step with political changes during the American colonial and republican periods. The city is closely linked with foundational narratives of the state of Maryland and of the United States, which are both precious to and taken for granted by most of the visitors to the site. These interlinked meanings are quite firmly ensconced in tourists' consciousness, the archaeologists found. Thus, the "unmasking" of the material conditions in which present-day privileges and disadvantages had their beginnings had no apparent effect. Exposure to the archaeology and to the critical theory behind it largely failed to unseat the preconceptions of visitors to the Annapolis sites through the 1980's.

The discouragingly limited success of this phase of the project was regarded by the investigators as an important part of their data. A more receptive audience was found after Leone and his colleagues sought guidance from two African American historians also working in Maryland. They were introduced to communities in Annapolis who, unlike the middle-class white tourists, were quite vitally interested in learning about the origins of social privilege in the city. Moveover, black Annapolitans helped Leone's team to refocus their research agenda. Leone writes:

> Our dialogue partners had three questions for us as archaeologists. Was there, indeed, any way to tell whether archaeological material was associated with African Americans? Did they have a share in the record? What would an African American historical archaeology look like?[63]

These questions were framed in 1988, and led to a wealth of evidence showing how African Americans, both as free persons and as slaves, had dwelled in the neighborhoods of the city during colonial and republican times. The kind and distribution of their material culture indicated that while the African Americans had participated in the economic and cultural life of the city, they had also been able to offer significant real and symbolic resistances to it.

> A third of the population of Annapolis is and has been African American, and there are several neighborhoods where free black people lived both before and after the Civil War. Their artifacts show us that they used table settings no different from any others; that they made selective use of white-dominated markets; that they had an established African American cuisine; and that African Americans used some items in symbolic ways that were different from white usage. Archaeological evidence [reported from other sites] shows negotiation between classes about class identities through the different use of knickknacks, wild foods, and mass-produced national brands. *Integration into and resistance to the market occurred simultaneously. A persecuted people strategically maintained cultural integrity.* Thus there is now some history of African Americans in a city where their historical presence has long been implicitly denied.[64]

In reading the past, Leone is writing the future. More precisely, he is intervening

---

[63]See Leone (1995: 262).

[64]See Leone (1995: 262), emphasis added.

in the replication of a myth of the "good old days" of the American Revolution, persuasively presented in the various historical monuments of old Annapolis. The nostalgic "official" history of the Chesapeake region not only flatters white middle-class tourists. Left alone, it also accomplishes the astonishing hegemonic feat of hiding one-third of the city's people--even from themselves. Leone and his colleagues, however, have revealed the African American citizens of the city: revealed them as agents, not victims. Black Annapolitan cultural literacy did not begin in 1988. It was already *selectively* following the scripts of seventeenth- and eighteenth-century colonial society, with creative variations. Leone enhanced his own scientific literacy by following the lead of persons different from himself. The creative, resistive, innovative choices of the past now point the way toward innovation in the future.

(b) Wylie's history of gender archaeology. Like the African Americans of Annapolis, women were invisible in the archaeological record until very recently. Alison Wylie, a philosopher with extensive archaeological field experience, has documented the rapid blooming of the new specialty of gender archaeology.[65] Her fine-grained epistemological studies show that contemporary feminist practice in archaeology is establshing a middle course between and beyond the two extremes of positivism and constructionism, where much of recent scholarship has run aground. As a historian of the science of archaeology, Wylie finds that the discipline has visited each of those extremes several times already in this century.[66]

In the most recent swing of the pendulum, the so-called New Archaeology of the 1960's and 1970's in North America aspired to value-free, rigorously scientific and "objective" data collection. Its practitioners regarded the cultures of the past as if they were systems blindly adapting to environmental influences such as topography, climate, and food supply. Those adaptations were understood to be natural processes governed by laws that could be discovered; in other words, they could be "modeled." The research program of this *processual* New Archaeology was explicitly positivist and deductivist. It aspired to be a science that could generate hypotheses from its models and then get to work confirming them. Failed hypotheses would lead to adjustments in one's model, enabling the model itself to evolve gradually toward the epistemological status of a cross-cultural universal--an

---

[65]The archaeology of gender, much of which is explicitly feminist, has developed rapidly during the last ten years. See Conkey and Spector (1984) and the collections of papers edited by Gero and Conkey (1991), Claassen (1992), Walde and Willows (1991), and de Cros and Smith (1993). For further bibliography, see Bacus (1993). Wylie has documented and discussed the rapid development of this subfield in several publications, and most recently in a position paper prepared for a workshop preceding the conference on the Women, Gender, and Science Question at the University of Minnesota in May 1995. See also Sawicki (1994b) for applications of contemporary gender anthropology to the archaeology of Israel.

[66]See Wylie (1995 and forthcoming). See also Sawicki (1994b) for discussion of the principal articulators of the New Archaeology and of the more recent postprocessual archaeology.

ideal never attained in practice. Having modeled social processes in this way, the New Archaeology could deduce explanations of its data from them.

However, as Wylie remarks, this method could not handle some of the most interesting cases in archaeology, where factors such as the uniqueness of a given culture or the unpredictability of a spontaneous human initiative often become the key to understanding the past. Cultural uniqueness and human initiative are the very factors that in principle must elude deductive explanation, because they cannot be deduced from universal explanatory models.[67]

While processual archaeology was producing diminishing returns, advances in general social theory by the early 1980's spurred an alternative methodological proposal: *postprocessual* archaeology. This movement went to the opposite extreme from the positivism of the New Archaeology, embracing instead a constructionist position with regard to archaeological data. On this account, the past is produced in the present, through a vicious circle of inquiry in which the data are largely or even entirely determined by the questions that the investigators ask and the models they employ, as well as by the investigators' own social locations and political commitments. From the extreme postprocessualist position, the quest for scientific rigor undertaken earlier by the New Archaeology would appear to be not only futile, but sinisterly complicit in a project of promoting the interests of today's elite classes.[68]

But Wylie detects a turn toward a more moderate mediating position in some very recent statements by representatives of postprocessualism. This turn is being led by practice, not theory. Feminist archaeological practice discloses that data are *not* entirely determined by the interests, hypotheses, and models that investigators have brought into the field and the lab, as the postprocessualists contended. Quite simply, Wylie shows, it does happen that material uncovered in the earth *surprises us* and disconfirms our expectations. (Of course this could not happen if postprocessual constructionism were correct, for then one would never disconfirm hypotheses.)

Some of the most productive feminist archaeology of gender conforms rather closely to the deductive or hypothesis-testing program proposed by the New Archaeology. Much feminist archaeology intends to be "better science," not anti-science. Through it, models of social process continue to be revised and improved --for example, by the addition of considerations of gender to a marxist socio-economic model. Feminist critics identify aspects of previous social modeling that have been "bad science" in that they have incorporated unfounded assumptions about the essential capacities of women.

Wylie suggests that the practices of feminist archaeology are currently leading the way toward a resolution of the standoff between positivism and constructionism in archaeological method. The key to this advance has been a

---

[67]As Brumfiel (1992) argues.

[68]Proponents of postprocessualism in archaeology draw on the same trends in literary theory that inspire Elisabeth Schüssler Fiorenza's a priori rejection of empirical historical-Jesus research.

recognition of the length and complexity of the inference chains that tie the reconstructed picture of the past onto the raw excavated earth. Because so many sorts of material and textual evidence are available for examination, these inference chains are multiple and they can stabilize one another. In our terms, these evidentiary chains involve the interbraiding of causality and motivation in the progressive constitution of objects in research. Thus, historical inferences that could not be supported wholly upon texts, or wholly upon skeletal remains, or wholly upon material artifacts and landscape, become much more reliable when many lines of reasoning converge from several different kinds of evidence.

Feminist archaeologists, then, for the most part have shunned the relativism and constructionism that many feminist critics of science have taken over from materialist and psychoanalytic literary criticism. Evidence for reconstructing women's lives is not as scant in the archaeological record as it seems to be in cultural and scientific texts. Thus there is less temptation for archaeologists to resort to extrapolating data from sociological models. However there is a dire need to correct those models when they omit gender, a principle of economic organization in all known human societies that was a key factor in the assignment of civic and nutritional prerogatives in the past.

Like Leone's "recursive" archaeology, feminist archaeology of gender retrieves meanings that were lodged *intentionally* in artifacts and landscape. Those meanings motivated choices toward the realization of objectives intended as possible in terms of what had already been actualized. Because objects are never constituted the same way twice, the built environment can have multiple potentialities for creative use. The movement of live bodies through an ancient habitation site left traces that can still be read through empathy. This reading requires that *causality be distinguished from motivation*, so that *the bounded latitude of creative choices* can be recovered.

4. Proposal and conclusion.

The archaeological initiatives that Wylie describes and those of Leone and his colleagues exemplify science in a new mode of literacy. Such science reads the past and writes the future *realistically*, that is, with attention to the realizations of gender, race, and other social structurations. This mode of science is what I intend to support with my articulation of a realfeminist philosophy of science. The realism to which my work aspires is no mere naive empiricism. My proposal must distinguish itself from the movement that Sandra Harding has termed "feminist empiricism," whose adherents endorse the prevailing methodological norms of science while calling for more rigorous application of them. Feminist empiricists charge that the androcentric biases of research methods have simply been "bad science." But in their view, science can correct its own failings; it can increase its rigor and effectiveness simply by implementing remedial gender-inclusive measures.[69] The claims of feminist empiricists, however, can be contradictory. On one hand, they hold that the practices and findings of science are not determined

[69]See Harding (1986: 24-26 and 151).

by the individual personal characteristics of scientists. In this respect, "the scientist" envisioned by empiricism highly resembles Husserl's "transcendental i." On the other hand, feminist empiricists promise that science will be better when it is done by women and when its practices take account of personal descriptors such as gender. Contradictory claims are made about the role of personal determinants within science.

By contrast, my realfeminist philosophy of science rests on the premise that no two scientists *ever* constitute *the same* object. The variance of objects is owing to their dependence upon causal factors that cannot be egoically inhabited, but only registered; factors such as:

- impulses;
- the differential sensitivity of individual human bodies;
- the sequence of states through which body and psyche may have passed;
- "types" imported into consciousness without undergoing constitution;
- the "life-feeling" that happens to imbue any soul;
- the inclinations that accompany motive-constitution.

Conscious registrations of these causal factors, while they may be quite clear and distinct, are not open to empathy; and this fact insures that no act of object-constitution can be perfectly replicated. Moreover, object-constitution always includes valuation. Objects, having invariably been idiosyncratically constituted as motives, uniquely attract or repel whoever constituted them. Thus we cannot precisely *know* the comforts of life at Mount Vernon that prompted Washington to hand the army back to Congress. We cannot *know* the motherhood of a neolithic woman by counting the ridges etched on her pelvic bones by successive births.

Another's *objects* are off-limits to me. But another's *acts* are not; they are quite available to lead my egoic following as I inwardly feel along with them in *Einfühlung*. I can understand whatever I can empathically feel myself into. Understandable, *followable* acts include:

- valuations, which are acts of motive-constitution;
- the motivation of inclination by valuation;
- strokes of will in regard to impulses and inclinations.

The empathizing of such acts is what makes science *possible*. What makes it *science* is that two crucial distinctions occur. As scientist I distinguish between causal sequences and motivated acts, since I know that I can empathically inhabit only acts. But while inhabiting and following the acts of another, I also detect the other's impulses, inclinations, "life-feelings," and other caused registrations--not empathically, to be sure, but as opaque causes conditioning the emergence of each motivated act. Thus I distinguish the limits of empathy itself: my own live experience registers with me as *determined* by what was free in the other's experience. The other had options bounded by causal factors. Within those bounds, the other's motivated act-path unfolded. As I follow it, I *scientifically* can distinguish the options from the obligations.

Commonplace archaeological examples will underscore this point. Why did Paleo-indians bury stillborn infants beneath the mother's bedding in times of famine? (The little bodies *could have* been buried in any soft soil, or been exposed as carrion. The selection of just these resting places was motivated, not caused, and so it is accessible to knowledge through empathy.) Why did complex hierarchies of dishware come into use within working households in the eighteenth-century Chesapeake region, supplanting the undifferentiated but servicable crockery of earlier eras? (Nothing *forced* people to upgrade their table settings. The inclination to stratify and regiment domestic practices can be understood in terms of motivation, but not as a mental cause that produced a social effect.) What one understands is the creativity at work in the realization of one option out of several. The scientist understands how realization redounds recursively onto the causal factors that conditioned it, and may very well alter them. Really. The "mechanisms" of resistance and liberation turn out not to be causal mechanisms at all; rather, they are motivated choices that have been constrained by material conditions without being determined by them.

Thus the intricate braid of causality and motivation within human life can be skillfully combed out, then braided back together in a different pattern. What exactly must realfeminist philosophy of science change, if it wants to change the world? It must wean people away from deficient modes of reading: the modes that we have termed docility, echoing, and chiseling. Perhaps the most important first step is to illuminate the ontological status of gender itself. In Steinian terms, gender is a set of "types" imported into consciousness whole without having been originarily constituted. That is, the gender types are "echoed." These types then figure *causally* into the constitution of further value-objects. Phenomenological inquiry into the origination of gender typologies will disclose that their apparent "natural" status is an achievement partly determined by biological and social causes and partly created by acts of valuation, expression, and choice.

Materialist and psychoanalytic critiques of science stop once they have descriptively unveiled the constructed, non-necessary character of gender systems. Having shredded the mystique of naturalness, they consider the job done. No further guidance is forthcoming. In comparison, a realfeminist critique grasps gendering valuations as creative adaptations to specific sets of conditioning circumstances. Although it discloses causal constraints on the construction of gender in past times, it does not explain away their enabling role in human projects of realization. On the contrary, the ingenuity of gendering is appreciated against the background of cultural causal constraints. In this way, individual innovative choices that were made possible by gendering can be seen as the creative achievements that they were.

Stein's amendment to Husserl's account of object-constitution gives realfeminist philosophy of science the means to indicate various nuances in "reality." Husserl, for his part, had meant to exclude any notion of "being" from intentional objects. *Whether or not it existed* was not to be counted into the noematic sense of a constituted object. Stein, however, held that the sense that a particular item *really does exist right here* can very well be an inseparable component of the noematic content of the item in question, as constituted. But there are various ways for objects to be or not to be. Consider the meanings represented in Figure 6.2

FIGURE 6.2: MODES OF BEING FOR MEANT-OBJECTS

(1) X is because it must be.
(2) X is, but it did not have to be.
(3) X is, but can change under conditions Y.

(4) Z is not, because it cannot be.
(5) Z is not, but only because conditions Y have not yet occurred.
(6) Z is not, but should be.

(7) Y is something that I can accomplish.

For Stein, the "contents" of an act of object-constitution may include indices of the object's existence and its possibility, as well as its desirability or value. Stein's meant-objects are fatter than Husserl's. Their sense has a thicker noematic content. Regarding the meanings in Figure 6.2, Stein would not count three noemata or meant-objects: X, Y, and Z. She would instead count seven, because she includes the (meant) mode of something's being *as an intrinsic component of* that being itself as meant. Therefore Stein's objects are not only noematically fatter; there are more of them as well.[70] For example, she distinguishes between an inclination that I resist, and an inclination to which I acceed. The difference between these two objects lies in an intentional motivated act--and this difference can make the difference between blind "reinscription" of cultural practices and perspicacious subversive variations of them.

The subtle distinctions of which a poietic hermeneutic like Stein's is capable can provide better scientific access to phenomena that are simply lumped together as "desire" by other sciences of science. Desire is not always a cause, although it is assumed to be such in both the psychoanalytic and the materialist traditions. While bodily impulses are causes, attitudes and the social structurations (or "types") of gender and race are mere inclinations that do not causally determine behaviors. They resemble literary genres in that they provide the framework and

[70]In principle the number of Stein's meant-objects would be equal to the number of Husserl's, *multiplied by* the number of modes of existence and possibility, *multiplied by* the number of persons there are. To illustrate: Let Z be academic careers for women. Husserl would not distinguish among cases 4, 5, and 6, as Stein does; nor would he allow that object Z has been constituted differently by Stein than by himself.

means of production for particular performances. But those performances require interpretation. They have no single unitary "explanation."

Moreover, we recognize causes at all *only by default*. Causality can be defined only negatively, as that coherence between sequential states of affairs for which we can detect no motivation, no poietic human performance. Caused occurrences can be described only by statements having form (1). "Cause" says "must." Physical science aspires to make statements only in form (1), and to convert statements in form (4) into statements in form (1).

What of the other forms of meant-objects? Surely statements of forms (2), (3), and (5) belong to practical sciences like engineering and politics, by implication. But suppose that an object or a state of affairs is constituted for consciousness as both nonexistent and impossible, as in (4). In that case, *it cannot be willed* even though inclinations may arise concerning it. The i issues no "fiat" in regard to such an object. For a "fiat," the object must be constituted as both nonexistent and possible, as in (5) plus (7). Forms (6) and (7) pertain to human opportunity, responsibility, and achievement; that is, to the dynamics of *realization* through decision.

But it is now clear that *physical science is a particular motivated aspiration within the general repertoire of poietico-hermeneutical understanding*. The "desire" that constitutes natural science does not do so causally; much less is that desire either a male gender defense mechanism (as alleged by feminist object-relations theorists) or an oppressive project of patriarchy (as alleged by feminist materialists). The desire constituting physical sciences is not "blind," but perspicacious. It is the motivation to detect the precise material constraints operating upon human action so that, over against them, we can understand the creative achievements of the past as well as the range of possibility open in the future. Physical science is a subsidiary of the *Geisteswissenschaften*. Its causal explanations serve first to clarify what i's have to work with, and then to provide it.

Thus a realfeminist philosophy of science can describe how creative choices are possible and are actualized. In other words, it can *recover historical agency*. For the natural sciences, this feature points the way toward research design and technological innovation in future scientific practices. For the historical and cultural sciences, it permits the recovery of human initiatives from the past. Because "the past" is where scientific findings accumulate, along with other cultural accomplishments, realfeminist philosophy of science offers a novel paradigm for the progress of scientific enterprises.

I have insisted that choice is a constituting factor within the *practices* of all kinds of science. Cultural sciences (*Geisteswissenschaften*) also take choices as the *objects* of their investigations. Therefore their practices are said to be "recursive" as they go about interpreting past social actions through the overlay of present scientific action. This recursive overlay is what I have termed a "reading," and I have discussed several modes in which it can be accomplished: docility, echoing, chiseling, and receptive adaptation. To recognize these modes is to *give oneself a choice* of approach to the meant-objects of science. That choice does not include, however, the option to replicate past intention exactly. Every reading is an interpretation that varies from "the original." (The present study of Stein's phenomenology is no exception. My reading has been a receptive adaptation. I have selected from among her texts, re-argued and augmented her positions, filled in lacunae, criticized certain elements and left others entirely out of consideration. My egoic following of Stein was made possible by the egoic pathways that Stein's texts opened up, but I was not constrained to follow her reasoning exactly. Insofar as I have understood Stein's philosophy, I have done so by following the motivations of her logical and interpretive acts. Her most personal choices, however, remain beyond my comprehension. Even if she had not been murdered seven years before my birth, even if I had been able to hold her hands and had spoken with her, I would not understand her. Persons are not available to one another directly. We share what we can, and we cannot share ourselves.)

But what of the physical sciences? Their objects are meant as beings that "must be" independently of human intentions. There would seem to be no possibility of docility, echoing, chiseling, or any reading whatsoever in regard to physical things and processes, for they involve no choices. Natural processes are not creative, and admit of no interpretation. The physical sciences, then, actually require a kind of *subtractive literacy*. Confronted with the data of sensory impressions, we must first "read out" anything that can be attributed to human choice of any kind. The so-called "biases" of culture, trends in the data betraying the influence of subjective factors, "static" introduced by gender or racial expectations--such flaws are recognized and subtracted from consideration by physical science *only to the extent that they allow themselves to be "read," that is, interpreted.* Literacy's most basic capability is to distinguish between what is readable and what is not. Thus the isolation of "pure" non-arbitrary and non-readable material for analysis by physical science is an achievement that depends upon literacy. "Better science," the motto of feminist empiricism, requires a preliminary interpretive step that purges from the data any "contaminating" artifacts of choice, including artifacts of the experimental procedure and observation itself. This decontamination, as it were, can be done only as a practice of *un*-natural, *cultural* science. (Without a sociological broom, no rocket science.)

The necessity of this data-sorting procedure for any physical science indicates that the cultural sciences--the *Geisteswissenschaften* or sciences of reading--are by no means the softer, weaker, derivative imitations of true, hard, physical science. The physical sciences, rather, are a specialized moment within a science whose general practices comprise a literacy of egoic leading and following.

With this thesis, I avoid the two extremes of constructionism and positivism. Unlike standpoint theorists and object-relations theorists, I do not relativistically reduce all science to desire, ideology, biography, or literary interpretation. In orienting my realism toward *realization*, I retain an account of the human creativity that has escaped positivistic and empiricist philosophies of science. *The logical priority of literacy to physical science does not undermine science, but rather guarantees its purchase upon reality.*

Let us conclude with a reminder that science is more than *Einfühlung*, egoic following. The i is not enough; to do science, you need a body as well.

Paradoxically, physical things escape our direct knowledge just as surely as persons do, but in the opposite direction. We know things and persons only by their traces. The person casts its intellectual shadow as character, mediating its free acts into the regions where they find expression: the soul, the psyche, and the sentient body at work in the real world. By the same token, events in the real world impact the body and reverberate back through those same regions. Stein coined the term "blending" for this multi-level, bi-directional bodily mediation of influence. The body creates by blending, but it understands creativity by "tending"--Stein's term for the i's inherent proclivity to flow forward into the completion of various conscious processes. Understanding, finally, creates anew as it invariably variantly follows.

Appendix One

Dissertations and Theses on Edith Stein

Doctoral dissertations and other academic theses treating the work of Edith Stein are listed here in chronological order.

Verbillion, June M.
1960 *A Critical Analysis of the Educational Theories of Edith Stein.* Ed.D. dissertation, Loyola University.

Baseheart, Mary Catharine
1960 *The Encounter of Husserl's Phenomenology and the Philosophy of St. Thomas in Selected Writings of Edith Stein.* Ph.D. dissertation, University of Notre Dame.

Bajsič, Aloisius
1961 *Begriff einer "christlichen" philosophie bei Edith Stein.* Doctoral dissertation in philosophy, Bozen.

Madden, Anselm Mary
1962 *Edith Stein and the Education of Women: Augustinian Themes.* Ann Arbor, MI: UMI. Doctoral dissertation, St. Louis University.

Bernbeck, Eligius
1966 *Individuum und Gemeinschaft bei Edith Stein.* Pontifica Studiorum Universitat a Thoma Aq. in Urbe. Thesis ad lauram in philosophia. Doctoral dissertation, Rome.

Höfliger, Anton
1968 *Das Universalienproblem in Edith Steins Werk "Endliche und ewiges Sein."* Studia Friburgensa 46. Freiburg Schweiz: Universitätsverlag.

Salmen, J.
1968 *Personverständnis bei Edith Stein.* Mödling, 1973. Dissertation at Rome.

Ruf, Josef
1973 *Das Abbild der Dreifaltigkeit in der Schöpfung in Edith Steins Buch: Endliches und ewiges Sein.* Doctoral dissertation in philosophy, München.

Oben, Freda Mary
1979 *An Annotated Edition of Edith Stein's Papers on Woman.* Ann Arbor, MI: UMI. Doctoral dissertation at The Catholic University of America.

Barukinamwo, Matthieu
1982 *Edith Stein: Pour une ontologie dynamique, overte à la transcendence totale.* European University Studies, series 23: Theology, volume 169. Frankfurt: Peter Lang. A Roman doctoral dissertation.

Gooch, Augusta Spiegelman
1982 *Metaphysical Ordination: Reflections on Edith Stein's Endliches und ewiges Sein.* Ann Arbor, MI: UMI. Doctoral dissertation at the University of Dallas.

Manshausen, Udo Theodore
1983 *Die Biographie der Edith Stein: Beispiel einer Mystagogie.* Frankfurt: Peter Lang, 1984. Diplom-Hausarbeit in pastoral theology, Münster.

Fidalgo, António Carreto
1985 *Der Übergang zur Objectiven Welt: Eine kritische Erörterung zum Problem der Einfühlung bei Edith Stein.* Doctoral dissertation, Würtzburg.

Bejás, Andrés E.
1987 *Edith Stein: Von der Phänomenologie zur Mystik. Eine Biographie der Gnade.* Disputationes Theologicae 17. Frankfurt: Peter Lang, 1987. Diplomarbeit, Bonn.

Imhof, Beat Walter
1986 *Edith Stein's philosophische Entwicklung: Leben und Werk.* Band 1. Basler Beitrage zur Philosophie und ihre Geschichte 10. Basel: Birkhauser Verlag, 1987. Doctoral dissertation, Basel.

Gottwald, Uwe
1987 *Die Pädagogik Edith Steins.* Diplomarbeit, Würzburg.

Otto, Elisabeth
1989 *Welt, Person, Gott: Eine Untersuchung zur theologischen Grundlege der Mystik bei Edith Stein.* Vallender Schonstatt: Patris-Verlag, 1990. Doctoral dissertation, Tübingen.

Schandl, Felix M.
1990 *"Ich sah aus meinem Volk die Kirche wachsen!" Jüdische Bezüge und Strukturen in Leben und Werk Edith Steins (1891-1942).* Sinziger theologische Texte und Studien 9. Sinzig: Sankt Meinrad Verlag für Theologie, 1990. Diplomarbeit.

Miles, Judy A.
1991 *Simone de Beauvoir and Edith Stein: A Philosophical Analysis of Feminism.* Ann Arbor, MI: UMI. Doctoral dissertation, Claremont Graduate School, Department of Philosophy.

Müller, Andreas Uwe
1991 *Grundzüge der Religionsphilosophie von Edith Stein.* Freiburg: Alber, 1993. Dissertation in theology at Freiburg.

Appendix Two

Critique of the Empathy Theory Presented in Susan Bordo's *The Flight to Objectivity*

In her 1987 book of essays, Susan Bordo interprets "Cartesianism" as a quest for clarity, detachment, and certainty in scientific knowing; that is, as a quest for objectivity of a particular kind. While her interpretation is directed first at the text of René Descartes' *Meditations*, Bordo enlarges it into a "psychocultural reading" of certain intellectual responses to transitions occurring in European cultural life between the years 1400 and 1600 and commonly associated with the rise of modern science. This interpretation imposes a "before and after" structuration upon history. The era of "before" is designated as the Middle Ages and the Renaissance, with "the Greeks" thrown in for good measure. The era of "after" dawns with Copernicus, the rise of merchant capitalism, and the development of perspective in painting. Descartes articulated the epistemological program of this era, which continues into the present. Bordo holds that pre-Cartesian, medieval epistemology resembles the cognitive habits of small children in that both presume a "connected" stance toward the object of knowledge. In both cases the knower is embodied and non-separate from the natural world. Bordo further asserts that the estrangement from nature that allegedly occurred in early modern history was as traumatic for human culture as the separation from the mother is for the small child. In both cases, Bordo says, the "wound" of separation is denied even while we struggle to assert control over the mothering presence that has left us. Thus "Cartesian" science--science in the mode of detachment, clean certitude, and disembodied clarity--is a cultural defense mechanism.

Bordo's thesis was rejected in chapter six of this work on grounds that it is historically and phenomenologically flawed. A fuller account of those flaws is given here. I then point to Bordo's own hermeneutical *practice* as an illustration of what is wrong with her *theory* of empathy.

1. Phenomenological criticisms.

A *historical* mistake is an error in reporting some event that was available in principle for anyone to observe. A *phenomenological* mistake is an error in apprehending the status of some content of consciousness. For example, it would be a *historical* mistake to believe that Sigmund Freud lived in ancient Athens. (He lived in modern Vienna.) But it would be a *phenomenological* mistake to believe that separation from the mother "causes" anxiety. (Separation may occasion anxiety if complex motivational processes also come into play; however anxiety is not "caused.") In science, phenomenological mistakes occur most often in regard to apprehensions of the status of scientific theories and methods. This is the sort of phenomenological mistake that one finds in Bordo's work. Four such mistakes will be examined here.

(a) Application of Piaget's theories. Born in 1896, Jean Piaget began publishing short articles on topics in natural history when he was 11 years old. For the next 70 years he went on to publish foundational works in biology, philosophy, education, and "genetic epistemology"--his favored label for his own academic discipline.[1] Piaget's child psychology became widely known in the United States only after World War II. His research method was to observe children in his home or in laboratory schools, and to engage them in individual conversations. In this way, Piaget developed a structural theory of stages of cognitional ability. Each stage builds upon the former stage, and the stages are separated by crucial and relatively brief threshold periods. Cognitive structure is conceived as an adaptive equilibration that the organism achieves through assimilating stimuli input from the environment and accommodating itself to them.

Piaget derived his notion of cumulative developmental stages from two related sources: evolutionary biology, and a nineteenth-century classical view of the progress of civilization. With his contemporaries in the early twentieth century Piaget shared a common belief in the *evolution* of life forms from algae to complex plants, animals, and human beings, and in the *progress* of civilization from savage to tribal (including Jewish) societies, to city-states, to industrialized Western nations. Philosophy as Piaget learned it was a story about the gradual, inexorable perfection of Western thought. This story was so well known to his readers that Piaget was able to use it as a familiar analogue to explain the behaviors that he observed in children. Piaget accounted for his "unknown"--the development of logic in the child--by referring to something "known"--the development of logic in Western philosophy. In other words, the story of Western philosophy as a "progress" away from participatory knowing *was a constitutive ingredient* in Piaget's account of child development. Piaget saw infant cognition as merged with the body and the mother *precisely because* he saw "savage" and "tribal" cognition in that way, according to the common wisdom of his day.

But here we see the danger of making a phenomenological mistake. If Piaget modeled the child's development after the presumed cultural development of the savage (as he did), then the validity of his account is diminished whenever that earlier account of progressive civilization is called into question (as has happened in recent decades, of course). In any event, Piaget's account must not be cited to support the very theory that it uses for its *own* support. Yet this is exactly what Bordo attempts to do, in the effort to establish her developmental scheme for the periodization of history. She takes Piaget's version of child cognition as if it were the secure, "known" analogue, and then draws parallels from it to concerns in Descartes' texts. (For example, see Bordo 1987: 30, 46-48, 57, 61.) Bordo is delighted to discover a resemblance between the structuration of infantile logical development, on one hand, and the transition from medieval to modern thinking, on the other. But that resemblance actually is owing to the way in which Piaget imported cultural presuppositions into the framing of his theory in the first place.

---

[1]For an overview of Piaget's work, see the comprehensive anthology edited by Gruber and Vonèche (1977).

Despite some strategic and shallow disclaimers, Bordo mistakes Piaget's descriptions for pure, distinct, untheorized *objective* scientific data.

(b) Application of psychoanalytic theories. Bordo mistakes the works of Freud and his disciples for factual data as well. She writes as though psychoanalytic terms like "anxiety" and "defense mechanism" could pick out discrete elements of the natural world. But they cannot do so. Even the psychoanalytic term "object" names an *achievement* of the young child; there are no "objects" lying around in nature apart from the psychosexual developmental careers of individuals. Like Piaget, Freud built his theory by analogies with known cultural motifs, such as Greek myths (Oedipus, for example) and hydraulic engineering (drives and repression, for example). His empirical database consisted in his clinical conversations with patients, along with his own dreams. He *interpreted* these contents by referring to mythic themes and mechanical laws. He did not literally *observe* clear and distinct "oedipal impulses" or "repressive mechanisms" within the intellectual content that he analyzed. Rather, Freud's discernment of patterns to which he gave such names had the character of hypothesis, diagnosis, heuristic wager. Did such wagers pay off? Freud answers that in a fraction of his clinical cases, disclosure of his interpretive hypothesis to the patient brought about the cessation of the person's neurotic symptoms.

The apparent efficacy of psychoanalytic therapy on the *individual* scale is owing to its hermeneutic character. What heals is the meaning-making capacity of individual consciousness. But when psychoanalytic interpretation is applied to the reading of *cultural* texts, however, one is at a loss to say what it is that will do the healing and what will undergo the healing. Furthermore, as I pointed out in chapter six, a psychoanalytic reading of cultural phenomena (such as Bordo attempts) must necessarily reduce those phenomena to the status of texts. Psychoanalytic interpretation disregards any *real* correlates that may pertain to the psychic content of the dreams, stories, fantasies, and conversations that it analyzes at the individual level. When raised to the level of cultural analysis, then, a psychoanalytic reading has no way to put forth claims about *real, political* repressions or defenses. The substantive correlates of the psychoanalytic terms--if any--are irrelevant to the analysis; the terms are mere literary fictions. Just as it is psychoanalytically meaningful to speak of a patient's oedipal impulses *whether or not he has ever really engaged in a parricidal struggle with his own father*, so it may be meaningful to speak of cultural parturitional anxiety *whether or not there are social practices of violence against women.* The meaning is entirely intertextual. The texts have no explanatory relation to reality.

Bordo, however, glosses over these important distinctions between theory and fact, between text and reality, and between individual and society. She is not the first to do so. In chapter six I discussed feminist object-relations theory, which derives from the psychoanalytic work of Freud. In particular, I pointed out that Nancy Chodorow's 1974 *hypothesis* that gender arises from culture-specific childrearing practices had never undergone the cross-cultural empirical testing and validation that Chodorow herself was calling for. Nevertheless a generation of feminist theorists now has accepted Chodorow's thesis as dogma. This dogma is the source that Bordo relies upon when she associates the masculine "Cartesian" objectivity of Western science with the enforcement of rigid separation between knower and known. This dogma undergirds the analogy that Bordo poses between individual parturitional anxiety and a cultural anxiety that spurs the practices of Western science.

(c) Creeping causality. Analogy is the engine of the imagination, and imagination makes dreams come true. But not without work. Bordo neglects to do the work that would be required to construct a path from analogy to identity in regard to her applications of psychological theories. Her argument just skips back and forth between two claims: (1) the quest for objectivity in Western science *is like* an anxious defense mechanism, and (2) that quest *is* an anxious defense mechanism. The relation between those two claims is heuristic, or suggestive of avenues for further research. The second claim cannot be established through question-begging, that is, through *describing* historical events *as if* they were concrete instantiations of the phenomena named by the psychoanalytic terminology. In the absence of an alternative justification, one must conclude that the second claim cannot be validated at all.

Bordo's own attempts to justify her use of this analogy indicate that she is quite aware of the danger of question-begging that is lurking here. The connection that Bordo intends between the analogues--infantile object-formation, and the historical transition to "Cartesianism"--is described in various ways:

- open up the imagination (page 48)
- an illuminative framework for understanding (49)
- provocative correspondences (55)
- a tool for exploring (57)
- reverberation of the psychological on the cultural level (58)
- give psychocultural coherence (100)
- place in a striking new perspective (106)

These avowals of a merely heuristic intention occur as futile gestures toward reining in the presumption rampant throughout the book: that the quest for scientific objectivity *is* a cultural mechanism to reduce the anxiety owing to our estrangement from nature. This presumption is operative in many passages. For example, Bordo writes:

> The particular genius of Descartes was to have philosophically transformed what was first experienced as estrangement and loss--the sundering of the organic ties between the person and world--into a requirement for the growth of human knowledge and progress. And at this point, we are in a better position to flesh out the mechanism of *defense* involved here. Cartesian objectivism and mechanism, I will propose, should be understood as a *reaction-formation*--a denial of the "separation anxiety" described above, facilitated by an aggressive intellectual *flight* from the female cosmos and "feminine" orientation towards the world. (Bordo 1987: 100)

Misunderstanding the status of Freud's findings, Bordo regards anxiety as *causing* denials and other defense mechanisms (rather than motivating them). Thus the analogical transition from the individual to the cultural level here produces a *causal explanation* of the practices of science. Bordo means both to explain and to indict scientific practices in the same stroke. Modern science is explained as being a result of a neurotic cultural need to deny the wound of the loss of intimacy with the natural world. This explanation overlooks the real efficacy of some scientific practices, the failure of others, and the grounds of the difference between the two. The very question of efficacy itself is regarded as a part of the pathological anxiety. A shadow of suspicion falls across the motivation constituting science as *labor toward consensus*.

One curious outcome of this maneuver is that it renders itself invulnerable to any criticism made on scientific grounds. Bordo need not answer to objective criteria, once she has discredited objectivity itself as a cultural pathology with a causal explanation.

(d) Complicity in canon construction. The fourth phenomenological mistake to be noted in Bordo's 1987 book is its complicity in constructing the canons of Western art and literature. A canon is a list; in this case, the list of works chosen to represent "the" intellectual expression of an age. The error that one is liable to make with a canon is to forget the activity of selection that composed it, and consequently to mistake the canon for the whole universe. The act of compiling a canon always entails leaving out some works and, indeed, whole genres and media of expression. Bordo's periodization of Western history necessarily relies upon a commonly recognized list of the greatest and most significant literary, philosophical, and artistic works of Antiquity, the Middle Ages, and the Renaissance. But in mistaking that *selection* for a *representative sample*, Bordo infers beyond what the evidence can support.

In graphic arts, for example, Bordo contrasts the anachronisms and spatial disproportion of early medieval paintings with the depth perspectives achieved by Renaissance artists. She interprets the former as characteristic of a childlike immersion in myth and in nature. Supposedly, a self-aware subjectivity dawns only with the Renaissance, as the seer is offered a proper perspective and a proportional place in a landscape. (See Bordo 1987: 62-64.) This canonical selection produces the desired "before and after" contrast between pre-objective childlike immersion, and mature subject-object distanciation. Yet if the representation of *subjectivity* and the visual are to be investigated, then why leave out the earlier forms of Roman portraiture and Byzantine iconography? Large expressive eyes are the most striking feature of funerary portraits from Roman imperial Egypt. Some centuries later, in Greek icons the eyes of saints become enormous and mystical, offering portals into another world and means of presence between the divine and the human. The viewer of these paintings is invited to see what the icon sees. More importantly, the subsequent *suppression* of these visual emblems of subjectivity by the aniconic artistic traditions of Islam and the iconoclastic traditions of Western Christianity certainly has some bearing upon the cultural history of subjectivity as a visual phenomenon. If ancient portraits, icons, and iconoclast decorative designs are included in the history of the portrayal of subjectivity, then it becomes impossible to maintain the simple periodization upon which Bordo's argument hinges.

More significant, perhaps, is the paucity of medieval, modern, and contemporary *religious* theorists among the authors whose works Bordo surveys. The medieval cognitive experience, which Bordo correctly describes as participation of knower with known object, was primarily a *religious* experience. Jewish and Christian mystical authors were the chief architects of that epistemology. Yet they are strangely absent from Bordo's history of subjectivity. Even Augustine is mentioned only to have his "rare" introspective passages discounted.[2] In fact, the sixteenth century was a period of struggle between conflicting religious accounts of how human beings can claim to know the truth of created and Uncreated realities. The distrust of the mind that Bordo attributes to Descartes is more plausibly laid at the door of Jean Calvin. That Protestant reformer argued that sin had ruined human reason and rendered it incapable of performing the epistemological functions for which it was designed by its divine Creator. His Catholic opponents maintained that reason, though diminished by sin, was rehabilitated by grace and revelation.

The very transitional centuries that Bordo means to interpret--1400 through 1600--are precisely the era when religious repression overtook toleration, when Jews and Christian "heretics" were *noticed* (and then hunted and killed), and when the Reformation challenged earlier assumptions about the terms under which human beings could know truth and goodness. There were many sides to these conflicts, and the differences were epistemological to a very large extent. Thus Descartes had much more to worry about than Copernicus.

By leaving Catholic and Jewish works out of consideration, Bordo has bought into a secular Protestant hegemonic reading of intellectual history. The religious sensibilities of Catholics and Jews are submerged in the "before" category: they are designated childlike, medieval, and *no longer extant.* The term "we" in Bordo's book stands for the intellectual heirs of Calvin. It presumes that there are no intellectuals or scientists *today* who have the "medieval" experience of participatory knowing.[3]

2. Historical criticisms.

Not all pre-Reformation religious thinkers were "medieval," of course. In Antiquity one encounters numerous authors whose sense of estrangement from their world is quite striking, and seems almost "modern." I mentioned the names of some of them in chapter six: Qohelet (author of the biblical book known as Ecclesiastes), Epictetus, Augustine, Beruriah (a woman in the rabbinic circle in second-century Tiberias, mentioned in the Talmud, to whom tradition ascribed

[2]See Bordo (1987: 55). For a contrasting opinion about Augustine's place within a discussion of the subjective experiences of temporality and narrative, see Ricoeur (1984).

[3]In her last chapter, Bordo suggests that feminist epistemology is recovering the capacity for such connected knowing. Technically, this assertion revokes her earlier acceptance of Owen Barfield's claim that "we can no longer duplicate, and can only imagine" an experience of being at one with the universe. See Bordo (1987: 60). At the conclusion of the work, however, Bordo seems to position "women's ways of knowing" as the desired model for *future* science, while earlier book she positioned childlike cognitive modes as the model for a nostalgically remembered *"pre-scientific" past.*

some of the Stoic teachings of Epictetus), and the early rabbis whose teachings come down to us in the Talmud. Any of these figures provides a decisive counter-example to Bordo's assertion that "before" the time of Descartes, human subjectivity had not yet been set at odds with the objective world. I also mentioned Boethius and Schopenhauer as two thinkers who ascribe decidedly feminine characteristics to philosophy or to reason. They count against the feminist claim that reason is always masculinized in Western culture, a claim that Bordo endorses in the final chapter of her 1987 book.

Besides these, there are numerous other mistakes of historical fact to be found in Bordo's work. Many of these inaccuracies are owing to Bordo's reliance on secondary sources and her uncritical acceptance of the hegemonic secular cultural canon:

Bordo's assertions:	**contrary evidence:**
The Greeks regarded the senses as ideally suited to provide information about the world. (37)	Parmenides, a Greek, taught that appearances provide no access to reality. Plato agreed.
The medievals also regarded the senses and the body as a reliable epistemological guide. (45)	Scholastic sacramental theology held that faith was needed to disclose realities that elude the senses.
With Descartes begins the problem of "subjectivity"; i.e., the notion of influences from "within" the human being which can affect how the world is perceived. (50)	Pauline epistle and commentaries on earthly vision "in a mirror, darkly" (1 Cor 13: 12). Comparable rabbinic themes.
After Descartes, the mind is regarded as an untrustworthy inner space that deceives me about the outer world. (51)	The *Monadology* of Leibniz. The German hermeneutical tradition of Schleiermacher, Dilthey, and Lipps.
No one before Descartes thought that the teachings of the past were a source of infectious error. (52)	Plato's distrust of the rhetors. Christian patristic critique of mythology and polemic against the teachings of Judaism and participation in its festivals. Abelard's *Sic et Non.*
The notion of an inner influence upon subjectivity is a new idea in modern times, unknown to the medievals. (53)	Patristic and medieval theologies of divine grace. Talmudic descriptions of how human beings make decisions.
The distinction between self and world is first constructed by Descartes. (57)	Pauline epistles and commentaries on the themes of flesh versus spirit and spirit versus world.

The Greeks did not exclude the body from the soul's quest for eternal truth. (94)	Neoplatonism.

It is not at all surprising to find a few factual errors in a work of this scope. However Bordo's historical mistakes stem from her phenomenological mistakes, which led her systematically to exclude religious works from her survey of Western cultural development. My own ability to spot these errors is not owing to any superior familiarity with the history of philosophy; far from it. I simply read a larger canon.

3. Malpractice of empathy.

The exclusion of religious thinkers, particularly Jews and Catholics, from the history of philosophy is no mere incidental flaw in Bordo's work. It is the keystone of her whole project, and provides the premise for constructing the analogy upon which the book is built.

To understand Descartes, Bordo argues, "we" must enter into the cultural mentality of his contemporaries. But there is a problem: today there are no medieval people around for "us" to interrogate. This is the reason why "we" need to invoke the analogy between the stages of infantile cognitive development and the medieval-modern cultural transition. For Bordo, "medievals" are people who experience no estrangement between their subjectivity and the world around them. They share the "childlike" confidence that their bodies are at home in the world and that their senses deliver reliable knowledge of reality. They participate in what the Scholastics called the *analogia entis*, the similarity of being among all things and their Creator. Bordo writes:

> The medievals *had* no "problem of knowledge" (at least, not as we conceive it, in terms of certifying a correspondence between ideas and external world). Nor was the self/world dichotomy a characteristic way of talking about the universe. Rather . . . the dominating ontological metaphor was of the universe as a single "organism," whose domains (although hierarchically ordered) were characterized by interdependence and interconnection rather than mutual exclusivity. . . . [T]his was not simply a philosophical or ideological superstructure but a mode of experiencing being human and being-in-the-world that permeated ordinary language, and even the most basic levels of perception. The human being of the Middle Ages . . . perceived itself in continuity with the rest of the universe, in a way that we can no longer precisely duplicate, and can only imagine. (1987: 60, citing Carolyn Merchant and Owen Barfield)

But *all of these medievals are dead.* Bordo looks to children to "stand in" for medievals, because she denies that any adult intellectuals today can represent that sort of subjectivity. She presumes that the hegemonic cultural history that she tells is the only one that could be told.[4]

The absence of "medievals" is the absence of Catholic and Jewish

---

[4]For an alternative, see Daniel Boyarin's (1993) account of the body in the talmudic tradition--an intellectual tradition that is alive and well today.

intellectuals who theorize out of their living experiences of the creaturely status of their bodies and the world. Of course, this is not a *real* absence; it is an artifice constructed by the hegemonic Western canon and by Bordo's text.[5] In effect, Bordo inserts the child (as constructed by developmental psychology) into the gap opened up by the ejection of religious intellectuals. Then she targets that child for empathy. The substitution of the available child for the unavailable "medieval" is explicitly selected as the means for entering the world of Descartes "empathically" (Bordo 1987: 31). This is the thesis of Bordo's book. Thus the reader is given to understand that this work is produced through the practice of empathetic understanding. Bordo's writing is offered as a demonstration of the *practice* of empathy.

Bordo formulates an empathy *theory* as well: a formalized epistemology of empathy or sympathy (the terms are synonymous for her). This epistemology "claims a natural foundation for knowledge, not in detachment and distance, but in . . . 'sympathy': in closeness, connectedness, and empathy." According to this theory, sympathetic understanding of an object entails merging with it, uniting with it. Knowing is "placing oneself within" the object and "allowing it to speak."[6] Bordo's theory of empathy as "moving in" complements her practice of empathy as displacement.

When one moves in on the other, there is always the danger that the other will be crowded out. When one speaks "from the place of another," whose voice is it that is heard? What *one* calls "sympathy," the *other* may experience as displacement, eviction, colonization, and stifling. What *one* calls "sympathetic thinking" may be dummying up and ventriloquism to the *other*. An epistemology of empathy has to provide protection for the integrity of the other in the face of interrogation. Bordo's *theorization* of empathetic knowing has neglected to fulfill this duty. Her theory leaves open the possibility that the other will be displaced and effaced during "sympathetic understanding." And, as I have shown, Bordo's *practice* of historical periodization has actualized that possibility. Bordo has replaced historical and contemporary religious thinkers with children.

Bordo's empathy theory supports writing as a practice of effacement, displacement, substitution, and ventriloquism. It supports reading as a practice of echoing the hypotheses of authoritative scholars so as to transform those hypotheses into dogmas without checking the evidence. *If I were to read Bordo as she reads modern philosophy*--that is, in a diagnostic search for the desires and anxieties that have *caused* her assertions--then I would conclude that the author of *Flight to Objectivity* yearns to identify and merge herself with secular intellectual authorities

[5]For a discussion of the living Catholic tradition of the *analogia entis*, see Tracy (1981). Tracy finds that contemporary forms of Thomist philosophy pursue "the liberating intellectual ideal of the Scholastic tradition: 'to distinguish without separation in order to unite without confusion'" (1981: 414). In earlier work, I have discussed Catholic theories of *symbol* as a vehicle for experiencing the human status of created creator and as a means of political agency; see Sawicki (1984, 1988, 1992).

[6]Bordo cites Carol Gilligan and Evelyn Fox Keller as advocates of "sympathetic thinking." See Bordo (1987: 102-103, 112).

while severing herself from her religious community. This author is trying to pass as a WASP intellectual. Yet anxiety over the loss of mystical religious union with creation and Creator drives her quest for a secular mysticism. She fantasizes a feminist epistemological community, with her own text as its talmud.[7] *However I do not read Bordo in that way.* Instead I have read her text by a Steinian poietic hermeneutic. Thus I have criticized Bordo's historical, phenomenological, and practical mistakes by following what I could follow, while indicating where and why I could not follow. I conclude that this text does not lead into the future realization of scientific or political objectives. It has lost its way in the textual forest of theory.

[7]The most distinctive literary feature of the Talmuds--and of their historical core, the Mishnah--is their cacophony. They preserve the lively conversation of many voices in disagreement. Bordo's text retains many "voices" in its scholarly citations and footnotes; however, her sages are coerced into harmony by selective suppression of disagreement and of contrary evidence.

References

Works cited have been grouped by chapter. For primary sources, the works of each author are listed in chronological order of composition. When a first edition or a manuscript was not available to me for consultation, its date is enclosed in brackets. Translations are placed just beneath original texts and their dates are marked with an asterisk. "Husserliana" is the multi-volume critical edition of Edmund Husserl's works published at the Hague by Martinus Nijhoff, continued by Kluwer.

FOR CHAPTER ONE: THE GENESIS OF PHENOMENOLOGY

PRIMARY TEXTS

Dilthey, Wilhelm

[1870] *Leben Schleiermachers*, Vol. 1. Edited by Martin Redeker. Gesammelte Schriften 13/1 and 13/2. Göttingen: Vandenhoeck & Ruprecht, 1970.

[various] *Schleiermachers System als Philosophie. Leben Schleiermachers*, Vol. 2,1. Edited by Martin Redeker. Gesammelte Schriften 14,1. Göttingen: Vandenhoeck & Ruprecht, 1966. (This edition is the first publication of these manuscripts.)

1877 "Über die Einbildungskraft der Dichter," *Zeitschrift für Völkerpsychologie* 10: 42-104.

[1883] *Einleitung in die Geisteswissenschaften (Versuch einer Grundlegung für der Studium der Gesellschaft und der Geschichte)*. Gesammelte Schriften 1. Göttingen: Vandenhoech & Ruprecht, 1966.

*1988 *Introduction to the Human Sciences: An Attempt to Lay a Foundation for the Study of the Human Sciences*. Translated by Ramon Betanzos. Detroit: Wayne State University Press.

[1887] *Die Einbildungskraft des Dichters: Baustein für eine Poetik*. Gesammelte Schriften 6: 103-241. Edited by Herman Nohl. Göttingen: Vandenhoeck & Ruprecht, 1958.

*1985 "The Imagination of the Poet: Elements for a Poetics." Trans. by Louis Agosta and Rudolf A. Makkreel. *Poetry and Experience*, 29-173. Selected Works 5: 29-173. Edited by Rudolf A. Makkreel and Frithjof Rodi. Princeton: Princeton University Press.

[1894] *Ideen über eine beschreibende und zergliedernde Psychologie*. Gesammelte Schriften 5/1: 139-240. Edited by Georg Misch. Göttingen: Vandenhoeck & Ruprecht, 1957.

*1985 "Ideas Concerning a Descriptive and Analytic Psychology." *Selected Works* 2. Princeton: Princeton University Press. Edited by Rudolf A. Makkreel and Frithjof Rodi. Princeton: Princeton University Press.

[1895-6] *Beiträge zum Studium der Individualität*. Gesammelte Schriften 5,1: 241-316. Edited by Georg Misch. Göttingen: Vandenhoeck & Ruprecht, 1957.

[1900] *Die Entstehung der Hermeneutik*. Gesammelte Schriften 5: 317-338. Edited by Georg Misch. Göttingen: Vandenhoeck & Ruprecht, 1957.

*1990 "The Rise of Hermeneutics." Translated by Fredric Jameson. *The Hermeneutic Tradition From Ast to Ricoeur*, 101-114. Albany: State University of New York Press.

[1907-10] "Awareness, Reality: Time." [Excerpts from "Draft for a Critique of Historical Reason."] *The Hermeneutics Reader: Texts of the German Tradition from the Enlightenment to the Present*, 149-164. Edited by Kurt Mueller-Vollmer. New York: Continuum, 1985.

Lipps, Theodor
1893 *Grundzüge der Logik.* Hamburg and Leipzig: Leopold Voss.
[1903a] *Leitfaden der Psychologie.* Second, revised edition. Leipzig: Wilhelm Engelmann, 1906.
[1903b] *Ästhetik: Psychologie des Schönen und der Kunst.* Vol. 1. *Grundlegung der Ästhetik.* Second, unchanged edition. Leipzig: Leopold Voss, 1914.
1906 *Ästhetik: Psychologie des Schönen und der Kunst.* Vol. 2. *Die ästhetische Betrachtung und die bildende Kunst.* Leipzig: Leopold Voss.
1907a "Das Ich und die Gefühle." *Psychologische Untersuchungen* 1: 641-693. Edited by Theodor Lipps. Leipzig, Wilhelm Engelmann.
1907b "Das Wissen von fremden Ichen." *Psychologische Untersuchungen* 1: 694-722. Edited by Theodor Lipps. Leipzig, Wilhelm Engelmann.

Pfänder, Alexander
[1899] *Phänomenologie des Wollens: Eine psychologische Analyse.* Leipzig: Barth, 1930.
*1967 "Introduction" to the *Phenomenology of Willing.* Translated by Herbert Spiegelberg. In *Phenomenology of Willing and Motivation and Other Phaenomenologica,* 3-11. Evansville: Northwestern University Press.
[1911] "Motive und Motivation." Reprinted from *Münchener Philosophische Abhandlungen. Theodor lipps zu seinem sechzigsten Geburtstag gewidmet von früheren Schülern,* 163-95. Leipzig: Barth, 1930.
*1967 "Motives and Motivation." Translated by Herbert Spiegelberg. In *Phenomenology of Willing and Motivation and Other Phaenomenologica,* 12-40. Evansville: Northwestern University Press.
1929 Review of Theodor Celms, *Der phänomenologische Idealismus Husserls. Deutsche Literaturzeitung* 43: 2048-2050.

Reinach, Adolf
[1911] "Zur Theorie des negativen Urteils." *Münchener philosophische Abhandlungen: Th. Lipps zu seinem 60. Geburtstage gewidmet von früheren Schülern,* 196-254. Edited by Alexander Pfänder. Reprinted in *Gesammelte Schriften,* 56-102. Edited by his students. Halle: Neimeyer, 1921. Critical edition, *Sämtliche Werke,* 95-278. Edited by Karl Schuhmann and Barry Smith. Munich: Philosophia Verlag, 1989.
*1982 "On the Theory of the Negative Judgment." Translated by Barry Smith. *Parts and Moments,* 315-378. Edited by Barry Smith. Munich: Philosophia Verlag, 1982.
[1913] "Einleitung in die Philosophie." Lecture course at Göttingen. *Sämtliche Werke,* 369-514. Edited by Karl Schuhmann and Barry Smith. Munich: Philosophia Verlag, 1989.
[1913-4] "Über das Wesen der Bewegung." Seminar at Göttingen, winter semester. Ausarbeitung by Edith Stein. *Gesammelte Schriften,* 407-461. Edited by his students. Halle: Neimeyer, 1921. Critical edition. *Sämtliche Werke,* 551-588. Edited by Karl Schuhmann and Barry Smith. Munich: Philosophia Verlag, 1989.
[1914] "Über Phänomenologie." Lecture delivered in Marburg in January 1914. *Gesammelte Schriften,* 379-405. Edited by his students. Halle: Neimeyer, 1921. Critical edition. *Sämtliche Werke,* 531-550. Edited by Karl Schuhmann and Barry Smith. Munich: Philosophia Verlag, 1989.
*1969 "Concerning Phenomenology." Translated by Dallas Willard. *The Personalist* 50: 194-221.

Scheler, Max
[1912] "Über Selbsttäuschungen" in *Zeitschrift für Pathopsychologie* 1: 87-163. (This article was revised and expanded, and published in 1915 as "Die Idole der Selbsterkenntnis.")
1913a *Zur Phänomenologie und Theorie der Sympathiegefühle und von Liebe und Hass.* Halle: Max Niemeyer.
[1913b] "Der Formalismus in der Ethik und die materiale Wertethik." Part 1. *Jahrbuch für Philosophie und phänomenologische Forschung* 1. (Printed

simultaneously in a separate edition with the subtitle *Versuch der Grundlegung eines ethischen Personalismus*, Halle: Max Niemeyer. Pages are cited from the unchanged second edition of the latter.)

1915 "Die Idole der Selbsterkenntnis." *Abhandlungen und Aufsätze* 2: 5-168. Leipzig: Weissen Bücher.

*1973a "The Idols of Self-Knowledge." *Selected Philosophical Essays, 3-97*. Translated by David R. Lachterman. Evanston: Northwestern University Press.

[1916] "Der Formalismus in der Ethik und die Materiale Wertethik." Part 2. *Jahrbuch für Philosophie und phänomenologische Forschung* 2. (This material was written at the same time as part 1. The two parts were printed together in a separate edition with the subtitle *Versuch der Grundlegung eines ethischen Personalismus*, Halle: Max Niemeyer. Pages are cited from the unchanged third edition of the latter.)

*1973b *Formalism in Ethics and Non-Formal Ethics of Values: A New Attempt toward the Foundation of an Ethical Personalism.* Translated by Manfred S. Frings and Roger L. Funk. Evanston: Northwestern University Press.

1923 *Wesen und Formen der Sympathie.* Second, enlarged edition of *Zur Phänomenologie und Theorie der Sympathiegefühle und von Liebe und Hass*. Bonn: Friedrich Cohen.

*1954 *The Nature of Sympathy.* Translated by Peter Heath. Introduced by W. Stark. London: Routledge & Kegan Paul.

Schleiermacher, Friedrich D.E.

[1805-10] "The Aphorisms on Hermeneutics from 1805 and 1809/10." Translated by Roland Hass and Jan Wojcik. *The Hermeneutic Tradition from Ast to Ricoeur*, 57-84. Edited by Gayle L. Ormiston and Alan D. Schrift. Albany: State University of New York Press, 1990. (These were unpublished until 1959.)

[1819] "The Hermeneutics: Outline of the 1819 Lectures." Translated by Jan Wojcik and Roland Haas. *The Hermeneutic Tradition from Ast to Ricoeur*, 85-100. Edited by Gayle L. Ormiston and Alan D. Schrift. Albany: State University of New York Press, 1990. (A version of this material was in print in 1838.)

[1828] "Foundations: General Theory and Art of Interpretation." Translated by J. Duke and J. Forstman. *The Hermeneutics Reader: Texts of the German Tradition from the Enlightenment to the Present*, 73-97. Edited by Kurt Mueller-Vollmer. New York: Continuum, 1985. (A version of this material was in print in 1838.)

[1838] *Hermeneutik.* Edited by Friedrich Lücke. Schleiermachers Werke 4: 135-206. Aalen: Scientia Verlag, 1967. (This material was compiled after Schleiermacher's death from his manuscripts and from his students' lecture notes.) Second edition (including manuscripts not used by Friedrich Lücke), edited by Heinz Kimmerle. Heidelberg: Carl Winter, 1974.

[1862] *Psychologie*. Edited by L. George. Schleiermachers Werke 4, 1-80. Aalen: Scientia Verlag, 1967. (This material was compiled after Schleiermacher's death from his manuscripts and from his students' lecture notes.)

SECONDARY SOURCES

Ames, Van Meter

1943 "On Empathy." *The Philosophical Review* 52: 490-494.

Avé-Lallemant, Eberhard

1975a "Die Antithese Freiburg-München in der Geschichte der Phänomenologie." In Kuhn, *Die Münchener Phänomenologie*, 19-38.

1975b "Bio-Bibliographischer Anhang." *Max Scheler im Gegenwartsgeschehen der Philosophie*, 267-284. Edited by Paul Good. Bern: Franke.

1982 "Hedwig Conrad-Martius (1888-1966): Phenomenology and Reality." Translated by Herbert Spiegelberg. *The Phenomenological Movement: A Historical Introduction,* 212-222. By Herbert Spiegelberg. Third edition. The Hague: Nijhoff.

Barber, M.D.
1993 *Guardian of Dialogue: Max Scheler's Phenomenology, Sociology of Knowledge, and Philosophy of Love.* Lewisburg: Bucknell University Press.
Bershady, Harold J.
1992 "Introduction." *On Feeling, Knowing, and Valuing*, 1-46. By Max Scheler. Edited by Harold J. Bershady. Chicago: University of Chicago Press.
Betanzos, Ramon J.
1988 "Wilhelm Dilthey: An Introduction." *Introduction to the Human Sciences: An Attempt to Lay a Foundation for the Study of the Human Sciences*, 9-63. Translated by Ramon Betanzos. Detroit: Wayne State University Press.
Brettler, Lucinda Ann VanderVort
1973 *The Phenomenology of Adolf Reinach: Chapters in the Theory of Knowledge and Legal Philosophy.* Ph.D. Dissertation, McGill University. Microfiche.
Briggs, Sheila
1992 "Schleiermacher and the Construction of the Gendered Self." *Schleiermacher and Feminism: Sources, Evaluations, and Responses*, 87-94. Schleiermacher: Studies-and-Translations 12. Edited by Iain G. Nichol. Lewiston, NY: Edwin Mellen Press.
Bulhof, Ilse N.
1980 *Wilhelm Dilthey: A Hermeneutic Approach to the Study of History and Culture.* The Hague: Martinus Nijhoff.
Conrad-Martius, Hedwig
1921 "Einleitung." *Gesammelte Schriften*, i-xxxvii. By Adolf Reinach. Edited by his students. Halle: Neimeyer.
Crosby, John F.
1983 "A Brief Biography of Reinach." *Aletheia* 3: ix-x.
Crosby, John F., and Josef Seifert
1981 "Introduction" to "A Contribution Toward the Theory of the Negative Judgment." *Aletheia* 2: 9-14.
Duke, James
1977 "Schleiermacher: On Hermeneutics." *Hermeneutics: The Handwritten Manuscripts*, 1-15. By Friedrich Schleiermacher. Edited by Heinz Kimmerle. Translated by James Duke and Jack Forstman. American Academy of Religion Texts and Translations 1. Missoula, MT: Scholars Press.
Ellison, Julie
1990 *Delicate Subjects: Romanticism, Gender, and the Ethics of Understanding.* Ithaca: Cornell University Press.
Frings, Manfred S.
1973 "Nachwort des Herausgebers." "Berichtungen und Ergänzungen des Verfassers in der zweiten Auflage des Sympathiebuches." *Wesen und Formen der Sympathie.* By Max Scheler. Gesammelte Werke 7: 333-339, 342-348. Edited by Manfred S. Frings. Munich: Franke.
Gadamer, Hans-Georg
1960 *Truth and Method.* Translator unknown. New York: Seabury Press, 1975.
Geiger, Moritz
1933 "Alexander Pfänders methodische Stellung." *Neue Münchener philosophische Abhandlung: Festschrift für Alexander Pfänder*, 1-16. Leipzig: Barth.
Hart, James G.
1972 *Hedwig Conrad-Martius' Ontological Phenomenology.* Ph.D. dissertation, University of Chicago. Microfilm.
Hering, Jean
1926 *Phénoménologie et philosophie religieux: Étude sur la théorie de la connaissance religieuse.* Paris: Alcan.
Husserl, Edmund
[1914] *Vorlesungen über Ethik und Wertlehre 1908-1914.* Husserliana XXVIII. Edited by Ullrich Melle. Dordrecht: Kluwer, 1988.
1919 Obituary notice for Adolf Reinach. *Kant-Studien* 13: 147-149.
*1983 "Reinach as a Philosophical Personality." Translated by John F. Crosby. *Altheia* 3: xi-xiv.
1994 *Briefwechsel.* Husserliana Dokumenta 3. Ten volumes. Edited by Elisabeth Schuhmann and Karl Schuhmann. Dordrecht: Kluwer.

Ingarden, Roman
[1967] *Einführung in die Phänomenologie Edmund Husserls: Osloer Vorlesungen 1967*. Gesammelte Werke 4. Edited by Gregor Haefliger. Tübingen: Niemeyer, 1992.

Jager, Bernd
1968 "Pfänder on Motivation." *Humanitas* 3: 285-92.

Kalinowski, Georges
1991 *La Phénoménologie de l'homme chez Husserl, Ingarden, et Scheler*. Paris: Editions Universitaires.

Kesserling, Michael
1962 "Theodor Lipps (1851-1914): Ein Beitrag zur Geschichte der Psychologie." *Psychologische Beiträge* 7: 73-100.

Kimmerle, Heinz
1967 "Hermeneutical Theory or Ontological Hermeneutics." Trans. by Friedrich Seifert. *History and Hermeneutic*, 107-121. Edited by Robert W. Funk. New York: Harper & Row.

Kockelmans, Joseph J.
1967 *Edmund Husserl's Phenomeneological Psychology: A Historical-Critical Study*. Pittsburgh: Duquesne University Press.

Koebner, [ ]
1922 Review of *Gesammelte Schriften*, by Adolf Reinach. *Literarisches Zentralblatt*, 16 December 1922, 964-965.

Kuhn, Helmut, Eberhard Avé-Lallement and Reinhold Gladiator, eds.
1975 *Die Münchener Phänomenologie*. Vorträge des internationalen Kongresses in München 13.-18. April 1971. Phaenomenologica 65. The Hague: Nijhoff.

Kuhn, Helmut
1975 "Phänomenologie und *wirkliche Wirklichkeit*." In Kuhn et al., *Die Münchener Phänomenologie*, 1-7.

Lachterman, David R.
1973 "Translator's Introduction." *Selected Philosophical Essays*, xi-xxxviii. By Max Scheler. Evanston: Northwestern University Press.

Lossky, Nicholas
1948 "Perception of Other Selves." *The Personalist* 29: 149-162.

Luther, A.R.
1972 *Persons in Love: A Study of Max Scheler's Wesen und Formen der Sympathie*. The Hague: Nijhoff.

Makkreel, Rudolf A., and Frithjof Rodi
1985 "Introduction to Volume V." *Poetry and Experience*, 3-26. By Wilhelm Dilthey. Selected Works 5. Edited by R. A. Makkreel and F. Rodi. Princeton: Princeton University Press.

Makkreel, Rudolf A., and John Scanlon
1987 "Preface." *Dilthey and Phenomenology*, vii-xi. Current Continental Research 6. Edited by R. A. Makkreel and J. Scanlon. Washington: Center for Advanced Research in Phenomenology and University Press of America.

Mallgrave, Harry Francis, and Eleftherios Ikonomou
1994 "Introduction." *Empathy, Form, and Space: Problems in German Aesthetics 1973-1893*, 1-85. Santa Monica, CO: The Getty Center for the History of Art and the Humanities.

Marbach, Eduard
1974 "Husserls Stellungnahme zum Problem des Ich in der Auseinandersetzung mit der zeitgenössischen Psychologie." In *Das Problem des Ich in der Phänomenologie Husserls*, 218-246. Phaenomenologica 59. The Hague: Nijhoff.

McGill, V.J.
1942 "Scheler's Theory of Sympathy and Love." *Philosophy and Phenomenological Research* 2: 273-291.

Melle, Ullrich
1988 "Einteitung des Herausgebers." *Vorlesungen über Ethik und Wertlehre 1908-1914*, xiii-xlix. Husserliana XXVIII. By Edmund Husserl. Dordrecht: Kluwer.

Mensch, James R.
1981 *The Question of Being in Husserl's* Logical Investigations. Phaenoomenologica 81. The Hague: Nijhoff.
Meyer, Hans
1957 "Zur Ontologie der Gegenwart." *Philosophisches Jahrbuch der Görres-Gesellschaft* 65: 251-93.
Misch, Georg
1923 "Vorbericht des Herausgebers." *Die Geistige Welt: Einleitung in die Philosophie des Lebens*, vii-cxvii. *Gesammelte Schriften* Vol. 5. Erste Hälfte. Abhandlungen zur Grundlegung der Geisteswissenschaften. By Wilhelm Dilthey. Göttingen: Vandenhoeck & Ruprecht, 1957.
Mueller-Vollmer, Kurt
1963 *Towards a Phenomenological Theory of Literature: A Study of Wilhelm Dilthey's "Poetik."* Stanford Studies in Germanics and Slavics 1. The Hague: Mouton & Co.
1985 "Language, Mind, and Artifact: An Outline of Hermeneutic Theory Since the Enlightenment." *The Hermeneutics Reader: Texts of the German Tradition from the Enlightenment to the Present*, 1-53. Edited by Kurt Mueller-Vollmer. New York: Continuum, 1985.
Mulligan, Kevin
1987 "Promisings and Other Social Acts: Their Constituents and Structure." *Speech Act and Sachverhalt*, 29-90. Edited by Kevin Mulligan. Dordrecht: Nijhoff.
Mulligan, Kevin, editor
1987 *Speech Act and Sachverhalt: Reinach and the Foundations of Realist Phenomenology*. Dordrecht: Nijhoff.
Mundt, Ernest K.
1959 "Three Aspects of German Aesthetic Theory." *Journal of Aesthetics & Art Criticism* 17: 287-310.
Niebuhr, Richard R.
1964 *Schleiermacher on Christ and Religion: A New Introduction*. New York: Charles Scribner's Sons.
1967 "Friedrich Daniel Ernst Schleiermacher." *The Encyclopedia of Philosophy* Vol. 7: 316-319. New York: Macmillan Publishing Co. & The Free Press.
Oppenheimer, Oscar
1967 "Streben, Wollen, Motivieren: Zu Pfänders Phänomenologie des Wollens." *Zeitschrift für philosophische Forschung* 21: 538-53.
Ormiston, Gayle L., and Alan D. Schrift
1990 "Editors' Introduction." *The Hermeneutic Tradition From Ast to Ricoeur*, 1-35. Edited by Gayle L. Ormiston and Alan D. Schrift. Albany: State University of New York Press.
Palmer, Richard E.
1969 *Hermeneutics: Interpretation Theory in Schleiermacher, Dilthey, Heidegger, and Gadamer*. Evanston: Northwestern University Press.
Perrin, K.
1991 *Max Scheler's Concept of the Person: An Ethics of Humanism*. New York: St. Martin's Press.
Peters, R.S.
1960 *The Concept of Motivation*. Second edition. London: Routledge & Kegan Paul.
Plantinga, Theodore
1980 *Historical Understanding in the Thought of Wilhelm Dilthey*. Toronto: University of Toronto Press.
Rang, Bernhard
1973 *Kausalität und Motivation: Untersuchungen zum Verhältnis von Perspektivität und Objektivität in der Phänomenologie Edmund Husserls*. Phaenomenologica 53. The Hague: Nijhoff.
Richardson, Ruth Drucilla
1991 *The Role of Women in the Life and Thought of the Early Schleiermacher (1768-1806): An Historical Overview*. Schleiermacher: Studies and Translations 7. Lewiston, NY: Edwin Mellen Press.
1992 "Schleiermacher's 1800 'Versuch über die Schaamhaftigkeit': a Contribution

Toward a Truly Human Ethic." *Schleiermacher and Feminism: Sources, Evaluations, and Responses*, 49-85. Schleiermacher: Studies and Translations 12. Edited by Iain G. Nichol. Lewiston, NY: Edwin Mellen Press.

Rickman, Hans Peter

1961 "General Introduction." In Wilhelm Dilthey, *Pattern & Meaning in History: Thoughts on History & Society*, 11-63. Edited by H. P. Rickman. New York: Harper & Row, 1962.

1967 "Wilhelm Dilthey." *The Encyclopedia of Philosophy* Vol. 2: 403-407. Edited by Paul Edwards. New York: Macmillan Publishing Co. & The Free Press

1979 *Wilhelm Dilthey: Pioneer of the Human Sciences*. Berkeley and Los Angeles: University of California Press.

1988 *Dilthey Today: A Critical Appraisal of the Contemporary Relevance of His Work*. Contributions in Philosophy No. 35. New York: Greenwood Press.

Ricoeur, Paul

1949 *La Philosophie de la Volonté*. Volume 1: Le Volontaire et l'Involontaire. Paris: Aubier, Éditions Montaigne.

*1966 *Freedom and Nature: The Voluntary and the Involuntary*. Translated by Erazim V. Kohak. Evanston: Northwestern University Press.

1952 "Méthode et taches d'une phénomélogie de la volonté." *Problèmes actuels de la phénoménologie*, 113-140. Actes du Colloque International de Phénoménologie, Brussels, April 1951. Edited by H. L. Van Breda. Louvain: Desclée de Brouwer.

1975 "Phänomenologie des Wollens und Ordinary Language Approach." Translated by Alexandre Métraux. In Kuhn, *Die Münchener Phänomenologie*, 105-124.

*1982 "Phénoménologie du vouloir et approche par le langage ordinaire." *Pfänder-Studien*, 79-96. Edited by Herbert Spiegelberg and Eberhard Avé-Lallemant. The Hague: Nijhoff

Rosenberg, Alfred

1930 *Der Mythus des 20. Jahrhunderts*. One hundred seventh edition. Munich: Hoheneichen, 1937.

Schäfer, Michael

1978 *Zur Kritik von Schelers Idolenlehre: Ansätze einer Phänomenologie der Wahrnehmungstäuschungen*. Bonn: Bouvier Verlag Herbert Grundmann.

Schingnitz, Wilhelm, and Joachim Schondorff, editors

1943 *Philosophisches Wörterbuch*. Founded by Heinrich Schmidt. Tenth revised edition. Stuttgart: Alfred Kröner.

Schuhmann, Karl

1973 *Die Dialektik der Phänomenologie I: Husserl über Pfänder*. Phaenomenologica 56. The Hague: Nijhoff.

1977 "Ein Brief Husserls an Theodor Lipps." *Tijdschrift voor Filosofie* 39: 141-150.

1987 "Husserl and Reinach." *Speech Act and Sachverhalt*, 239-256. Edited by Kevin Mulligan. Dordrecht: Nijhoff.

1989 "Kommentar" and "Textkritik" for "Über das Wesen der Bewegung." *Sämtliche Werke*, 775-785. By Adolf Reinach. Edited by Karl Schuhmann and Barry Smith. Munich: Philosophia Verlag.

Schuhmann, Karl, and Barry Smith

1987 "Adolf Reinach: An Intellectual Biography." *Speech Act and Sachverhalt*, 3-27. Edited by Kevin Mulligan. Dordrecht: Nijhoff.

Schutz, Alfred

1942 "Scheler's Theory of Intersubjectivity and the General Thesis of the Alter Ego." *Philosophy and Phenomenological Research* 2: 323-347.

Shuster, George N.

1942 "Introductory Statement" for "Symposium on the Significance of Max Scheler for Philosophy and Social Science." *Philosophy and Phenomenological Research* 2: 269-272.

Smid, Reinhold Nikolaus

1982 "'Münchener Phänomenologie'--zur Frühgeschichte des Begriffs." In Spiegelberg, *Pfänder-Studien*, 109-153.

1983 "Ähnlichkeit als Thema der Münchener Lipps-Schule." *Zeitschrift für philosophische Forschung* 37: 606-616.

Smith, Barry
1982 "Introduction to Adolf Reinach on the Theory of the Negative Judgment." *Parts and Moments*, 289-313. Edited by Barry Smith. Munich: Philosophia Verlag.
1987 "Adolf Reinach: An Annotated Bibliography." *Speech Act and Sachverhalt*, 299-332. Edited by Kevin Mulligan. Dordrecht: Nijhoff.
Sokolowski, Robert
1965-66 Review of new editions of three works by Pfänder. *The Modern Schoolman* 43: 292-296.
Spiegelberg, Herbert
1941 "Notes: Alexander Pfänder [Obituary]." *Philosophy and Phenomenological Research* 2: 263-5.
1967 "Translator's Introduction" to Pfänder, *Phenomenology of Willing and Motivation*, xv-xxvii.
1972 *Phenomenology in Psychology and Psychiatry: A Historical Introduction.* Evanston, IL: Northwestern University Press.
1973 "Is the Reduction Necessary for Phenomenology? Husserl's and Pfänder's Replies." *Journal of the British Society for Phenomenology* 4: 3-15.
1974a "'Epoche' Without Reduction: Some Replies to My Critics." *Journal of the British Society for Phenomenology* 5: 256-261.
1974b "Neues Licht auf die Beziehungen zwischen Husserl und Pfänder: Bemerkungen und Ergänzungen anlässlich von Karl Schuhmanns *Husserl über Pfänder*." *Tijdschrift voor Filosofie* 36: 565-573.
1982a *The Phenomenological Movement: A Historical Introduction.* Third edition, revised and enlarged. The Hague: Nijhoff.
1982b "Aus der Diskussion (zu W. Trillhaas und P. Ricoeur)." In Spiegelberg, *Pfänder-Studien*, 97-106.
Spiegelberg, Herbert and Eberhard Avé-Lallemant, editors
1982 *Pfänder-Studien.* Phaenomenologica 84. The Hague: Nijhoff.
Stark, W.
1954 "Editor's Introduction." *The Nature of Sympathy*, ix-xlii. By Max Scheler. London: Routledge & Kegan Paul.
Stein, Edith
[1933-5] *Aus dem Leben einer jüdischen Familie. Das Leben Edith Steins: Kindheit und Jugend.* Edith Steins Werke 7. Freiburg: Herder, 1965. Reprinted as *Aus meinem Leben. Mit einer Weiterführung über die zweite Lebenshälfte von Maria Amata Neyer OCD.* Freiburg: Herder, 1987.
*1986 *Life in a Jewish Family: Her Unfinished Autobiographical Account.* Collected Works of Edith Stein 1. Translated by Josephine Koeppel. Washington: ICS Publications.
Stern, Paul
1898 *Einfühlung und Association in der neueren Ästhetik: Ein Beitrag zur Psychologischen Analyse der ästhetischen Anschauung.* Hamburg: Voss.
Trillhaus, Wolfgang
1975 "'Selbst leibhaftig gegeben': Reflexion einer phänomenologischen Formel nach Alexander Pfänder." In Kuhn, *Die Münchener Phänomenologie*, 8-18.
Vacek, Edward
1987 "Scheler's Evolving Methodologies." *Morality Within the Life- and Social World: Interdisciplinary Phenomenology of the Authentic Life in the "Moral Sense," 165-183.* Analecta Husserliana 22. Edited by Anna-Teresa Tymieniecka. Dordrecht: Reidel, Kluwer.
Wenzl, Aloys
1947 "Hundert Jahre philosophische Tradition in München." *Geistige Welt* 11: 41-48.

FOR CHAPTER TWO: HUSSERL ON INTERSUBJECTIVITY

PRIMARY TEXTS

(Husserl's manuscripts cannot be arranged in chronological order. It was his custom to return to them again and again to rework them.)

Husserl, Edmund

1900 *Logische Untersuchungen.* Part 1. Prolegomena zur reinen Logik. Halle: Niemeyer.

*1970 *Logical Investigations.* Volume 1. Prolegomena to Pure Logic, Investigations I and II. Translated by J.N. Findlay. New York: Humanities Press.

1901 *Logische Untersuchungen.* Part 2. Untersuchungen zur Phänomenologie und Theorie der Erkenntnis. Halle: Niemeyer.

*1970 *Logical Investigations.* Volume 2. Investigations III, IV, V, and VI. Translated by J.N. Findlay. New York: Humanities Press.

[1904] "Meine Stellung zum Psychologismus." *Briefwechsel.* Volume 2. Munich Phenomenologists. Husserliana Dokumenta 3. Edited by Elisabeth Schuhmann and Karl Schuhmann. Dordrecht: Kluwer, 1994.

[various] Research papers on intersubjectivity. *Zur Phänomenologie der Intersubjektivität: Texts aus dem Nachlass. Erster Teil: 1905-1920.* Husserliana XIII, 1973.

[1911] "Philosophie als strenge Wissenschaft." *Logos* 1: 289-341. Reprinted in Husserliana XXV, 1987.

*1965 "Philosophy as Rigorous Science." Translated by Quentin Lauer. *Phenomenology and the Crisis of Philosophy*, 71-147. New York: Harper & Row.

[1913] *Ideen zu einer reinen Phänomenologie und phänomenologischen Philosophie.* Book 1. Allgemeine Einführung in die reine Phänomenologie. *Jahrbuch für Philosophie und phänomenologische Forschung* 1: 1-323. Reprinted in Husserliana III, 1950.

*1982 *Ideas Pertaining to a Pure Phenomenology and to Phenomenological Philosophy.* Book 1. General Introduction to a Pure Phenomenology. Translated by F. Kersten. Dordrecht: Kluwer.

[1928] "Vorlesungen zur Phänomenologie des inneren Zeitbewusstseins." Edited by Martin Heidegger. *Jahrbuch für Philosophie und phänomenologische Forschung* 10 (1928): 367-498. Reprinted in Husserliana X, 1966.

*1991 *On the Phenomenology of the Consciousness of Internal Time (1893-1917).* Translated by John Barnett Brough. Dordrecht: Kluwer.

1952a *Ideen zu einer reinen Phänomenologie und phänomenologischen Philosophie.* Book 2. Phänomenologische Untersuchungen zur Konstitution. Edited by Marly Biemel. Husserliana IV.

*1989 *Ideas Pertaining to a Pure Phenomenology and to Phenomenological Philosophy.* Book 2. Studies in the Phenomenology of Constitution. Translated by Richard Rojcewicz and André Schuwer. Dordrecht: Kluwer.

1952b *Ideen zu einer reinen Phänomenologie und phänomenologischen Philosophie.* Book 3. Die Phänomenologie und die Fundamente der Wissenschaften. Edited by Marly Biemel. Husserliana V.

*1980 *Ideas Pertaining to a Pure Phenomenology and to Phenomenological Philosophy.* Book 3. Phenomenology and the Foundations of the Sciences. Translated by Ted E. Klein and William E. Pohl. The Hague: Martinus Nijhoff.

1987 "Natur und Geist." *Aufsäze und Vorträge (1911 - 1921).* Edited by Thomas Nenon and Hans Rainer Sepp. Husserliana XXV.

SECONDARY SOURCES

Bell, David
1991 *Husserl*. London: Routledge.

Bernet, Rudolf, Iso Kern, and Eduard Marbach
1993 *An Introduction to Husserlian Phenomenology*. Evanston: Northwestern.

Biemel, Marly
1952 "Einleitung des Herausgebers." *Ideen zu einer Phänomenologie und phänomenologischen Philosophie*. Book 2. Phänomenologische Untersuchungen zur Konstitution, xiii-xx. Edited by Marly Biemel. Husserliana IV.

Biemel, Walter
1981 "Introduction to the Dilthey-Husserl Correspondence." Translated by Jeffner Allen. *Husserl: Shorter Works*, 198-201. Edited by Peter McCormick and Frederick A. Elliston. Notre Dame, IN: University of Notre Dame Press.
1989 "Dank an Löwen. Erinnerung an die Zeit von 1945-1952." *Profile der Phänomenologie: Zum 50. Todestag von Edmund Husserl*, 236-268. Phänomenologische Forschungen 22. Freiburg: Karl Alber.

Bruzina, Ronald
1995 "Translator's Introduction." *Sixth Cartesian Meditation: The Idea of a Transcendental Theory of Method*, vii-xcii. By Eugen Fink. Translated by Ronald Bruzina. Bloomington: Indiana University Press.

Carr, David
1974 *Phenomenology and the Problem of History: A Study of Husserl's Transcendental Philosophy*. Evanston: Northwestern University Press.

Gibson, W.R. Boyce
1971 "Excerpts From a 1928 Freiburg Diary." *Journal of the British Society for Phenomenology* 2: 63-76.

Herbstrith, Waltraud
[1985] *Edith Stein: A Biography*. Translated by Barnard Bonowitz. New York: Harper and Row. Republished in San Francisco by Ignatius Press, 1992.

Ingarden, Roman
1962 "Edith Stein on her Activity as an Assistant of Edmund Husserl." Translated by Janina Makota. *Philosophy and Phenomenological Research* 23: 155-175.
[1967] *Einführung in die Phänomenologie Edmund Husserls: Oslöer Vorlesungen 1967*. Gesammelte Werke 4. Edited by Gregor Haefliger. Tübingen: Neimeyer, 1992.

Kern, Iso
1973 "Einleitung der Herausgebers." *Zur Phänomenologie der Intersubjektivität: Texts aus dem Nachlass. Erster Teil: 1905-1920*, xvii-xlviii. Husserliana 13. By Edmund Husserl. The Hague: Nijhoff.

Kockelmans, Joseph
1967 *Phenomenology: The Philosophy of Edmund Husserl and Its Interpretation*. Garden City, NY: Doubleday, Anchor Books.

Lauer, Quentin
1965 "Introduction." *Phenomenology and the Crisis of Philosophy*, 1-68. By Edmund Husserl. Translated by Quentin Lauer. New York: Harper & Row.

McCormick, Peter
1981 "Husserl on Philosophy as Rigorous Science." *Husserl: Shorter Works*, 161-165. Edited by Peter McCormick and Frederick A. Elliston. Notre Dame, IN: University of Notre Dame Press.

Mensch, James Richard
1988 *Intersubjectivity and Transcendental Idealism*. Albany: State University of New York Press.

Ricoeur, Paul
1967 *Husserl: An Analysis of His Phenomenology*. Translated by Edward G. Ballard and Lester E. Embree. Evanston: Northwestern University Press.

Rojcewicz, Richard, and André Schuwer
1989 "Translators' Introduction." *Ideas Pertaining to a Pure Phenomenology and to a Phenomenological Philosophy*. Book 2. Studies in the Phenomenology of Constitution, xi-xvi. By Edmund Husserl. Translated by Richard Rojcewicz and André Schuwer. Dordrecht: Kluwer.

Schuhmann, Elisabeth, and Karl Schuhmann
1994 *Briefwechsel.* By Edmund Husserl. 10 volumes. Dordrecht: Kluwer.
Schuhmann, Karl
1977 *Husserl-Chronik: Denk- und Lebensweg Edmund Husserls.* Husserliana Dokumente 1. The Hague: Martinus Nijhoff.
Smith, Barry
1995 "Common Sense." *The Cambridge Companion to Husserl*, 394-437. Edited by Barry Smith and David Woodruff Smith. Cambridge: Cambridge University Press.
Société Thomiste
1933 "La Phénoménologie, Juvisy, 112 Septembre 1932." *Journées d' Études de la Société Thomiste.* Juvisy: Les Éditions du Cerf.
Stein, Edith
[1917] *Zum Problem der Einfühlung.* Halle: Buchdruckerei des Waisenhauses. Reprinted in Munich: Kaffke, 1980.
*1989 *On the Problem of Empathy.* Translated by Waltraut Stein. Third revised edition. The Collected Works of Edith Stein 3. Washington: ICS Publications.
Stein, Waltraut
[1962] "Translator's Introduction." *On the Problem of Empathy*, xvi-xxiii. By Edith Stein. Third revised edition. Washington: ICS Publications, 1989.
Strasser, Stephan
1989 "Monadologie und Teleologie in der Philosophie Edmund Husserls." *Profile der Phänomenologie: Zum 50. Todestag von Edmund Husserl*, 217-253. Phänomenologische Forschungen 22. Freiburg: Karl Alber.
Ströker, Elisabeth
1993 *Husserl's Transcendental Phenomenology.* Translated by Lee Hardy. Stanford: Stanford University Press.
Waldenfels, Bernhard
1989 "Erfahrung des Fremden in Husserls Phänomenologie." *Profile der Phänomenologie: Zum 50. Todestag von Edmund Husserl*, 39-62. Phänomenologische Forschungen 22. Freiburg: Karl Alber.

FOR CHAPTER THREE: EDITH STEIN'S HERMENEUTIC THEORY

Bernet, Rudolf, Iso Kern, and Eduard Marbach
1993 *An Introduction to Husserlian Phenomenology.* Evanston: Northwestern.
Husserl, Edmund
1900 *Logische Untersuchungen.* Part 1. Prolegomena zur reinen Logik. Halle: Niemeyer.
*1970a *Logical Investigations.* Volume 1. Prolegomena to Pure Logic, Investigations I and II. Translated by J.N. Findlay. New York: Humanities Press.
1901 *Logische Untersuchungen.* Part 2. Untersuchungen zur Phänomenologie und Theorie der Erkenntnis. Halle: Niemeyer.
*1970b *Logical Investigations.* Volume 2. Investigations III, IV, V, and VI. Translated by J.N. Findlay. New York: Humanities Press.
[1911] "Philosophie als strenge Wissenschaft." *Logos* 1: 289-341. Reprinted in Husserliana XXV, 1987.
*1965 "Philosophy as Rigorous Science." Translated by Quentin Lauer. *Phenomenology and the Crisis of Philosophy*, 71-147. New York: Harper & Row.
[1913] *Ideen zu einer reinen Phänomenologie und phänomenologischen Philosophie.* Book 1. Allgemeine Einführung in die reine Phänomenologie. *Jahrbuch für Philosophie und phänomenologische Forschung* 1: 1-323. Reprinted in Husserliana III, 1950
*1982 *Ideas Pertaining to a Pure Phenomenology and to Phenomenological Philosophy.* Book 1. General Introduction to a Pure Phenomenology. Translated by F. Kersten. Dordrecht: Kluwer.

1952a *Ideen zu einer reinen Phänomenologie und phänomenologischen Philosophie.* Book 2. Phänomenologische Untersuchungen zur Konstitution. Edited by Marly Biemel. Husserliana IV.

*1989 *Ideas Pertaining to a Pure Phenomenology and to Phenomenological Philosophy.* Book 2. Studies in the Phenomenology of Constitution. Translated by Richard Rojcewicz and André Schuwer. Dordrecht: Kluwer.

1952b *Ideen zu einer reinen Phänomenologie und phänomenologischen Philosophie.* Book 3. Die Phänomenologie und die Fundamente der Wissenschaften. Edited by Marly Biemel. Husserliana V.

*1980 *Ideas Pertaining to a Pure Phenomenology and to Phenomenological Philosophy.* Book 3. Phenomenology and the Foundations of the Sciences. Translated by Ted E. Klein and William E. Pohl. The Hague: Martinus Nijhoff.

1987 "Natur und Geist." *Aufsäze und Vorträge (1911 - 1921).* Edited by Thomas Nenon and Hans Rainer Sepp. Husserliana XXV.

Scheler, Max

1912 "Über Selbsttäuschungen" in *Zeitschrift für Pathopsychologie* 1: 87-163. (This article was revised and expanded, and published in 1915 as "Die Idole der Selbsterkenntnis.")

1913 *Zur Phänomenologie und Theorie der Sympathiegefühle und von Liebe und Hass.* Halle: Max Niemeyer.

1915 "Die Idole der Selbsterkenntnis." *Abhandlungen und Aufsätze* 2: 5-168. Leipzig: Weissen Bücher.

*1973 "The Idols of Self-Knowledge." *Selected Philosophical Essays, 3-97.* Translated by David R. Lachterman. Evanston: Northwestern University Press.

1923 *Wesen und Formen der Sympathie.* Second, enlarged edition of *Zur Phänomenologie und Theorie der Sympathiegefühle und von Liebe und Hass.* Bonn: Friedrich Cohen.

*1954 *The Nature of Sympathy.* Translated by Peter Heath. Introduced by W. Stark. London: Routledge & Kegan Paul.

Stein, Edith

[1917] *Zum Problem der Einfühling.* Halle: Buchdruckerei des Waisenhauses. Reprinted in Munich: Kaffke, 1980.

[1932] *Des hl. Thomas con Aquino Untersuchungen über die Wahrheit.* Two volumes. Translated by Edith Stein. Breslau: Otto Borgmeyer. Second edition, with Latin-German glossary, 1934.

1965 *Aus dem Leben einer jüdischen Familie. Das Leben Edith Steins: Kindheit und Jugend.* Edith Steins Werke 7. Freiburg: Herder. Reprinted as *Aus meinem Leben. Mit einer Weiterführung über die zweite Lebenshälfte von Maria Amata Neyer OCD.* Freiburg: Herder, 1987.

*1986 *Life in a Jewish Family: Her Unfinished Autobiographical Account.* Collected Works of Edith Stein 1. Translated by Josephine Koeppel. Washington: ICS Publications.

FOR CHAPTER FOUR: EDITH STEIN'S HERMENEUTIC PRACTICES

PRIMARY TEXTS

Stein, Edith

[1914] "Husserls Exzerpt aus der Staatsexamensarbeit von Edith Stein." Edited by Karl Schuhmann. *Tijdschrift voor Filosofie* 53 (1991) 686-699.

[various] *Selbstbildnis in Briefen.* Part 1: 1916-1934. Part 2: 1934-1942. Edith Steins Werke 8 and 9. Freiburg: Herder, 1976 and 1977. (Includes some letters to Ingarden and all letters to Conrad-Martius, although these have also been published separately.)

*1993 *Self-Portrait in Letters 1916-1942.* Collected Works of Edith Stein 5. Translated by Josephine Koeppel. Washington: ICS Publications.

[various] *Breife an Roman Ingarden 1917-1939.* Edith Steins Werke 14. Freiburg: Herder, 1991.

[various] *Briefe an Hedwig Conrad-Martius*. Munich: Kösel, 1960.

[1917a] *Zum Problem der Einfühlung*. Halle: Buchdrucheri des Waisenhauses. Reprinted Munich: Kaffke, 1980.

*1989 *On the Problem of Empathy*. Translated by Waltraut Stein. Third revised edition. The Collected Works of Edith Stein 3. Washington: ICS Publications.

[1917b] "Zur Kritik an Theodor Elsenhans und August Messer." Worked out by Edith Stein. *Aufsätze und Vorträge (1911-1921)*, 226-248. Husserliana 25. By Edmund Husserl. Edited by Thomas Nenon and Hans Rainer Sepp. Boston: Nijhoff, 1987.

[1917c] "Zu Heinrich Gustav Steinmanns Aufsatz ‚Zur systematischen Stellung der Phänomenologie'." *Aufsätze und Vorträge (1911-1921)*, 253-266. Husserliana 25. By Edmund Husserl. Edited by Thomas Nenon and Hans Rainer Sepp. Boston: Nijhoff, 1987.

1920 Review of *Naturerlebnis und Wirklichkeitsbewußtsein*, by Gertrud Kutznizky. *Kant-Studien* 24/4: 402-405.

1921 "Vorwort" and commentary. *Adolf Reinach: Gesammelte Schriften*, 406 and passim. Halle: Neimeyer.

1922a "Psychische Kausalität." Beiträge zur philosophischen Begründung der Psychologie und der Geisteswissenschaften, Erste Abhandlung. *Jahrbuch für Philosophie und phänomenologische Forschung* 5: 1-116. Reprinted Tübingen: Niemeyer, 1970.

*1956 (Partial translation) "Motivation as a Fundamental Law of the Life of the Mind." *Writings of Edith Stein*, 177-197. Translated by Hilda Graef. Westminster, Md.: Newman.

[1922b] "Individuum und Gemeinschaft." Beiträge zur philosophischen Begründung der Psychologie und der Geisteswissenschaften, Zweite Abhandlung. *Jahrbuch für Philosophie und phänomenologische Forschung* 5: 116-283. Reprinted Tübingen: Niemeyer, 1970.

[1925] "Eine Untersuchung über den Staat. *Jahrbuch für Philosophie und phänomenologische Forschung* 7: 1-123. Reprinted Tübingen: Niemeyer, 1970.

[1928] *John H. Kardinal Newman: Briefe und Tagebücher 1801-1845*. Translated by Edith Stein. Munich: Theatinerverlag.

1929 "Husserls Phänomenologie und die Philosophie des hl. Thomas v. Aquino: Versuch einer Gegenüberstellung." *Jahrbuch für Philosophie und phänomeno-logische Forschung*. Ergänzungsband, 315-338.

[1928-33] *Die Frau: Ihre Aufgabe nach Natur und Gnade*. Edith Steins Werke 5. Freiburg: Herder, 1959.

*1987 *Essays on Woman*. The Collected Works of Edith Stein 2. Translated by Freda Mary Oben. Washington: ICS Publications.

[1930] *Einführung in die Philosophie*. Edith Steins Werke 13. Freiburg: Herder, 1991.

[1931] "Potenz und Akt." Typescript, later expanded into *Endliches und ewiges Sein*. Publication planned by "Archivum Carmelitanum Edith Stein" in the series Edith Steins Werke.

[1932] *Des hl. Thomas von Aquino Untersuchungen über die Wahrheit (Quaestiones disputatae de veritate)*. Two volumes. Translated by Edith Stein. Breslau: Otto Borgmeyer. Second edition, with Latin-German glossary, 1934. Reprinted in Edith Steins Werke 3 and 4. Freiburg: Herder, 1952 and 1955.

[1932-33] *Der Aufbau der menschlichen Person*. Lectures delivered in Münster. Edith Steins Werke 16. Freiburg: Herder, 1994.

1933a Review of *Die Abstraktionslehre des hl. Thomas von Aquin*, by L.M. Habermehl. *Philosophisches Jarhbuch der Görres-Gesellschaft* 46: 502-3.

1933b "Karl Adam's Christusbuch." *Die Christliche Frau* (Münster) 31 (March 1933) 84-89.

[1933-35] *Aus dem Leben einer jüdischen Familie. Das Leben Edith Steins: Kindheit und Jugend*. Edith Steins Werke 7. Freiburg: Herder, 1965. Reprinted as *Aus meinem Leben. Mit einer Weiterführung über die zweite Lebenshälfte von Maria Amata Neyer OCD*. Freiburg: Herder, 1987.

*1986 *Life in a Jewish Family: Her Unfinished Autobiographical Account.*

Collected Works of Edith Stein 1. Translated by Josephine Koeppel. Washington: ICS Publications.

1934 "Die deutsche Summa." *Die christliche Frau* (Münster) 32 (August-September 1934) 245-252 and (October 1934) 276-281.

[1935-6] *Endliches und ewiges Sein: Versuch eines Aufstieges zum Sinn des Seins.* Edith Steins Werke 2. Freiburg: Herder, 1950. Second edition, 1962. Third edition, 1986.

*1956 (Partial translation) "Life of the Mind and Motivation." *Writings of Edith Stein, 198-199.* Translated by Hilda Graef. Westminster, Md.: Newman.

[1935-36] "Martin Heideggers Existentialphilosophie." (Originally an appendix to *Endliches und ewiges Sein,* but deleted by the publisher.) *Welt und Person: Beitrag zum Christlichen Wahrheitsstreben,* 69-135. Edith Steins Werke 6. Freiburg: Herder, 1962.

[various] *Welt und Person: Beitrag zum Christlichen Wahrheitsstreben.* Edith Steins Werke 6. Freiburg: Herder, 1962.

1937 Review of "La crise de la science st de la philosophie transcendentale. Introduction à la philosophie phénoménologique," by Edmund Husserl. *Revue Thomiste* (May-June 1973) 327-329.

[various] "Zwei Betrachtungen zu Edmund Husserl." *Welt und Person: Beitrag zum Christlichen Wahrheitsstreben,* 33-38. Edith Steins Werke 6. Freiburg: Herder, 1962.

[1935-41] *Verborgenes Leben: Hagiographische Essays, Meditationen, geistliche Texte.* Edith Steins Werke 11. Herder: Freiburg, 1987.

*1992 *The Hidden Life: Hagiographic Essays, Meditations, Spiritual Texts.* Translated by Waltraut Stein. The Collected Works of Edith Stein 4. Washington: ICS Publications.

[various] *Ganzheitliches Leben: Schriften zur religiösen Bildung.* Edith Steins Werke 12. Freiburg: Herder, 1990.

[1942] *Kreuzeswissenschaft: Studie über Joannes a Cruce.* Edith Steins Werke 1. Freiburg: Herder, 1954. Second edition, 1983.

*1960 *The Science of the Cross: A Study of St. John of the Cross.* Translated by Hilda Graef. London: Burns & Oates, 1960.

1946 "Ways to Know God." Translated by M. Rudolf Allers. *The Thomist.*

SECONDARY SOURCES

Baseheart, Mary Catharine

1989 "Edith Stein's Philosophy of Woman and of Women's Education." *Hypatia* 4: 120-131.

Baseheart, Mary Catharine, Linda Lopez McAlister and Waltraut Stein

1995 "Edith Stein." *A History of Women Philosophers,* 157-187. Volume 4. Edited by Mary Ellen Waithe. Boston: Kluwer.

Bello, Angela Ales

1993 "Edith Stein und Hedwig Conrad-Martius: eine menschliche und intellektuelle Begegnung." *Studien zur Philosophie von Edith Stein: Internationales Edith Stein-Symposion, Eichstätt 1991,* 256-284. Phänomenologische Forschungen 26/27. Edited by Reto Luzius Fetz, Matthias Rath, and Peter Schulz. Freiburg: Karl Alber.

Berendsohn, Walter A.

1972 "Lion Feuchtwanger and Judaism." *Lion Feuchtwanger: The Man, his Ideas, his Work,* 25-32. Edited by John M. Spalek. Los Angeles: Hennessey & Ingalls.

Bergmann, Ernst

1932 *Erkenntnisgeist und Muttergeist: Eine Soziosophie der Geschlechter.* Breslau: Ferdinand Hirt.

1933 *Fichte und der Nationalsozialismus.* Breslau: Ferdinand Hirt.

Berndt, Wolfgang

1972 "The Trilogy *Der Wartesaal.*" *Lion Feuchtwanger: The Man, his Ideas, his Work,* 131-156. Edited by John M. Spalek. Los Angeles: Hennessey & Ingalls.

Biemel, Marly
1952 "Einleitung des Herausgebers." *Ideen zu einer reinen Phänomenologie und phänomenologischen Philosophie*, xiii-xx. Second book. Phänomenologische Untersuchungen zur Konstitution. Husserliana 4. By Edmund Husserl. Edited by Marly Biemel. The Hague: Nijhoff.

Boehm, Rudolf
1966 "Einleitung des Herausgebers." *Zur Phänomenologie des inneren Zeitbewussyseins (1893-1917)*, xiii-xliii. Husserliana 10. By Edmund Husserl. Edited by Rudolf Boehm. The Hague: Nijhoff.

Brough, John Barnett
1991 "Translator's Introduction." *On the Phenomenology of the Consciousness of Internal Time (1893-1917)*, xi-lvii. By Edmund Husserl. Dordrecht: Kluwer.

Carmely, Klara Pomeranz
1981 *Das Identitätsproblem jüdischer Autoren im deutschen Sprachraum: Von der Jahrhundertwende bis Hitler.* Königstein: Monographien Literaturwissenschaft.

Cecil, Robert
1972 *The Myth of the Master Race: Alfred Rosenberg and Nazi Ideology.* New York: Dodd Mead & Co.

Chandler, Albert R.
1945 *Rosenberg's Nazi Myth.* Ithaca: Cornell University Press.

Conrad-Martius, Hedwig
1958 "Edith Stein." *Hochland.* 51: 38-46. (Reprinted as Introduction to *Briefe an Hedwig Conrad-Martius.* By Edith Stein. Munich: Kösel, 1960.)

Elders, Leo J.
1991 "Edith Stein und Thomas von Aquin." *Edith Stein: Leben, Philosophie, Vollendung. Abhandlungen des internationalen Edith-Stein-Symposiums, Rolduc, 2.-4. November 1990*, 253-271. Edited by Leo Elders. Würzburg: Naumann.

Faulhaber, Uwe Karl
1972 "Lion Feuchtwanger's Theory of the Historical Novel." *Lion Feuchtwanger: The Man, his Ideas, his Work*, 67-81. Edited by John M. Spalek. Los Angeles: Hennessey & Ingalls.

Feuchtwanger, Lion
1933 *Die Geschwister Oppermann.* Berlin: Aufbau-Verlag, 1957.
*1934 *The Oppermanns.* New York: Viking.

Gelber, Lucy
1955 "Nachwort zur kritischen Neuausgabe." *Des hl. Thomas von Aquino Untersuchungen über die Wahrheit*, 449-467. Part 2. Translated by Edith Stein. Edith Steins Werks 4. Freiburg: Herder.

Gerl, Hanna-Barbara
1991 "Ganz offenes Auge." Afterword to *Einführung in die Philosophy*, 265-278. By Edith Stein. Freiburg: Herder.

Gilman, Sander L.
1986 *Jewish Self Hatred: Anti-Semitism and the Hidden Language of the Jews.* Baltimore: Johns Hopkins.
1991 *Inscribing the Other.* Lincoln: University of Nebraska Press.
1992 "The Jewish Body." *People of the Body: Jews and Judaism From an Embodied Perspective.* Edited by Howard Eilberg-Schwartz. Albany: State University of New York Press.

Hedwig, Klaus
1991 "Über den Begriff der Einfühlung in der Dissertationsschrift Edith Steins." *Edith Stein: Leben, Philosophie, Vollendung. Abhandlungen des internationalen Edith-Stein-Symposiums, Rolduc, 2.-4. November 1990*, 239-251. Edited by Leo Elders. Würzburg: Naumann.

Heidegger, Martin
[1925] *History of the Concept of Time: Prolegomena.* Translated by Theodore Kisiel. Bloomington: Indiana University Press, 1992.

Herbstrith, Waltraud
[1985] *Edith Stein: A Biography.* Trans. B. Bonowitz from the fifth German edition. New York: Harper and Row, 1985; San Francisco: Ignatius Press, 1992.

1987 *Das wahre Gesicht Edith Steins*. Sixth edition. Aschaffenburg: Kaffke.
1989 "Hans Lipps im Blick Edith Steins." *Dilthey-Jahrbuch für Philosophie und Geschichte der Geisteswissenschaften* 6: 31-51.
1991 "Das philosophische Denken Edith Steins." *Denken im Dialog: Zur Philosophie Edith Steins*, 23-41. Edited by Waltraud Herbstrith. Tübingen: Attempto Verlag.

Honecker, Raimund
1991 "Wegkreuzungen: Edith Stein und Martin Honecker." *Ein Leben für die Wahrheit: Zur geistigen Gestalt Edith Steins*, 37-76. Edited by Lina Börsig-Hover. Fridingen a.D.: Börsig-Verlag.

Husserl, Edmund
[1917] "Phänomenologie und Psychologie" and "Phänomenologie und Erkenntnislehre." *Aufsätze und Vorträge (1911 - 1921)*, 82-225. Husserliana XXV. Edited by Thomas Nenon and Hans Rainer Sepp. The Hague: Martinus Nijhoff, 1987.

Husserl, Edmund and Malvine Husserl
[various] *Briefe an Roman Ingarden: Mit Erläuterungen und Erinnerungen an Husserl*. Edited by Roman Ingarden. The Hague: Nijhoff, 1968.

Imhof, Beat Walter
1987 *Edith Steins philosophische Entwicklung: Leben und Werk*. Volume 1. Basler Beitrage zur Philosophie und ihre Geschichte 10. Basel: Birkhauser.

Ingarden, Roman
1918 "The Letter to Husserl about the VI [Logical] Investigation and 'Idealism'." *Ingardiana: A Spectrum of Specialized Studies Establishing the Field of Research*. Analecta Husserliana 4. Edited by Anna-Teresa Tymieniecka. Boston: Reidel, 1976.
1931 *Das Literarische Kunstwerke*. Second editon. Tübingen: Niemeyer, 1960.
*1973 *The Literary Work of Art: An Investigation on the Borderlines of Ontology, Logic, and Theory of Literature*. Translated by George G. Grabowicz. Evanston: Northwestern University Press.
1962 "Edith Stein on Her Activity as an Assistant of Edmund Husserl." *Philosophy and Phenomenological Research* 23: 155-175.
1986a "Meine Erinnerungen an Edmund Husserl und Erläuterungen zu den Briefen Husserls." *Briele an Roman Ingarden*, 105-184. Edited by Roman Ingarden. The Hague: Nijhoff.
1986b "Über die philosophischen Forschungen Edith Steins." *Edith Stein: Eine grosse Glaubenzeugin*, 203-229. Edited by Waltraut Herbstrith. Essen: Ploger Verlag Annweiler.
1991 "Zu Edith Steins Analyse der Einfühlung und des Aufbaus der menschlichen Person." *Denken im Dialog: Zur Philosophie Edith Steins*, 72-82. Edited by Waltraud Herbstrith. Tübingen: Attempto Verlag.

Jaegerschmid, Adelgundis
1981a "Edith Stein: Ein Lebensbild." *Internationale katholische Zeitschrift "Communio"* 10: 465-478.
1981b "Gespräche mit Edmund Husserl." *Stimmen der Zeit* 199: 48-58, 129-138.
1987 "So erbebte ich Edith Stein." *Edith Stein: Wege zur inneren Stille*. Edited by Waltraut Stein. Aschaffenburg: Kaffke.

Kern, Iso
1973 "Einleitung der Herausgebers." *Zur Phänomenologie der Intersubjektivität: Texts aus dem Nachlass. Erster Teil: 1905-1920*, xvii-xlviii. Husserliana 13. By Edmund Husserl. The Hague: Nijhoff.

Kersten, F.
1982 "Translator's Note." *Ideas Pertaining to a Pure Phenomenology and a Phenomenological Philosophy*. First Book: General Introduction to Pure Phenomenology, xiii-xvi. By Edmund Husserl. Translated by F. Kersten. Boston: Kluwer.

Koepcke, Cordula
1991 *Edith Stein: Ein Leben*. Würzburg: Echter Verlag.

Koeppel, Josephine
1986 "Chronology 1916-1942," "Translator's Afterword," and "Notes." *Life in a Jewish Family*, 415-512. By Edith Stein. Translated by Josephine Koeppel.

Washington: ICS Publications.

1993 "Translator's Preface." *Self-Portrait in Letters 1916-1942*, vii-viii. By Edith Stein. Translated by Josephine Koeppel. Washington: ICS Publications.

Kühn, Rolf

1991 "Leben aus dem Sein: Zur philosophischen Grundintuition Edith Steins." *Denken im Dialog: Zur Philosophie Edith Steins*, 118-132. Edited by Waltraud Herbstrith. Tübingen: Attempto Verlag.

Leuven, Romanus

1983 *Heil im Unheil. Das Leben Edith Steins: Reife und Vollendung*. Edith Steins Werke 10. Freiburg: Herder, 1983.

Müller, Andreas Uwe

1993 *Grundzüge der Religionsphilosophie von Edith Stein*. Freiburg: Alber, 1993.

Nenon, Thomas, and Hans Reiner Sepp

1987 "Einleitung der Herausgeber." *Aufsätze und Vorträge (1911-1921)*, xi-xxxix. Husserliana 25. By Edmund Husserl. Edited by Thomas Nenon and Hans Rainer Sepp. Boston: Nijhoff, 1987.

Neuss, Wilhelm

1947 *Der Kampf gegen den Mythus*. Dokumente zur Zeitgeschichte 4. Cologne: J. P. Bachem.

Nota, Jan [John]

1991a "Die frühe Phänomenologie Edith Steins." *Denken im Dialog: Zur Philosophie Edith Steins*, 57-71. Edited by Waltraud Herbstrith. Tübingen: Attempto Verlag.

1991b "Edith Stein und Martin Heidegger." *Denken im Dialog: Zur Philosophie Edith Steins*, 93-116. Edited by Waltraud Herbstrith. Tübingen: Attempto Verlag.

1991c "Edith Stein--Max Scheler--Martin Heidegger." *Edith Stein: Leben, Philosophie, Vollendung. Abhandlungen des internationalen Edith-Stein-Symposiums, Rolduc, 2.-4. November 1990*, 227-237. Edited by Leo Elders. Würzburg: Naumann.

Nova, Fritz

1986 *Alfred Rosenberg: Nazi Theorist of the Holocaust*. New York: Hippocrene Books.

Ott, Hugo

1987 "Edith Stein (1891-1942) und Freiburg: Ein Beitrag anlasslich der Seligsprechung am 1. Mai 1987." *Freiburger Diozesan-Archiv* 107: 253-274.

1988 "Die Weltanschauungsprofessuren (Philosophie und Geschichte) an der Universität Freiburg: Besonders im Dritten Reich." *Historisches Jarhbuch* 108: 155-173.

1993 "Edith Stein und Freiburg." *Studien zur Philosophie von Edith Stein: Internationales Edith Stein-Symposion, Eichstätt 1991*, 107-139. Phänomenologische Forschungen 26/27. Edited by Reto Luzius Fetz, Matthias Rath, and Peter Schulz. Freiburg: Karl Alber.

Ott, Hugo, editor

1993 "Die Randnotizen Martin Honeckers zur Habilitationsschrift ‚Potenz und Akt'." *Studien zur Philosophie von Edith Stein: Internationales Edith Stein-Symposion, Eichstätt 1991*, 140-145. Phänomenologische Forschungen 26/27. Edited by Reto Luzius Fetz, Matthias Rath, and Peter Schulz. Freiburg: Karl Alber.

Panzer, Ursula

1984 "Einleitung der Herausgeberin." *Logische Untersuchungen*. Volume 2, Part 1. Untersuchungen zur Phänomenologie und Theorie der Erkenntnis, xix-lxv. By Edmund Husserl. Edited by Ursula Panzer. Boston: Nijhoff.

Rath, Matthias

1993 "Die Stellung Edith Steins im Psychologismusstreit." *Studien zur Philosophie von Edith Stein: Internationales Edith Stein-Symposion, Eichstätt 1991*, 197-225. Phänomenologische Forschungen 26/27. Edited by Reto Luzius Fetz, Matthias Rath, and Peter Schulz. Freiburg: Karl Alber.

Reinach, Adolf

[1913-4] "Über das Wesen der Bewegung." *Gesammelte Schriften*, 407-461. Edited by his students. Halle: Niemeyer, 1921. Critical edition, *Sämtliche Werke*,

551-588. Edited by Karl Schuhmann and Barry Smith. Munich: Philosophie Verlag, 1989.

Rojcewicz, Richard, and André Schuwer

1989 "Translators' Introduction." *Ideas Pertaining to a Pure Phenomenology and a Phenomenological Philosophy.* Second Book. Studies in the Phenomenology of Constitution, xi-xix. By Edmund Husserl. Translated by Richard Rojcewicz and André Schuwer. Boston: Kluwer.

Rosenberg, Alfred

1930 *Der Mythus des 20. Jahrhunderts.* One hundred seventh edition. Munich: Hoheneichen, 1937.

1935 *An die Dunkelmänner unserer Zeit: Eine Antwort auf die Angriffe gegen den „Mythus des 20. Jahrhunderts."* Thirtieth edition. Munich: Hoheneichen.

Rosenberg, Arthur

1936 *A History of the German Republic.* Translated by Ian F.D. Morrow and L. Marie Sieveking. London: Methuen.

Schuhmann, Karl

1976 "Einleitung des Herausgebers." *Ideen zu einer reinen Phänomenologie und zu phänomenologische Philosophie*, xiii-lvii. Book 1. Allgemeine Einführung in die reine Phänomenologie. By Edmund Husserl. Edited by Karl Schuhmann. The Hague: Nijhoff.

1993 "Edith Stein und Adolf Reinach." *Studien zur Philosophie von Edith Stein: Internationales Edith Stein-Symposion, Eichstätt 1991*, 53-88. Phänomenologische Forschungen 26/27. Edited by Reto Luzius Fetz, Matthias Rath, and Peter Schulz. Freiburg: Karl Alber.

Schulz, Peter

1993 "Die Schrift ‚Einführung in die Philosophie'." *Studien zur Philosophie von Edith Stein: Internationales Edith Stein-Symposion, Eichstätt 1991*, 228-255. Phänomenologische Forschungen 26/27. Edited by Reto Luzius Fetz, Matthias Rath, and Peter Schulz. Freiburg: Karl Alber.

Stallmach, Josef

1991 "Edith Stein--von Husserl zu Thomas von Aquin." *Denken im Dialog: Zur Philosophie Edith Steins*, 42-56. Edited by Waltraud Herbstrith. Tübingen: Attempto Verlag.

Stockhausen, Alma von

1991 "Edith Stein und die Phänomenologie von Edmund Husserl." *Edith Stein: Leben, Philosophie, Vollendung. Abhandlungen des internationalen Edith-Stein-Symposiums, Rolduc, 2.-4. November 1990*, 213-226. Edited by Leo Elders. Würzburg: Naumann.

Waldo, Hilda

1972 "Lion Feuchtwanger: A Biography (July 7, 1884-December 21, 1958)." *Lion Feuchtwanger: The Man, his Ideas, his Work*, 1-24. Edited by John M. Spalek. Los Angeles: Hennessey & Ingalls.

FOR CHAPTER FIVE: INTERPRETATIONS OF EDITH STEIN

Ales Belo, Angela

1981 "Le problème de l'être dans la phénoménologie de Husserl." *Analecta Husserilana* 11: 41-50.

1993 "Edith Stein und Hedwig Conrad-Martius: eine menschliche und intellektuelle Begegnung." *Studien zur Philosophie von Edith Stein*, 256-284. Internationales Edith Stein-Symposion, Eichstatt 1991. Edited by Reto Luzius Fetz, Matthias Rath, and Peter Schulz. Munich: Karl Alber.

Baaden, James Raphael

1987 "A Question of Martyrdom." *The Tablet* 31 (January 31, 1987) 107-8.

Barukinamwo, Matthieu

1982 *Edith Stein: Pour une ontologie dynamique, overte à la transcendence totale.* European University Studies, series 23: Theology, volume 169. Frankfurt: Peter Lang.

1990 *Edith Stein: Un Don de Dieu à l'Humanité*. Bujumbura, Burundi: Les Presse Lavigerie.

Baseheart, Mary Catharine

1960 *The Encounter of Husserl's Phenomenology and the Philosophy of St. Thomas in Selected Writings of Edith Stein.* Ph.D. dissertation, University of Notre Dame.

1981 "Infinity in Edith Stein's *Endliches und ewiges Sein.*" *Proceedings of the American Catholic Philosophical Association* 55: 126-134.

1987 "Edith Stein's Philosophy of Person." *Carmelite Studies* 4: 34-49.

1989 "Edith Stein's Philosophy of Woman and of Women's Education." *Hypatia* 4: 120-131.

Baseheart, Mary Catharine, Linda Lopez McAlister, and Waltraud Stein

1995 "Edith Stein (1892-1942)." *A History of Women Philosophers.* Volume 4. Contemporary Women Philosophers 1900-Today, 157-187. Edited by Mary Ellen Waithe. Boston: Kluwer

Batzdorff, Susanne

1987a "A Martyr of Auschwitz." *The New York Times*, April 12, 1987, 52-55, 70. Reprinted in *Edith Stein: Selected Works*, 103-113. Edited by Susanne M. Batzdorff. Springfield, IL: Templegate Publishers, 1991.

1987b "Watching Tante Edith Become Teresa, Blessed Martyr of the Church." *Moment* (September 1987) 46-53.

1989 "Catholics and Jews: Can We Bridge the Abyss?" *America* 160 (1989) 223-224, 230.

1990a "Was bedeutet die Seligsprechung Edith Steins für ihre Familie?" *Erinnere dich--vergiß es nicht*, 31-39. Edited by Waltraut Stein. Essen: Ploger Verlag Annweiler.

1990b "Aus dem Leben einer judischen Familie--Tante Ediths Vermachtnis as ihre Nachkommen." *Erinnere dich--vergiß es nicht*, 41-63. Edited by Waltraut Stein. Essen: Ploger Verlag Annweiler, 1990.

1990c *Edith Stein: Selected Writings. With Comments, Reminiscences and Translations of her Prayers and Poems.* Springfield, IL: Templegate Publishers.

1991 "Edith Stein aus der Sicht der Verwandten." Afterword to *Edith Stein und das Judentum*, 130-139. 2nd edition. By Matthias Böchel. Ramstein: Paqué, 1991.

Bejás, Andrés E.

1987 *Edith Stein: Von der Phänomenologie zur Mystik. Eine Biographie der Gnade*. Disputationes Theologicae 17. Frankfurt: Peter Lang.

Bergmann, Ernst

1932 *Erkenntnisgeist und Muttergeist: Eine Soziosophie der Geschlecter*. Breslau: Ferdinand Hirt.

Böckel, Matthias

1991 "Edith Stein aus jüdischer Sicht." *Edith Stein und das Judentum*, 114-129. Second edition. Ramstein: Paqué.

Collins, James

1942 "Edith Stein and the Advance of Phenomenology." *Thought* 17: 685-708.

1952 Review of *Endliches und ewiges Sein. Modern Schoolman* 29: 139-145.

Conrad-Martius, Hedwig

1958 "Edith Stein." *Hochland* 51: 38.46. Reprinted in *Edith Stein: Briefe an Hedwig Conrad-Martius*, 61-83. Munich: Kösel-Verlag, 1960.

Constantini, Elio

1981 "Einfühlung und Intersubjektivität bei Edith Stein und bei Husserl." *Analecta Husserliana* 11: 335-339.

Cross, Nancy M.

1989 "A Higher Middle Ground: Blessed Edith Stein's Feminism." *Review for Religious* 48: 86-94.

Dempf, Alois

1934 Review of *Untersuchungen über die Wahrheit,* Edith Stein's translation of the *De veritate* of Thomas Aquinas. *Blätter für Deutsche Philosophie* 8.

1953 Review of *Endliches und ewiges Sein. Philosophisches Jahrbuch der Görresgesellschaft* 62: 201-204.

Drahos, Mary
1953 "Eternal Sabbath." A biographical drama in three acts. Privately printed typescript. Presented at Blackfriars Theater, New York, October 1953.

Draï, Raphaël
1989 *Lettre ouverte au Cardinal Lustiger sur l'autre révisionnisme.* Aix-en-Provence: Alinea.

Dubois, Marcel-Jacques
1973 "L'Itineraire philosophique et spirituel d'Edith Stein," *Revue Thomiste* 73: 181-210.

Eichmann-Leutenegger, B.
1984 "Edith Stein, Menorah und Christuskreuz. Jüdische und christliche Einflussbereiche als Studien einer inneren Biografie." *Judaica Zürich* 40: 176-189.

Elders, Leo
1991 "Edith Stein und Thomas Aquinas." *Edith Stein: Leben, Philosophie, Vollendung.* Proceedings of the International Edith Stein Symposium, Rolduc, 2-4 November 1990, 253-271. Edited by Leo Elders. Wurzburg: Naumann.

Elders, Leo, editor
1991 *Edith Stein: Leben, Philosophie, Vollendung.* Proceedings of the International Edith Stein Symposium, Rolduc, 2-4 November 1990. Würzburg: Nauman.

Endres, Elisabeth
1987 *Edith Stein: Christliche Philosophin und jüdische Märtyrerin.* Munich: Piper.

Erlich, Ernst L.
1984 "Katholische Kirke und Judentum zur Zeit des Nationalsozialismus: Eine geschichtliche Erfahrung und Herausforderung an uns." *Judaica* 40: 145-158.

Faber, Marvin
1930 "A Review of Recent Phenomenological Literature." *Journal of Philosophy* 27: 141-160.

Fetz, Reto Luzius
1993 "Ich, Seele, Selbst: Edith Steins Theorie personaler Identität." *Studien zur Philosophie von Edith Stein*, 286-319. Internationales Edith Stein-Symposion, Eichstatt 1991. Edited by Reto Luzius Fetz, Matthias Rath, and Peter Schulz. Munich: Karl Alber.

Fetz, Reto Luzius, Matthias Rath, and Peter Schulz, editors
1993 *Studien zur Philosophie von Edith Stein.* Internationales Edith Stein-Symposion, Eichstatt 1991. Munich: Karl Alber.

Fidalgo, António Carreto
1985 *Der Übergang zur objektiven Welt: Einer kritische Erörterung zum Problem der Einfühlung bei Edith Stein.* Doctoral dissertation, University of Würzburg. Typescript.
1993 "Edith Stein, Theodor Lipps und die Einfühlungsproblematik." *Studien zur Philosophie von Edith Stein*, 90-106. Internationales Edith Stein-Symposion, Eichstatt 1991. Edited by Reto Luzius Fetz, Matthias Rath, and Peter Schulz. Munich: Karl Alber.

Fuchs-Kreimer, Nancy
1991 "Sister Edith Stein: A Rabbi Reacts." *Lilith* 16/1: 6-7, 28.

Gaboriau, Florent
1988 "Edith Stein philosophe." *Revue Thomiste* 88: 87-107, 256-277, 440-459, 589-619. Reprinted Paris: FAC, 1989.

Geiger, L.B.
1954 Review of *Endliches und ewiges Sein. Revue des sciences philosophiques et théologiques* 38: 275-277.

Gelber, Lucy
1959 "Vorwort der Herausgeber." *Die Frau: Ihre Aufgabe nach Natur und Gnade*, v-xxxix. Edith Steins Werke 5. Edited by Lucy Gelber. Freiburg: Herder.
*1987 "Editors' Introduction." *Essays on Women*, 1-40. The Collected Works of Edith Stein, volume two. Translated by Freda Mary Oben. Washington: ICS, 1987.

Gerl, Hanna-Barbara
1989 "Edith Stein und die Frauenfrage." *Edith Stein. Keine Frau is ja nur Frau:*

Texte zur Frauenfrage, 5-22. Edited by Hanna-Barbara Gerl. Freiburg: Herder.

1991a *Unerbittliches Licht. Edith Stein: Philosophie, Mystik, Leben.* Mainz: Matthias-Grünewald-Verlag. Three of the fifteen chapters offer philosophical analysis.

1991b "Ganz offenes Auge." Nachwort to *Einführung in die Philosophie*, 265-278. By Edith Stein. Freiburg: Herder.

Graef, Hilda

1955 *The Scholar and the Cross: The Life and Work of Edith Stein.* London: Longmans and Green, 1955; Westminster, MD: Newman Press, 1956.

Graef, Hilda, editor and translator

1956 *Writings of Edith Stein.* London: P. Owen; Westminster, MD: Newman Press, 1956.

Guilead, Reuben

1974 *De la Phenomenologie à la Science de la Croix: L'Itineraire d'Edith Stein.* Philosophes Contemporains: Textes et Etudes 17. Louvain: Editions Nauwelaerts.

Hauke, Johanna, and Gabriele Dick

1984 "Edith-Stein-Forschung 1984." *Archiv für schlesische Kirkengeschichte* 42: 215-236.

Hedwig, Klaus

1991 "Über den Begriff der 'Einfühlung' in der Dissertationsschrift Edith Steins." *Edith Stein: Leben, Philosophie, Vollendung*, 239-251. Proceedings of the Internationalen Edith Stein Symposium, Rolduc, 2-4 November 1990. Edited by Leo Elders. Wurzburg: Naumann.

1993 "Edith Stein und die analogia entis." *Studien zur Philosophie von Edith Stein*, 320-352. Internationales Edith Stein-Symposion, Eichstatt 1991. Edited by Reto Luzius Fetz, Matthias Rath, and Peter Schulz. Munich: Karl Alber.

Heinemann, F.H.

1951 "Philosophical Survey: German Philosophy." *Philosophy* 26: 358-360.

Herbstrith, Waltraud

[1971] *Das wahre Gesicht Edith Steins.* 6th edition. Aschaffenburg: Kaffke, 1987. First edition, 1971. The fifth edition was translated by Barnard Bonowitz as *Edith Stein: A Biography.* New York: Harper and Row, 1985; San Francisco: Ignatius Press, 1992.

1988 "Edith Stein (1891-1942)." *Christliche Philosophie im katholischen Denken des 19. und 20. Jahrhunderts.* Volume 2: Rückgriff auf scholastisches Erbe. Edited by Emerich Coreth *et al.* Cologne: Styria.

1991a *Edith Stein: Versohnerin zwischen Jüden und Christen.* Leutesdorf: Johannes-Verlag.

1991b *Edith Stein: Etappen einer leidenschaftlichen Suche nach der Wahrheit.* München: Verlag Neue Stadt.

1991c "Das philosophische Denken Edith Steins." *Denken im Dialog: Zur Philosophie Edith Steins*, 23-41. Edited by Waltraud Herbstrith. Tübingen: Attempto Verlag.

1993 *Edith Stein--Das eine Menschsein: Die Frau in Christentum.* München: Verlag J. Pfeiffer.

Herbstrith, Waltraud, editor

1988 *Edith Stein. Aus der Tiefe leben: Ausgewahlte Texte zu Fragen der Zeit, Mit zahlreichen erstveroffentlichten Texten.* München: Kosel-Verlag.

1990 *Erinnere dich--vergiß es nicht: Edith Stein, Christlich-judische Perspektiven.* Essen: Ploger Verlag Annweiler.

1991 *Denken im Dialog: Zur Philosophie Edith Steins.* Tübingen: Attempto Verlag.

Höfliger, Anton

1968 *Das Universalienproblem in Edith Steins Werk "Endliches und ewiges Sein."* Freiburg Schweiz: Univertätsverlag.

Honecker, Raimund

1991 "Wegkreuzungen: Edith Stein und Martin Honecker." In *Ein Leben für die Wahrheit: Zur Geistigen Gestalt Edith Steins*, 37-76. Edited by Lina Börsig-

Hover. Fridingen a.D.: Börsig-Verlag.

Hughes, J.
1985 "Edith Stein's Doctoral Dissertation on Empathy and the Philosophical Climate From Which It Emerged." *Teresianum* 36: 455-484.

Huning, Alois
1969 *Edith Stein und Peter Wust: Von der Philosophie zum Glaubenszeugnis.* Münster: Verlag Regensberg, 1969.

Imhof, Beat Walter
1987 *Edith Stein's philosophische Entwicklung: Leben und Werk.* Band 1. Basler Beitrage zur Philosophie und ihre Geschichte 10. Basel: Birkhauser Verlag, 1987.

Ingarden, Roman
1979 "Über die philosophischen Forschungen Edith Steins." *Freiburger Zeitschrift für Philosophie und Theologie* 26: 456-480. Reprinted in *Edith Stein: Eine grosse Glaubenzeugin*, 203-229. Edited by Waltraud Herbstrith. Essen: Ploger Verlag Annweiler, 1986.
1991 "Zu Edith Steins Analyse der Einfühlung und des Aufbaus der menschlichen Person." *Denken im Dialog: Zur Philosophie Edith Steins*, 72-82. Edited by Waltraud Herbstrith. Tübingen: Attempto Verlag.

Inglis, John A.
1993 *Aquinas and the Historiography of Medieval Philosophy: A Reappraisal.* Ph.D. dissertation, University of Kentucky.

Jaegerschmid, Adelgundis
1981a "Edith Stein: ein Lebensbild." *Internationale katholische Zeitschrift "Communio"* 10: 465-478.
1981b "Gespräche mit Edmund Husserl." *Stimmen der Zeit* 199: 48-58, 129-138.
1987 "So erbebte ich Edith Stein." In *Edith Stein: Wege zur inneren Stille.* Edited by Waltraud Herbstrith. Aschaffenburg: Kaffke.

Kalinowski, Georges
1984 "Edith Stein et Karol Wojtyla sur la personne." *Revue Philosophique de Louvain* 82: 545-561. Reprinted in *Autour de "Personne et Acte" de Karol Cardinal Wojtyla: Articles et Conferences sur une rencontre du thomisme avec la phénoménologie.* Aix-en-Provence: Presses Universitaires d'Aix-Marseille, 1987.

Kaufmann, Fritz
1952 Review of *Endliches und ewiges Sein. Philosophy and Phenomenological Research* 12: 572-577.

Knight, Helen
1927 "Philosophy in Germany." *Journal of Philosophical Studies* 2: 79-88.
1933 "Philosophy in Germany." *Philosophy* 8: 95-98.

Koepcke, Cordula
1985 *Edith Stein: Philosophin und Ordensfrau.* Hamburg: Freidrich Wettig Verlag.
1991 *Edith Stein: Ein Leben.* Würzburg: Echter Verlag.

Koeppel, Josephine
1990 *Edith Stein: Philosopher and Mystic.* The Way of the Christian Mystics, volume 12. Collegeville, MN: Liturgical Press.

Koyré, Alexandre
1932-3 Review of *Untersuchungen über die Wahrheit,* Edith Stein's translation of the *De veritate* of Thomas Aquinas. *Philosophie Médiévale 2.*

Kühn, Rolf
1988 "Leben aus dem Sein: Zur philosophischen Grundintuition Edith Seins." *Freiburger Zeitschrift für Philosophie und Theologie* 35: 159-173. Reprinted in *Denken im Dialog: Zur Philosophie Edith Steins*, 118-132. Edited by Waltraud Herbstrith. Tübingen: Attempto Verlag.

Lembeck, Karl-Heinz
1988 "Die Phänomenologie Husserls und Edith Stein." *Theologie und Philosophie* 63: 182-202.
1991 "Glaube im Wissen? Zur aporetischen Grundstruktur der Stätphilosophie Edith Steins." *Denken im Dialog: Zur Philosophie Edith Steins*, 156-175. Edited by Waltraud Herbstrith. Tübingen: Attempo.
1993 "Von der Kritik zur Mystik: Edith Stein und der Marburger

Neukantianismus." *Studien zur Philosophie von Edith Stein*, 170-196. Internationales Edith Stein-Symposion, Eichstatt 1991. Edited by Reto Luzius Fetz, Matthias Rath, and Peter Schulz. Munich: Karl Alber.

Leuven, Romaeus

1983 *Heil im Unheil. Das Leben Edith Steins: Reife und Vollendung.* Edith Steins Werke 10. Freiburg: Herder, 1983.

Linssen, Michael

1991 "Arbeitsbericht über die Edition der Werke Edith Steins durch das ‚Archivum Carmelitanum Edith Stein'." *Edith Stein: Leben, Philosophie, Vollendung, 291-294.* Proceedings of the International Edith Stein Symposium, Rolduc, 2-4 November 1990. Edited by Leo Elders. Wurzburg: Naumann.

MacIntyre, Alasdair

1993 "Edith Stein and Openness to Others." Lecture delivered at the Edith Stein Center, Spalding University, Louisville, KY. Typescript.

Madden, Anselm Mary

1962 *Edith Stein and the Education of Women: Augustinian Themes.* Ann Arbor, MI: UMI, 1962. Doctoral dissertation, St. Louis University.

Manshausen, Udo Theodore

1983 *Die Biographie der Edith Stein: Beispiel einer Mystagogie.* Frankfurt: Peter Lang, 1984.

Matzker, Reiner

1991 *Einfühlung: Edith Stein und Phänomenologie.* Frankfurt: Peter Lang.

McAlister, Linda Lopez

1989 "Feminist Saint? Edith Stein's Feminism." Florida International University Occasional Papers in Women's Studies.

1993 "Edith Stein: Essential Differences." *Philosophy Today* 37: 70-77. First presented as a lecture at the Edith Stein Center, Spalding University, 1991.

McInerny, Ralph M.

1966 *Thomism in an Age of Renewal.* Garden City, NY: Doubleday.

1987 "Edith Stein and Thomism." *Carmelite Studies* 4: 74-87.

Miles, Judy A.

1991 *Simone de Beauvoir and Edith Stein: A Philosophical Analysis of Feminism.* Ann Arbor, MI: UMI. Doctoral dissertation, Claremont Graduate School, Department of Philosophy.

Miribel, Elisabeth de ("une moniale française")

1954 *Edith Stein: 1891-1942.* Paris: Éditions du Seuil.

Moossen, Inge

1987 *Das unselige Leben der "seligen" Edith Stein: Eine dokumentarische Biographie.* Frankfurt: Haag & Herchen.

Naber, A.

1932 Review of *Untersuchungen über die Wahrheit,* Edith Stein's translation of the *De veritate* of Thomas Aquinas. *Gregorianum* 13.

Nota, Jan [John]

1987a "Misunderstanding and Insight About Edith Stein's Philosophy." *Human Studies* 10: 205-212.

1987b "Edith Stein and Martin Heidegger." *Carmelite Studies* 4: 50-73. Translated as ""Edith Stein und Martin Heidegger." *Denken im Dialog: Zur Philosophie Edith Steins*, 93-116. Edited by Waltraud Herbstrith. Tübingen: Attempto Verlag, 1991.

1991a "Die frühe Phänomenologie Edith Steins." *Denken im Dialog: Zur Philosophie Edith Steins*, 57-71. Edited by Waltraud Herbstrith. Tübingen: Attempto Verlag.

1991b "Edith Stein--Max Scheler--Martin Heidegger." *Edith Stein: Leben, Philosophie, Vollendung.* Proceedings of the International Edith Stein Symposium, Rolduc, 2-4 November 1990, 227-237. Edited by Leo Elders. Wurzburg: Naumann.

Oben, Freda M.

1979 *An Annotated Edition of Edith Stein's Papers on Woman.* Ann Arbor, MI: University Microfilms.

1990 "Edith Stein as Educator." *Thought* 65: 113-126.

Öesterreicher, John M.
1953 "Edith Stein: Witness of Love." *Walls Are Crumbling: Seven Jewish Philosophers Discover Christ*, 288-329. New York: Devin Adair Co.

Orth, Ernst Wolfgang
1993 "Richard Hönigswalds Neukantianismus und Edmund Husserls Phänomenologie als Hintergrund des Denkens von Edith Stein." *Studien zur Philosophie von Edith Stein*, 16-52. Internationales Edith Stein-Symposion, Eichstatt 1991. Edited by Reto Luzius Fetz, Matthias Rath, and Peter Schulz. Munich: Karl Alber.

Ott, Hugo
1987 "Edith Stein (1891-1942) und Freiburg: Ein Beitrag anlasslich der Seligsprechung am 1. Mai 1987." *Freiburger Diozesan-Archiv* 107: 253-274.
1988 "Die Weltanschauungsprofessuren (Philosophie und Geschichte) an der Universität Freiburg: Besonders im Dritten Reich." *Historisches Jahrbuch* 108: 155-173.
1991 "Edith Stein in Freiburg." Lecture for the Rotary Club of Freiburg-Zähringen, November 26, 1991. Typescript.

Otto, Elisabeth
1990 *Welt, Person, Gott: Eine Untersuchung zur theologischen Grundlege der Mystik bei Edith Stein.* Vallender Schonstatt: Patris Verlag, 1990.

Posselt, Teresia Renata de Spiritu Sancto
1954 *Edith Stein: Lebensbild einer Philosophin und Karmelitin.* Nurnburg: Glock und Lutz, 1948. 7th edition, 1954. Translated by C. Hastings and Donald Nicholl as *Edith Stein.* New York: Sheed & Ward, 1952.

Przywara, Erich
1931 Review of *Untersuchungen über die Wahrheit,* Edith Stein's translation of the *De veritate* of Thomas Aquinas. *Stimmen der Zeit* 121.
1933 "Germany." *The Modern Schoolman* 10: 91-92.

Rath, Matthias
1993 "Die Stellung Edith Steins im Psychologismusstreit." *Studien zur Philosophie von Edith Stein*, 197-225. Internationales Edith Stein-Symposion, Eichstatt 1991. Edited by Reto Luzius Fetz, Matthias Rath, and Peter Schulz. Munich: Karl Alber.

Reifenrath, Bruno H.
1985 *Erziehung im Licht des Ewigen: Die Pädagogik Edith Steins.* Frankfurt: Diesterweg.

Schandl, Felix M.
1990 *"Ich sah aus meinem Volk die Kirche wachsen!" Jüdische Bezüge und Strukturen in Leben und Werk Edith Steins (1891-1942)*. Sinziger theologische Texte und Studien 9. Sinzig: Sankt Meinrad Verlag für Theologie, 1990. Diplomarbeit, 1990.

Schlafke, Jakob
1980 "Works by and About Edith Stein in English." *Edith Stein: Documents Concerning Her Life and Death*, 42-44. Translated by Susanne M. Batzdorff. New York: Edith Stein Guild.

Schmidbauer, Robert
1980 "Edith Stein und die Entwicklung des europäischen Judentums seit dem 19 Jahrhundert." In *Zeugen der Zeit*, 131-136. Edited by Waltraud Herbstrith. München: 1980.
1984 "Die christologische Offenheit der Philosophie Edith Steins." In *Praesentia Christi: Festschrift Johannes Betz zum 70. Geburtstag*, 467-473. Edited by Lothar Lies. Düsseldorf: Patmos Verlag.

Schuhmann, Karl
1993 "Edith Stein und Adolf Reinach." *Studien zur Philosophie von Edith Stein*, 53-88. Internationales Edith Stein-Symposion, Eichstatt 1991. Edited by Reto Luzius Fetz, Matthias Rath, and Peter Schulz. Munich: Karl Alber.

Schulz, Peter
1993 "Die Schrift ‚Einführung in die Philosophie'." *Studien zur Philosophie von Edith Stein*, 228-255. Internationales Edith Stein-Symposion, Eichstatt 1991. Edited by Reto Luzius Fetz, Matthias Rath, and Peter Schulz. Munich: Karl Alber.

1994 *Edith Steins Theorie der Person: Von der Bewußtseinsphilosophie zur Geistmetaphysik.* Freiburg: Alber.

Secretan, Philibert

1979 "Essence et personne: Contribution a la connaissance d'Edith Stein." *Freiburger Zeitschrift für Philosophie und Theologie* 26: 481-504.

1981 "Edith Stein on the 'Order and Chain of Being'." *Analecta Husserliana* 11: 113-123.

1992 *Erkenntnis und Aufsteig: Einführung in die Philosophie von Edith Stein.* Innsbruck: Tyrolia Verlag.

1993 "Individuum, Individualität und Individuation nach Edith Stein und Wilhelm Dilthey." *Studien zur Philosophie von Edith Stein*, 148-169. Internationales Edith Stein-Symposion, Eichstatt 1991. Edited by Reto Luzius Fetz, Matthias Rath, and Peter Schulz. Munich: Karl Alber.

Stallmach, Josef

1991 "Edith Stein--von Husserl zu Thomas von Aquin." *Denken im Dialog: Zur Philosophie Edith Steins*, 42-56. Edited by Waltraud Herbstrith. Tübingen: Attempto Verlag.

Stockhausen, Alma von

1991 "Edith Stein und die Phänomenologie von Edmund Husserl." *Edith Stein: Leben, Philosophie, Vollendung, 213-226.* Proceedings of the International Edith Stein Symposium, Rolduc, 2-4 November 1990. Edited by Leo Elders. Wurzburg: Naumann.

Tilliette, Xavier

1990 "Edith Stein et la philosophie chretienne: À propos d'Être fini et Être eternel." *Gregorianum* 71: 97-113.

Verbillion, June M.

1960 *A Critical Analysis of the Educational Theories of Edith Stein.* Ed.D. dissertation, Loyola University. Typescript.

Ziegenaus, A.

1987 "Judentum und Christentum: Erwägungen in Blick auf Edith Stein." *Forum Katholische Theologie* 3: 253-268.

Zimmermann, Albert

1987 "Edith Stein als Philosophin." *Reden anläßlich der Vortragsveranstaltung "Edith Stein: Lebensweg und wissenschaftliches Werke" am 15. Mai 1987 in der Universität zu Köln*, 26-39. Privately printed.

1991 "Begriff und Aufgabe einer christlichen Philosophie bei Edith Stein." *Denken im Dialog: Zur Philosophie Edith Steins*, 133-140. Edited by Waltraud Herbstrith. Tübingen: Attempto Verlag.

FOR CHAPTER SIX: SCIENCE AS LITERACY

Bacus, Elisabeth A., et al., editors

1993 *A Gendered Past: A Critical Bibliography of Gender in Archaeology.* Technical Report 25. Ann Arbor: University of Michigan Museum of Anthropology Publications.

Benhabib, Seyla

1992 *Situating the Self: Gender, Community and Postmodernism in Contemporary Ethics.* New York: Routledge.

Bordo, Susan R.

1987 *The Flight to Objectivity: Essays on Cartesianism and Culture.* Albany: State University of New York Press.

Brumfiel, Elizabeth M.

1992 "Breaking and Entering the Ecosystem: Gender, Class, and Faction Steal the Show." *American Anthropologist* 94: 551-567.

Chodorow, Nancy

1974 "Family Structure and Feminine Personality." *Women, Culture, and Society*, 43-66. Edited by Michelle Zimbalist Rosaldo and Louise Lamphere. Stanford: Stanford University Press.

1978 *The Reproduction of Mothering: Psychoanalysis and the Sociology of Gender.* Berkeley: University of California.

Claassen, Cheryl, editor
1992 *Exploring Gender Through Archaeology: Selected papers From the 1991 Boone Conference*. Monographs in World Archaeology 11. Madison, WI: Prehistory Press.
Conkey, Margaret W., and Janet D. Spector
1984 "Archaeology and the Study of Gender." *Advances in Archaeological Method and Theory* 7: 1-38.
du Cros, Hilary, and Laurajane Smith, editors
1993 *Women in Archaeology: A Feminist Critique*. Canberra: Department of Prehistory, Australian National University.
Flax, Jane
1991 *Thinking in Fragments: Psychoanalysis, Feminism and Post-Modernism in the Contemporary West*. Berkeley: University of California Press.
Freud, Sigmund
1966 "Project for a Scientific Psychology." *Pre-Psycho-Analytic Publications and Unpublished Drafts*, 295-343. The Standard Edition of the Complete Psychological Works of Sigmund Freud, volume 1. Translated by James Strachey et al. London: Hogarth Press.
Haraway, Donna J.
1989 *Primate Visions: Gender, Race, and Nature in the World of Modern Science*. New York: Routledge.
1991 *Simians, Cyborgs, and Women: The Reinvention of Nature*. New York: Routledge.
Harding, Sandra
1986 *The Science Question in Feminism*. Ithaca: Cornell University Press.
Hartsock, Nancy
1983 "The Feminist Standpoint: Developing the Ground for a Specifically Feminist Historical Materialism." *Discovering Reality: Feminist Perspectives on Epistemology, Metaphysics, and Philosophy of Science*, 283-310. Edited by Sandra Harding and Merrill B. Hintikka. Dordrecht: Reidel.
Hennessy, Rosemary
1993 *Materialist Feminism and the Politics of Discourse*. New York: Routledge.
Keller, Evelyn Fox
1985 *Reflections on Gender and Science*. New Haven: Yale University Press.
Leone, Mark P.
1982 "Some Opinions About Recovering Mind." *American Antiquity* 47: 742-760.
1992 "The Productive Nature of Material Culture and Archaeology." *Historical Archaeology* 26: 130-133.
1995 "A Historical Archaeology of Capitalism: Considering Political Context." *American Anthropologist* 97: 251-268.
Millett, Kate
1969 *Sexual Politics*. New York: Doubleday and Company.
Potter, Parker B., Jr.
1992 "Critical Archaeology: In the Ground and on the Street." *Historical Archaeology* 26: 117-129.
Ricoeur, Paul
1965 "Le conflit des interprétations." *De l'interprétation: Essai sur Freud*, 29-44. Paris: Éditions du Seuil.
Sacks, Karen Brodkin
1979 *Sisters and Wives: The Past and Future of Sexual Equality*. Westport, CT: Greenwood Press.
1989 "Toward a Unified Theory of Class, Race, and Gender," *American Ethnologist* 16: 534-550.
Sawicki, Marianne
1994a *Seeing the Lord: Resurrection and Early Christian Practices*. Minneapolis: Fortress Press.
1994b "Archaeology as Space Technology: Digging for Gender and Class in Holy Land." *Method & Theory in the Study of Religion* 6: 319-348.
Scarry, Elaine
1985 *The Body in Pain: The Making and Unmaking of the World*. New York: Oxford University Press.

Schüssler Fiorenza, Elisabeth

1992 *But She Said: Feminist Practices of Biblical Interpretation.* Boston: Beacon Press.

1994a "The Bible, the Global Context, and the Discipleship of Equals," *Reconstructing Christian Theology*, 79-98. Edited by Rebecca S. Chopp and Mark Lewis Taylor. Minneapolis: Fortress Press.

1994b "The Rhetoricity of Historical Knowledge." *Religious Propaganda and Missionary Competition in the New Testament World: Essays Honoring Dieter Georgi.* Edited by Lukas Bormann et al. Leiden: Brill.

1995 *Jesus -- Miriam's Child, Sophia's Prophet: Critical Issues in Feminist Christology.* New York: Continuum.

Shackel, Paul A., and Barbara J. Little

1992 "Post-Processual Approaches to Meanings and Uses of Material Culture in Historical Archaeology." *Historical Archaeology* 26: 5-11.

Stein, Edith

[1917] *Zum Problem der Einfühlung.* Halle: Buchdruckerei des Waisenhauses. Reprinted in Munich: Kaffke, 1980.

[1922] "Psychische Kausalität." *Beiträge zur philosophischen Begründung der Psychologie und der Geisteswissenschaften*, 2-116. *Jahrbuch für Philosophie und phänomenologische Forschung* 5. Reprinted in Tübingen: Max Niemeyer Verlag, 1970.

1994 *Der Aufbau der Menschlichen Person.* Edith Steins Werke, volume 16. Greiburg: Herder.

Strachey, James.

1966 "Editor's Introduction" to "Project for a Scientific Psychology," by Sigmund Freud. *Pre-Psycho-Analytic Publications and Unpublished Drafts*, 283-293. The Standard Edition of the Complete Psychological Works of Sigmund Freud, volume 1. Translated by James Strachey et al. London: Hogarth Press.

Walde, Dale, and Noreen D. Willows, editors

1991 *The Archaeology of Gender.* Proceedings of the Twenty-Second Annual Conference of the Archaeological Association of the University of Calgary. Calgary: Archaeological Association.

Winnicott, D.W.

1965 *The Maturational Processes and the Facilitating Environment in the Theory of Emotional Development.* New York: International Universities Press.

Wylie, Alison

1995a "Angles of Vision: The Engendering of Archaeology Past and Present." Paper presented to a conference on "The Women and Gender Question in Science," Minneapolis.

1995b "The Constitution of Archaeological Evidence: Gender, Politics, and Science." *The Disunity of Science: Boundaries, Contexts, and Power.* Edited by Peter Galison and David J. Stump. Stanford: Stanford University Press.

FOR APPENDIX TWO: CRITIQUE OF BORDO'S EMPATHY THEORY

Bordo, Susan

1987 *The Flight to Objectivity: Essays on Cartesianism and Culture.* Albany: State University of New York Press.

Boyarin, Daniel

1993 *Carnal Israel: Reading Sex in Talmudic Culture.* Berkeley: University of California Press.

Gruber, Howard E., and J. Jacques Vonèche, editors

1977 *The Essential Piaget.* New York: Basic Books.

Ricoeur, Paul

1984 "The Aporias of the Experience of Time: Book 11 of Augustine's *Confessions.*" *Time and Narrative.* Volume 1, 5-30. Translated by Kathleen McLaughlin and David Pellauer. Chicago: University of Chicago Press.

Sawicki, Marianne
1984 *Aesthetic Catechetics: An Approach Via the Effective Symbol as Described by Schillebeeckx, Rahner, and Heidegger*. Ann Arbor: University Microfilms International.
1988 *The Gospel in History: Portrait of a Teaching Church*. New York: Paulist Press.
1994 *Seeing the Lord: Resurrection and Early Christian Practices*. Minneapolis: Fortress Press.
1997a "Spatial Management of Gender and Labor in Greco-Roman Galilee." *Archaeology and the World of Galilee: Texts and Contexts in the Roman and Byzantine Periods*. University of South Florida Studies in Judaica. Edited by Douglas R. Edwards and Thomas McCollough. Atlanta: Scholars Press.
1997b "Caste and Contact in the Galilee of Jesus: Research Beyond Positivism and Constructionism." *Galilean Archaeology and the Historical Jesus: The Integration of Material and Textual Remains*." Edited by J. Andrew Overman and Richard Horsley. Philadelphia: Trinity Press International.

Tracy, David
1981 *The Analogical Imagination: Christian Theology and the Culture of Pluralism*. New York: Crossroad.

Index
Of Subjects and Persons

(The names of the numerous commentators catalogued in Chapter Five have been omitted from this index, for the most part, in the interest of utility.)

Phaenomenologica

23. K. Held: *Lebendige Gegenwart.* Die Frage nach der Seinsweise des transzendentalen Ich bei Edumund Husserl, entwickelt am Leitfaden der Zeitproblematik. 1966
ISBN 90-247-0254-2
24. O. Laffoucrière: *Le destin de la pensée et 'La Mort de Dieu' selon Heidegger.* 1968
ISBN 90-247-0255-0
25. E. Husserl: *Briefe an Roman Ingarden.* Mit Erläuterungen und Erinnerungen an Husserl. Hrsg. von R. Ingarden. 1968 ISBN Hb: 90-247-0257-7; Pb: 90-247-0256-9
26. R. Boehm: *Vom Gesichtspunkt der Phänomenologie* (I). Husserl-Studien. 1968
ISBN Hb: 90-247-0259-3; Pb: 90-247-0258-5
For *Band II* see below under Volume 83
27. T. Conrad: *Zur Wesenslehre des psychischen Lebens und Erlebens.* Mit einem Geleitwort von H.L. van Breda. 1968 ISBN 90-247-0260-7
28. W. Biemel: *Philosophische Analysen zur Kunst der Gegenwart.* 1969
ISBN Hb: 90-247-0263-1; Pb: 90-247-0262-3
29. G. Thinès: *La problématique de la psychologie.* 1968
ISBN Hb: 90-247-0265-8; Pb: 90-247-0264-X
30. D. Sinha: *Studies in Phenomenology.* 1969
ISBN Hb: 90-247-0267-4; Pb: 90-247-0266-6
31. L. Eley: *Metakritik der formalen Logik.* Sinnliche Gewissheit als Horizont der Aussagenlogik und elementaren Prädikatenlogik. 1969
ISBN Hb: 90-247-0269-0; Pb: 90-247-0268-2
32. M.S. Frings: *Person und Dasein.* Zur Frage der Ontologie des Wertseins. 1969
ISBN Hb: 90-247-0271-2; Pb: 90-247-0270-4
33. A. Rosales: *Transzendenz und Differenz.* Ein Beitrag zum Problem der ontologischen Differenz beim frühen Heidegger. 1970 ISBN 90-247-0272-0
34. M.M. Saraïva: *L'imagination selon Husserl.* 1970 ISBN 90-247-0273-9
35. P. Janssen: *Geschichte und Lebenswelt.* Ein Beitrag zur Diskussion von Husserls Spätwerk. 1970 ISBN 90-247-0274-7
36. W. Marx: *Vernunft und Welt.* Zwischen Tradition und anderem Anfang. 1970
ISBN 90-247-5042-3
37. J.N. Mohanty: *Phenomenology and Ontology.* 1970 ISBN 90-247-5053-9
38. A. Aguirre: *Genetische Phänomenologie und Reduktion.* Zur Letztbegründung der Wissenschaft aus der radikalen Skepsis im Denken E. Husserls. 1970
ISBN 90-247-5025-3
39. T.F. Geraets: *Vers une nouvelle philosophie transcendentale.* La genèse de la philosophie de Maurice Merleau-Ponty jusqu'à la 'Phénoménologie de la perception.' Préface par E. Levinas. 1971 ISBN 90-247-5024-5
40. H. Declève: *Heidegger et Kant.* 1970 ISBN 90-247-5016-4
41. B. Waldenfels: *Das Zwischenreich des Dialogs.* Sozialphilosophische Untersuchungen in Anschluss an Edmund Husserl. 1971 ISBN 90-247-5072-5
42. K. Schuhmann: *Die Fundamentalbetrachtung der Phänomenologie.* Zum Weltproblem in der Philosophie Edmund Husserls. 1971 ISBN 90-247-5121-7
43. K. Goldstein: *Selected Papers/Ausgewählte Schriften.* Edited by A. Gurwitsch, E.M. Goldstein Haudek and W.E. Haudek. Introduction by A. Gurwitsch. 1971
ISBN 90-247-5047-4

Phaenomenologica

44. E. Holenstein: *Phänomenologie der Assoziation.* Zu Struktur und Funktion eines Grundprinzips der passiven Genesis bei E. Husserl. 1972 ISBN 90-247-1175-4
45. F. Hammer: *Theonome Anthropologie?* Max Schelers Menschenbild und seine Grenzen. 1972 ISBN 90-247-1186-X
46. A. Pažanin: *Wissenschaft und Geschichte in der Phänomenologie Edmund Husserls.* 1972 ISBN 90-247-1194-0
47. G.A. de Almeida: *Sinn und Inhalt in der genetischen Phänomenologie E. Husserls.* 1972 ISBN 90-247-1318-8
48. J. Rolland de Renéville: *Aventure de l'absolu.* 1972 ISBN 90-247-1319-6
49. U. Claesges und K. Held (eds.): *Perspektiven transzendental-phänomenologischer Forschung.* Für Ludwig Landgrebe zum 70. Geburtstag von seiner Kölner Schülern. 1972 ISBN 90-247-1313-7
50. F. Kersten and R. Zaner (eds.): *Phenomenology: Continuation and Criticism.* Essays in Memory of Dorion Cairns. 1973 ISBN 90-247-1302-1
51. W. Biemel (ed.): *Phänomenologie Heute.* Festschrift für Ludwig Landgrebe. 1972 ISBN 90-247-1336-6
52. D. Souche-Dagues: *Le développement de l'intentionnalité dans la phénoménologie husserlienne.* 1972 ISBN 90-247-1354-4
53. B. Rang: *Kausalität und Motivation.* Untersuchungen zum Verhältnis von Perspektivität und Objektivität in der Phänomenologie Edmund Husserls. 1973 ISBN 90-247-1353-6
54. E. Levinas: *Autrement qu'être ou au-delà de l'essence.* 2nd. ed.: 1978 ISBN 90-247-2030-3
55. D. Cairns: *Guide for Translating Husserl.* 1973 ISBN (Pb) 90-247-1452-4
56. K. Schuhmann: *Die Dialektik der Phänomenologie, I.* Husserl über Pfänder. 1973 ISBN 90-247-1316-1
57. K. Schuhmann: *Die Dialektik der Phänomenologie, II.* Reine Phänomenologie und phänomenologische Philosophie. Historisch-analytische Monographie über Husserls 'Ideen I'. 1973 ISBN 90-247-1307-2
58. R. Williame: *Les fondements phénoménologiques de la sociologie compréhensive: Alfred Schutz et Max Weber.* 1973 ISBN 90-247-1531-8
59. E. Marbach: *Das Problem des Ich in der Phänomenologie Husserls.* 1974 ISBN 90-247-1587-3
60. R. Stevens: *James and Husserl: The Foundations of Meaning.* 1974 ISBN 90-247-1631-4
61. H.L. van Breda (ed.): *Vérité et Vérification / Wahrheit und Verifikation.* Actes du quatrième Colloque International de Phénoménologie / Akten des vierten Internationalen Kolloquiums für Phänomenologie (Schwabisch Hall, Baden-Württemberg, 8.-11. September 1969). 1974 ISBN 90-247-1702-7
62. Ph.J. Bossert (ed.): *Phenomenological Perspectives.* Historical and Systematic Essays in Honor of Herbert Spiegelberg. 1975. ISBN 90-247-1701-9
63. H. Spiegelberg: *Doing Phenomenology.* Essays on and in Phenomenology. 1975 ISBN 90-247-1725-6
64. R. Ingarden: *On the Motives which Led Husserl to Transcendental Idealism.* 1975 ISBN 90-247-1751-5
65. H. Kuhn, E. Avé-Lallemant and R. Gladiator (eds.): *Die Münchener Phänomenologie.* Vorträge des Internationalen Kongresses in München (13.-18. April 1971). 1975 ISBN 90-247-1740-X

Phaenomenologica

66. D. Cairns: *Conversations with Husserl and Fink*. Edited by the Husserl-Archives in Louvain. With a foreword by R.M. Zaner. 1975 ISBN 90-247-1793-0
67. G. Hoyos Vásquez: *Intentionalität als Verantwortung*. Geschichtsteleologie und Teleologie der Intentionalität bei Husserl. 1976 ISBN 90-247-1794-9
68. J. Patočka: *Le Monde naturel comme problème philosophique*. 1976 ISBN 90-247-1795-7
69. W.W. Fuchs: *Phenomenology and the Metaphysics of Presence*. An Essay in the Philosophy of Edmund Husserl. 1976 ISBN 90-247-1822-8
70. S. Cunningham: *Language and the Phenomenological Reductions of Edmund Husserl*. 1976 ISBN 90-247-1823-6
71. G.C. Moneta: *On Identity*. A Study in Genetic Phenomenology. 1976 ISBN 90-247-1860-0
72. W. Biemel und das Husserl-Archiv zu Löwen (eds.): *Die Welt des Menschen - Die Welt der Philosophie*. Festschrift für Jan Patočka. 1976 ISBN 90-247-1899-6
73. M. Richir: *Au-delà du renversement copernicien*. La question de la phénoménologie et son fondement. 1976 ISBN 90-247-1903-8
74. H. Mongis: *Heidegger et la critique de la notion de valeur*. La destruction de la fondation métaphysique. Lettre-préface de Martin Heidegger. 1976 ISBN 90-247-1904-6
75. J. Taminiaux: *Le regard et l'excédent*. 1977 ISBN 90-247-2028-1
76. Th. de Boer: *The Development of Husserl's Thought*. 1978 ISBN Hb: 90-247-2039-7; Pb: 90-247-2124-5
77. R.R. Cox: *Schutz's Theory of Relevance*. A Phenomenological Critique. 1978 ISBN 90-247-2041-9
78. S. Strasser: *Jenseits von Sein und Zeit*. Eine Einführung in Emmanuel Levinas' Philosophie. 1978 ISBN 90-247-2068-0
79. R.T. Murphy: *Hume and Husserl*. Towards Radical Subjectivism. 1980 ISBN 90-247-2172-5
80. H. Spiegelberg: *The Context of the Phenomenological Movement*. 1981 ISBN 90-247-2392-2
81. J.R. Mensch: *The Question of Being in Husserl's* Logical Investigations. 1981 ISBN 90-247-2413-9
82. J. Loscerbo: *Being and Technology*. A Study in the Philsophy of Martin Heidegger. 1981 ISBN 90-247-2411-2
83. R. Boehm: *Vom Gesichtspunkt der Phänomenologie II*. Studien zur Phänomenologie der Epoché. 1981 ISBN 90-247-2415-5
84. H. Spiegelberg and E. Avé-Lallemant (eds.): *Pfänder-Studien*. 1982 ISBN 90-247-2490-2
85. S. Valdinoci: *Les fondements de la phénoménologie husserlienne*. 1982 ISBN 90-247-2504-6
86. I. Yamaguchi: *Passive Synthesis und Intersubjektivität bei Edmund Husserl*. 1982 ISBN 90-247-2505-4
87. J. Libertson: *Proximity*. Levinas, Blanchot, Bataille and Communication. 1982 ISBN 90-247-2506-2

Phaenomenologica

88. D. Welton: *The Origins of Meaning*. A Critical Study of the Thresholds of Husserlian Phenomenology. 1983 ISBN 90-247-2618-2
89. W.R. McKenna: *Husserl's 'Introductions to Phenomenology'*. Interpretation and Critique. 1982 ISBN 90-247-2665-4
90. J.P. Miller: *Numbers in Presence and Absence*. A Study of Husserl's Philosophy of Mathematics. 1982 ISBN 90-247-2709-X
91. U. Melle: *Das Wahrnehmungsproblem und seine Verwandlung in phänomenologischer Einstellung*. Untersuchungen zu den phänomenologischen Wahrnehmungstheorien von Husserl, Gurwitsch und Merleau-Ponty. 1983 ISBN 90-247-2761-8
92. W.S. Hamrick (ed.): *Phenomenology in Practice and Theory*. Essays for Herbert Spiegelberg. 1984 ISBN 90-247-2926-2
93. H. Reiner: *Duty and Inclination*. The Fundamentals of Morality Discussed and Redefined with Special Regard to Kant and Schiller. 1983 ISBN 90-247-2818-6
94. M. J. Harney: *Intentionality, Sense and the Mind*. 1984 ISBN 90-247-2891-6
95. Kah Kyung Cho (ed.): *Philosophy and Science in Phenomenological Perspective*. 1984 ISBN 90-247-2922-X
96. A. Lingis: *Phenomenological Explanations*. 1986 ISBN Hb: 90-247-3332-4; Pb: 90-247-3333-2
97. N. Rotenstreich: *Reflection and Action*. 1985 ISBN Hb: 90-247-2969-6; Pb: 90-247-3128-3
98. J.N. Mohanty: *The Possibility of Transcendental Philosophy*. 1985 ISBN Hb: 90-247-2991-2; Pb: 90-247-3146-1
99. J.J. Kockelmans: *Heidegger on Art and Art Works*. 1985 ISBN 90-247-3102-X
100. E. Lévinas: *Collected Philosophical Papers*. 1987 ISBN Hb: 90-247-3272-7; Pb: 90-247-3395-2
101. R. Regvald: *Heidegger et le Problème du Néant*. 1986 ISBN 90-247-3388-X
102. J.A. Barash: *Martin Heidegger and the Problem of Historical Meaning*. 1987 ISBN 90-247-3493-2
103 J.J. Kockelmans (ed.): *Phenomenological Psychology*. The Dutch School. 1987 ISBN 90-247-3501-7
104. W.S. Hamrick: *An Existential Phenomenology of Law: Maurice Merleau-Ponty*. 1987 ISBN 90-247-3520-3
105. J.C. Sallis, G. Moneta and J. Taminiaux (eds.): *The Collegium Phaenomenologium. The First Ten Years*. 1988 ISBN 90-247-3709-5
106. D. Carr: *Interpreting Husserl*. Critical and Comparative Studies. 1987. ISBN 90-247-3505-X
107. G. Heffernan: *Isagoge in die phänomenologische Apophantik*. Eine Einführung in die phänomenologische Urteilslogik durch die Auslegung des Textes der *Formalen und transzendenten Logik* von Edmund Husserl. 1989 ISBN 90-247-3710-9
108. F. Volpi, J.-F. Mattéi, Th. Sheenan, J.-F. Courtine, J. Taminiaux, J. Sallis, D. Janicaud, A.L. Kelkel, R. Bernet, R. Brisart, K. Held, M. Haar et S. IJsseling: *Heidegger et l'Idée de la Phénoménologie*. 1988 ISBN 90-247-3586-6
109. C. Singevin: *Dramaturgie de l'Esprit*. 1988 ISBN 90-247-3557-2

Phaenomenologica

110. J. Patočka: *Le monde naturel et le mouvement de l'existence humaine.* 1988 ISBN 90-247-3577-7
111. K.-H. Lembeck: *Gegenstand Geschichte.* Geschichtswissenschaft in Husserls Phänomenologie. 1988 ISBN 90-247-3635-8
112. J.K. Cooper-Wiele: *The Totalizing Act.* Key to Husserl's Early Philosophy. 1989 ISBN 0-7923-0077-7
113. S. Valdinoci: *Le principe d'existence.* Un devenir psychiatrique de la phénoménologie. 1989 ISBN 0-7923-0125-0
114. D. Lohmar: *Phänomenologie der Mathematik.* 1989 ISBN 0-7923-0187-0
115. S. IJsseling (Hrsgb.): *Husserl-Ausgabe und Husserl-Forschung.* 1990 ISBN 0-7923-0372-5
116. R. Cobb-Stevens: *Husserl and Analytic Philosophy.* 1990 ISBN 0-7923-0467-5
117. R. Klockenbusch: *Husserl und Cohn.* Widerspruch, Reflexion und Telos in Phänomenologie und Dialektik. 1989 ISBN 0-7923-0515-9
118. S. Vaitkus: *How is Society Possible?* Intersubjectivity and the Fiduciary Attitude as Problems of the Social Group in Mead, Gurwitsch, and Schutz. 1991 ISBN 0-7923-0820-4
119. C. Macann: *Presence and Coincidence.* The Transformation of Transcendental into Ontological Phenomenology. 1991 ISBN 0-7923-0923-5
120. G. Shpet: *Appearance and Sense.* Phenomenology as the Fundamental Science and Its Problems. Translated from Russian by Th. Nemeth. 1991 ISBN 0-7923-1098-5
121. B. Stevens: *L'Apprentissage des Signes.* Lecture de Paul Ricœur. 1991 ISBN 0-7923-1244-9
122. G. Soffer: *Husserl and the Question of Relativism.* 1991 ISBN 0-7923-1291-0
123. G. Römpp: *Husserls Phänomenologie der Intersubjektivität.* Und Ihre Bedeutung für eine Theorie intersubjektiver Objektivität und die Konzeption einer phänomenologischen. 1991 ISBN 0-7923-1361-5
124. S. Strasser: *Welt im Widerspruch.* Gedanken zu einer Phänomenologie als ethischer Fundamentalphilosophie. 1991 ISBN Hb: 0-7923-1404-2; Pb: 0-7923-1551-0
125. R. P. Buckley: *Husserl, Heidegger and the Crisis of Philosophical Responsibility.* 1992 ISBN 0-7923-1633-9
126. J. G. Hart: *The Person and the Common Life.* Studies in a Husserlian Social Ethics. 1992 ISBN 0-7923-1724-6
127. P. van Tongeren, P. Sars, C. Bremmers and K. Boey (eds.): *Eros and Eris.* Contributions to a Hermeneutical Phenomenology. Liber Amicorum for Adriaan Peperzak. 1992 ISBN 0-7923-1917-6
128. Nam-In Lee: *Edmund Husserls Phänomenologie der Instinkte.* 1993 ISBN 0-7923-2041-7
129. P. Burke and J. Van der Veken (eds.): *Merleau-Ponty in Contemporary Perspective.* 1993 ISBN 0-7923-2142-1
130. G. Haefliger: *Über Existenz: Die Ontologie Roman Ingardens.* 1994 ISBN 0-7923-2227-4
131. J. Lampert: *Synthesis and Backward Reference in Husserl's* Logical Investigations. 1995 ISBN 0-7923-3105-2
132. J.M. DuBois: *Judgment and Sachverhalt.* An Introduction to Adolf Reinach's Phenomenological Realism. 1995 ISBN 0-7923-3519-8

Phaenomenologica

133. B.E. Babich (ed.): *From Phenomenology to Thought, Errancy, ar ' . ɜsire.* Essays in Honor of William J. Richardson, S.J. 1995 ISBN 0-7923-3567-8
134. M. Dupuis: *Pronoms et visages. Lecture d'Emmanuel Levinas.* 1Ꞔ ˬ ISBN 0-7923-3655-0; Pb 0-7923-3994-0
135. D. Zahavi: *Husserl und die transzendentale Intersubjektivität.* Eine Antwort auf die sprachpragmatische Kritik. 1996 ISBN 0-7923-3713-1
136. A. Schutz: *Collected Papers, IV.* Edited with preface and notes by H. Wagner and G. Psathas, in collaboration with F. Kersten. 1996 ISBN 0-7923-3760-3
137. P. Kontos: *D'une phénoménologie de la perception chez Heidegger.* 1996 ISBN 0-7923-3776-X
138. F. Kuster: *Wege der Verantwortung.* Husserls Phänomenologie als Gang durch die Faktizität. 1996 ISBN 0-7923-3916-9
139. C. Beyer: *Von Bolzano zu Husserl.* Eine Untersuchung über den Ursprung der phänomenologischen Bedeutungslehre. 1996 ISBN 0-7923-4050-7
140. J. Dodd: *Idealism and Corporeity.* An Essay on the Problem of the Body in Husserl's Phenomenology. 1997 ISBN 0-7923-4400-6
141. E. Kelly: *Structure and Diversity.* Studies in the Phenomenological Philosophy of Max Scheler. 1997 ISBN 0-7923-4492-8
142. J. Cavallin: *Content and Object.* Husserl, Twardowski and Psychologism. 1997 ISBN 0-7923-4734-X
143. H.P. Steeves: *Founding Community.* A Phenomenological-Ethical Inquiry. 1997 ISBN 0-7923-4798-6
144. M. Sawicki: *Body, Text, and Science.* The Literacy of Investigative Practices and the Phenomenology of Edith Stein. 1997 ISBN 0-7923-4759-5

Previous volumes are still available

Further information about *Phenomenology* publications are available on request.

Kluwer Academic Publishers – Dordrecht / Boston / London